UAV and Sensors Applications for Navigation and Positioning

UAV and Sensors Applications for Navigation and Positioning

Guest Editors

Yongbo Zhang
Yuhang Li

MDPI

Basel • Beijing • Wuhan • Barcelona • Belgrade • Novi Sad • Cluj • Manchester

Guest Editors
Yongbo Zhang
School of Aeronautic Science and Engineering
Beihang University
Beijing
China

Yuhang Li
Institute of Solid Mechanics
Beihang University
Beijing
China

Editorial Office
MDPI AG
Grosspeteranlage 5
4052 Basel, Switzerland

This is a reprint of the Special Issue, published open access by the journal *Sensors* (ISSN 1424-8220), freely accessible at: https://www.mdpi.com/journal/sensors/special_issues/82F80BO47Y.

For citation purposes, cite each article independently as indicated on the article page online and as indicated below:

Lastname, A.A.; Lastname, B.B. Article Title. *Journal Name* **Year**, *Volume Number*, Page Range.

ISBN 978-3-7258-4137-0 (Hbk)
ISBN 978-3-7258-4138-7 (PDF)
https://doi.org/10.3390/books978-3-7258-4138-7

Contents

sensors

MDPI

Article

Scheduling Framework for Accelerating Multiple Detection-Free Object Trackers

Myungsun Kim [1,*], Inmo Kim [2], Jihyeon Yong [2] and Hyuksoo Kim [2]

1 Department of Applied Artificial Intelligence, Hansung University, Seoul 02876, Republic of Korea
2 Department of IT Convergence Engineering, Hansung University, Seoul 02876, Republic of Korea
* Correspondence: kmsjames@hansung.ac.kr; Tel.: +82-2-760-4045

Abstract: In detection-free tracking, after users freely designate the location of the object to be tracked in the first frame of the video sequence, the location of the object is continuously found in the following video frame sequence. Recently, technologies using a Siamese network and transformer based on DNN modules have been evaluated as very excellent in terms of tracking accuracy. The high computational complexity due to the usage of the DNN module is not a preferred feature in terms of execution speed, and when tracking two or more objects, a bottleneck effect occurs in the DNN accelerator such as the GPU, which inevitably results in a larger delay. To address this problem, we propose a tracker scheduling framework. First, the computation structures of representative trackers are analyzed, and the scheduling unit suitable for the execution characteristics of each tracker is derived. Based on this analysis, the decomposed workloads of trackers are multi-threaded under the control of the scheduling framework. CPU-side multi-threading leads the GPU to a work-conserving state while enabling parallel processing as much as possible even within a single GPU depending on the resource availability of the internal hardware. The proposed framework is a general-purpose system-level software solution that can be applied not only to GPUs but also to other hardware accelerators. As a result of confirmation through various experiments, when tracking two objects, the execution speed was improved by up to 55% while maintaining almost the same accuracy as the existing method.

Keywords: GPU scheduling; object tracking; multi-DNN; multi-threading, detection-free tracker

1. Introduction

In a wide range of AI (artificial intelligence)-enabled service fields such as human–computer interaction [1], traffic control [2], video surveillance [3], and augmented reality [4], object-tracking technology has drawn constant attention. Object tracking is largely divided into detection-free tracking and tracking-by-detection. Recent studies have used tracking-by-detection methodologies to realize MOT (multi-object tracking). However, this is a method of tracking classified objects in advance and the process of revealing the association between the detection results. Detection-free tracking, which allows users to track any object from the user point of view, can be a more useful technology for security and safety-related applications such as crime prevention and facility safety. VOT (visual object tracking) is a kind of detection-free tracking, which estimates the position of the user-defined target object in a series of video frames. In doing so, the estimated position in each frame is usually defined by the bounding box including the target object to be tracked.

Without loss of generality, in order to secure the improved inference accuracy, DNN (deep neural network) models are getting bigger and more complicated [5]. Trackers with the latest technologies are also equipped with DNN models, and the computational complexity is also very high [6,7]. Therefore, tracking two or more user-specified objects in a video frame is even more computationally complex and makes the system very slow.

The recent trend of VOT technology can be divided into Siamese-network-based and transformer-based studies. Siamese network structure tracks target object by computing the similarity between the target patch designated by the user and the search region of the video frames [6]. Transformer-based trackers conduct tracking by fusing the features of the target patch and search region of the frames using attention mechanism [7]. These two kinds of trackers have very different structures. Therefore, using the optimization technique of the same method is not meaningful to make up for the speed-performance deterioration incurred when tracking two or more objects.

In order to maximize the execution speed of DNN, hardware accelerators specialized in specific DNN modules have been released [5]. However, these specialized accelerators do not guarantee their performance even in the new DNN structure. Therefore, edge devices and servers usually use GPUs to accelerate DNN modules. The DNN modules used in object trackers are also dependent on the GPU-specific libraries used by deep learning frameworks such as TensorFlow [8] and PyTorch [9], and the libraries are not optimized for various kinds of GPU hardware structures. Furthermore, these deep learning frameworks do not provide optimization techniques for two or more DNN modules to run parallel in the GPU even when the GPU is experiencing under utilization, which may lead the tracker performance to be suboptimal.

To tackle the above-mentioned issues, we propose a software-based solution approach, which provides an efficient scheduling framework for the two well-performing object trackers running on edge devices and GPU-server computing systems. We first lay the groundwork for the proposed scheduling framework to optimally map workloads included in the tracker to computing units. To this end, an in-depth computational structure analysis is conducted on SiamRPN++, which epitomizes Siamese-network-based trackers, and CSWinTT, which best exemplifies transformer-based trackers. Particularly, we give most of our attention to the large-scale computational structure of MHA (multi-head attention), which transformer-based trackers have in common, from the DNN module perspective.

Second, the proposed scheduling framework improves the tracking speed of two or more trackers when they are running together. This means that the tracking performance is improved when two or more objects are simultaneously tracked in a detection-free manner. The proposed scheduling framework is a system-level acceleration technology designed to be independent of the different structures of GPUs. Additionally, the approach proposed in this study can be applied to hardware accelerators other than GPUs. This is possible with only a library provided by the accelerator manufacturer.

2. Background and Related Work

To provide an aid in understanding the remaining part of this paper, this section gives background knowledge and related object tracking studies that have been previously conducted. In addition, the two types of trackers targeted in this paper are technically described.

2.1. Object Tracking

Object tracking refers to the process of estimating the position of an object or several objects that move over time in video frames, and generally, object tracking is divided into two categories depending on employed tracking algorithms: VOT (visual object tracking) and MOT (multiple object tracking). Object tracking typically outputs a bounding box, which has the location information of the object in each video frame [6,10,11].

VOT tracks a single object and class-agnostic. In VOT, only the position (i.e., bounding box) of the object in the first frame is given without any other information. There is no detailed information about the object, but as long as the location information of the object in the first frame is provided, the object can be tracked continuously in consequent video frames. VOT falls under the category of detection-free tracking, which means that a manually initialized bounding box is required for the tracking target rather than the detection of the predefined target object.

Unlike VOT, MOT tracks objects in predetermined classes. MOT automatically identifies multiple objects in a video and shows them as a series of trajectories. MOT tracks multiple objects and is commonly known as detection-based tracking, thus performing object detection every frame and associating the results with tracking. In other words, connecting the detected location information of the current frame with the one of the previous frame. For example, if there is a video of several cars driving on the road, MOT tracks each car separately.

2.2. Detection-Free Tracking

Our target tracking systems are detection-free trackers, and thus the trackers in this paper aim to keep track of multiple objects designated by the user in the first frame with VOT as the default mechanism, not MOT. Representative trends of trackers using VOT technology are based on either the Siamese network or the transformer architecture.

2.2.1. Siamese-Network-Based Trackers

SiamFC [10] is a seminal study using a Siamese network for object tracking. A user creates an exemplar image z including a tracking target, and search image frames x means video frames that need to be inferred. x and z pass through the same CNN and their output tensors become the input of the cross-correlation operation. Thereafter, each component of the calculated similarity map corresponds to the similarity with z with respect to the inside x. The SiamRPN [11] adopts an RPN (region proposal network), which was used as a standard in image detection problems, in SiamFC, and performs bounding box regression to determine the location of tracked target. As a result, the size of the tracked object can be estimated more accurately than before, and at the same time, the iterative calculation due to the adoption of the image pyramid can be avoided. As a way to solve the problem of decreasing accuracy by padding inside CNN, SiamRPN++ [6] proposes a spatial-aware-sampling strategy to make the locations of tracked objects in the search image frame have a uniform distribution. In fact, by applying learning data collected by the strategy, a SiamRPN tracker that adopted ResNet-50 [12] as a backbone obtains a higher accuracy than a SiamRPN with AlexNet [13].

2.2.2. Transformer-Based Trackers

Transformer-based approaches have drawn great performance in various AI applications such as object detection, semantic segmentation, and image recognition. The success factors in such fields are from the fact that a cross-attention mechanism enables relevant reasoning between image patches [14]. Even in object tracking research works, transformer-based trackers have presented their excellent achievements by incorporating the pixel-level attention to mingle the features of the target object and tracked object in search image frames. TransT [15] introduces an attention mechanism to perform feature fusion of target object and search image frames. The designed feature fusion network is structured with two modules: ECA (ego-context augment) module and CFA (cross-feature augment) module. The two modules expedite bounding box regression and object localization. STARK [16] suggests an encoder–decoder structured transformer which takes both spatial and temporal information into account. The encoder with self-attention modules learns the relationship between the target object and the incoming video frames by analyzing feature dependencies. To achieve target position estimation, the decoder learns a query embedding. Swin Transformer [17] takes up a hierarchical structure and consists of transformers. To obtain an expanded receptive region, it gradually increases the size of image patches. CSWinTT [7] develops pixel-level attention into window-level attention while inheriting the structural advantage of Swin Transformer. Cyclic shifting has the effect of expanding the window area, which greatly improves the accuracy.

2.3. Structural Analysis of Detection-Free Object Trackers

In this subsection, we provide a concise description of the architecture of two representative detection-free object trackers.

2.3.1. SiamRPN++: Siamese-Network-Based

Figure 1 shows the overall workflow of SiamRPN++. For input data, the target patch with the object to be tracked and the video frames are used, and the video frames are generally called as search region or search image frames. In the first frame, the user sets the location information of the target patch containing the object to be tracked. The location information of the target patch consists of a total of four values, given the x and y coordinate values at the upper left, and the values of width and height based on them.

The target patch and search image frames are transmitted to ResNet-50-based backbones. Each backbone outputs three different feature maps; then, they are inputted to three RPN (region proposal network) blocks to perform a similarity check. To do so, two identical DW-XC (depth-wise cross-correlation) modules are applied [6]. Through the weighted sum operation, the three bounding box values and classification results from the three RPN modules derive one bounding box regression and a classification result, which are the final results of inference. The closer the value of classification result is to one, the higher the probability that the object is in the bounding box.

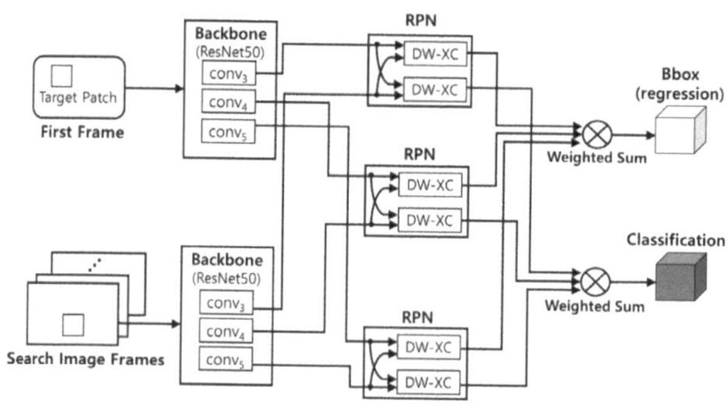

Figure 1. Workflow of SiamRPN++.

Equation (1) details the two weighted sum procedures shown in Figure 1. S_{all} and B_{all} represent the final classification and regression results, respectively [6]. S_l and B_l represent the classification and regression of each RPN, and l is one of 3, 4, and 5, indicating that it is the result from $conv3$, $conv4$, and $conv5$ in backbones. α_i and β_i are combination weights, and they are obtained after offline end-to-end optimization [6].

$$S_{all} = \sum_{l=3}^{5} \alpha_i * S_l, \quad B_{all} = \sum_{l=3}^{5} \beta_i * B_l. \tag{1}$$

2.3.2. CSWinTT: Transformer-Based

Basically, just as SiamRPN++ explained above, CSWinTT also receives the target patch and search image frames while using just one ResNet-50 backbone. Overall, it is constructed very similar to general transformer-based object trackers [14–17] and has a large computational complexity. CSWinTT has six-layer encoder and decoder blocks, and each encoder and decoder has MHA (multi-head attention). The core technology of CSWinTT centers around MHA incorporating window-partitioning and cyclic shifting, and the gray box in Figure 2 denotes it.

CSWinTT also sets the bounding box position as the final result and outputs confidence score every frame along with the bounding box. Confidence score is the probability of the presence or absence of the target object to be found in the bounding box, which is the result of inference about every single frame. The closer it is to the value of one, the higher the probability that the object is in that bounding box.

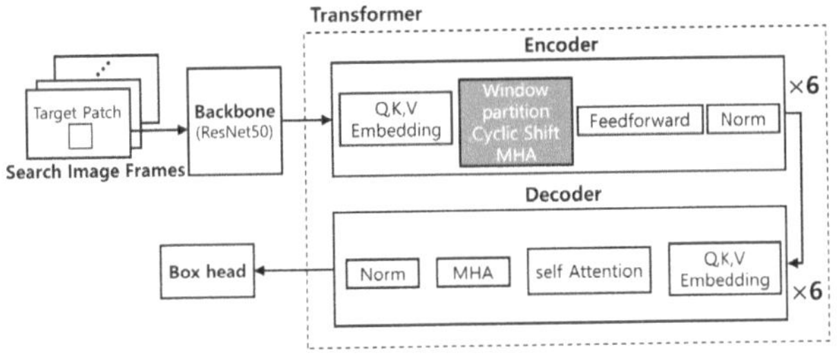

Figure 2. Workflow of CSWinTT.

3. Problem Settings

In this paper, we aim to show a mechanism that maximizes the execution speed while maintaining inference accuracy when multiple detection-free-based trackers are running on edge device or GPU-server systems to track multiple objects. There are some difficulties that must be noted in such an execution environment.

- Systems that run DNN models, such as embedded edge devices and server systems, are typically composed of many CPUs and a much smaller number of DNN accelerators. Thus, multiple tracker tasks hosted by the CPUs can throw DNN workloads required for object tracking into the accelerator independently of each other. In this situation, the performance of trackers may differ greatly depending on the scheduling policy imposed on DNN workloads delivered to the accelerator.
- For instance, a GPU is a representative DNN accelerator. As shown in Figures 1 and 2, backbone, RPN, and MHA blocks have different computational complexity. Here, the GPU is not always 100% used depending on which block is computed. For example, when there are two DNN workloads that occupy 30% of GPU utilization, they maintain 30% utilization if they are performed in order in a row. However, if the two workloads are on the GPU at the same time, they can have twice the execution speed with 60% utilization. However, it is not easy to double GPU utilization under real-world applications. Libraries such as cuDNN [18] and CUDA runtime [19] do not adaptively allocate DNN workloads to all different GPU hardware architectures.
- Since multiple CPUs are supported, multiple trackers may allocate DNN workloads to the accelerator such as the GPU. Moreover, with the help of commercial DNN frameworks such as TensorFlow [8] and PyTorch [9], we can easily design trackers using Python. However, if hardware blocks such as MPS (multi-process service) [20] are not supported in an embedded environment, occurred overhead is unavoidable due to the context-switching, where context means the virtual address boundary of processes. In addition, even if multi-threading is available within the same context (process), it is difficult to avoid the serialization problem caused by the GIL (global interpreter lock) policy of Python.
- SiamRPN++ and CSWinTT have very heterogeneous structures, as shown earlier in Figures 1 and 2. Thus, a uniform DNN workload scheduling scheme can lead to poor performance for some trackers in a way that can benefit some trackers.

The problem we want to solve is providing an effective software-based means to get out of the difficulties listed above.

4. Solution

In this section, to solve the aforementioned problems, we explain our solution approach with sufficient technical details. First, the overall solution architecture is given, and then workload scheduling techniques and parallelization methods preferable to the execution characteristics of each tracker are described in detail.

4.1. Overall Solution Approach

Figure 3 details the operational workflow of the proposed solution approach in this paper. Roughly, the approach consists of offline and run-time phases. Offline, first, the execution time of each function block constituting the target tracker is measured through profiling. Then, a basic scheduling unit is derived by comprehensively considering this result and the data dependency between each functional block. In Figure 3, each work in the work list becomes an instance scheduled on the CPU (i.e., scheduling unit), and the figure shows an example of eight CPUs and w works. When each scheduling instance is obtained, a computing unit suitable for each instance, either a CPU or a GPU, is defined. Finally, the work list to be executed at run-time phase is completed according to this offline procedure.

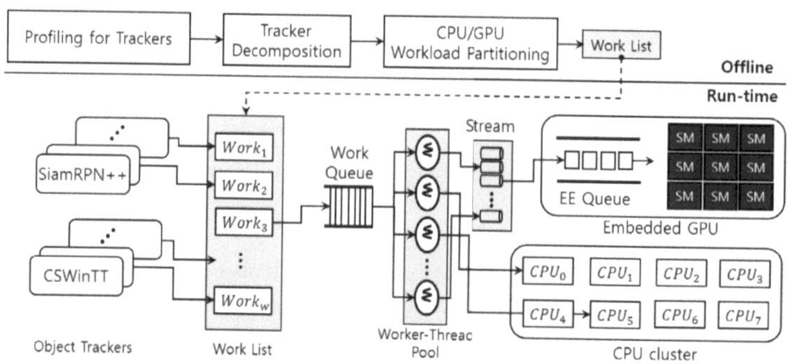

Figure 3. Overview of the scheduling framework for accelerating multiple detection-free object trackers.

In the run-time phase, multiple works in the work list are executed over the solution architecture, which consists largely of an offline defined work list, a work queue, and a worker-thread pool. A work $Work_i$ defined in the work list can be edited on a variety of scales. Particularly, it is possible from a small layer of a DNN model to the entire tracker. For example, $Work_0$ is the tracker task itself, and $Work_1$ and $Work_2$ are DNN workloads allocated by the tracker $Work_0$ to the GPU. First, $Work_0$ is mapped to one of threads in the worker-thread pool, and then $Work_0$ assigns $Work_1$ and $Work_2$ to the threads in the pool.

Object trackers can request their works ($Work_0, Work_1, \cdots, Work_7$) asynchronously to the work queue. One of threads in the worker-thread pool immediately extracts the work at the queue front of the work queue whenever it is in the idle state. Then the worker thread first determines whether the delivered work is GPU-side or CPU-side. If the work is a DNN workload it is mapped to one of the streams [5,21], and then enqueued to the EE (execution engine) queue of the GPU [5,21], if not, it is assigned to one of the CPUs directly.

4.2. Scheduling Works in the Work List

The detailed operation of each function block shown in Figure 3 is explained through Algorithm 1. In the algorithm, the function **worker_thread( )** is the pseudo code of each worker thread. Both **queue_pull( )** and **queue_push( )** are provisioned as tools for accessing the work queue, and **queue_pull( )** is only called by worker threads and **queue_push( )** is utilized by each work, i.e., *work* in the algorithm.

When the proposed scheduling framework starts, the all threads within the worker-thread pool wait indefinitely since there is no work in the work queue. Then, offline defined works in the work list arrive at the work queue through the function **queue_push( )**. At this point in time, *sig* is broadcasted to all the threads in the worker-thread pool. Then, one of threads in the pool receives *sig* and exits the blocked state, as shown in line 3. Immediately after this, the function **queue_pull( )** is used, and the work (*work*) at the front of the work queue is transferred as an argument to the function **execute( )**. Looking at lines 13 and 14,

queue_pull( ) also broadcasts *sig*. This allows worker threads to take the next work after the first work is exited when there are two or more works in the work queue.

Algorithm 1 Scheduling framework for multiple object trackers.

```
1:  function worker_thread( )
2:      A:
3:      wait_signal(sig)
4:      work ← queue_pull( )
5:      execute(work)
6:      goto A:
7:  end function
8:  function queue_push(work)
9:      insert work to the work queue
10:     broadcast(sig)
11: end function
12: function queue_pull( )
13:     if (number of works in the queue > 1) then
14:         broadcast(sig)
15:     end if
16:     return work at the queue front
17: end function
18: function execute(work)
19:     if (work is not a pure CPU workload) then
20:         i ← index of work
21:         set the stream index as i
22:     end if
23:     if (work is a pure DNN workload) then
24:         executing work
25:         synchronize host until the DNN workload in the i^th stream has completed
26:         return
27:     end if
28:     executing work
29: end function
```

The properties of the works in the work list can be expressed in three ways: pure CPU workloads, pure DNN workloads, and CPU–GPU mixed workloads. The parameter *work* passed to the function **execute( )** is one of these three. First, if it is not a pure CPU workload, it means that the GPU is used, so set the stream index for parallel processing in the GPU (lines 19 to 22).

If *work* is a pure DNN workload, *work* is transferred to the GPU's EE queue and executed (line 24). At this time, synchronization must be performed with the host CPU for the next operation. For instance, let us say we have $Work_1$, $Work_2$, and $Work_3$ in the work list. $Work_1$ and $Work_2$ are DNN workloads that use GPU, and *Work*3 needs to concatenate the operation results of $Work_1$ and $Work_2$. In this case, *Work*3 using the CPU must wait for the synchronization event that the GPU has finished all its assigned work (line 25).

In the case of a mix of CPU workload and DNN workload, *work* is simply executed as in line 28. At this time, the internal operation of *work* uses the GPU sporadically, but there is no synchronization process like in line 25. The reason for this is as follows. First, *work* is a decomposed internal tracker function. Therefore, in one function flow, the next workload that uses the CPU cannot be called until the GPU finishes its operation.

4.3. Execution Time Analysis for Decomposing Object Trackers

Tables 1 and 2 represent execution time profiles for each functional blocks of SiamRPN++ and CSWinTT trackers, respectively. For measurement, we used one NVIDIA RTX A6000

GPU [22] and randomly extracted 3350 images from TrackingNet [23] as a dataset. As we can see from the two tables, CSwinTT is a very large-sized tracker that takes more than twice the time when dealing with 3350 images compared to SiamRPN++.

Table 1. Execution time profile of SiamRPN++.

	Backbone	RPN	Others	Total
Exe. Time (s)	22.7	8.87	37.13	68.7
Ratio	33.0%	12.9%	54.1%	100%

Table 2. Execution time profile of CSWinTT.

	Backbone	Encoder	Decoder	Others	Total
Exe. Time (s)	22.6	100.69	11.41	36.83	171.53
Ratio	13.1%	58.7%	6.7%	21.5%	100%

If we look at closely the time required for each functional block, the execution time of using DNN workloads (backbone and RPN) is shorter than that of others, where others in the tables include image loading and pre-processing. In other words, it has a short use of GPU. Therefore, it is not suitable to apply parallelism on DNN workloads, and because the execution time is short as a whole, small-sized subdivided works are not efficient. Too small-sized works can only cause scheduling overhead.

On the contrary, CSWinTT has a long overall processing time and has a long time to use DNN workloads (backbone, encoder, and decoder). In particular, the encoder block using DNN workloads accounts for almost 60% of the total execution time, so if we apply parallelism to this part, considerable performance gain can be expected.

4.4. Placement of Works Constituting SiamRPN++

In Figure 4, we illustrate the operational flow of the proposed solution architecture with a walk-through example in a sequence of four works. As mentioned above, based on the computation analysis of SiamRPN++, each work is the entire SiamRPN++ tracker itself. D_j^i denotes a DNN workload of $Work_i$ where j does the operational sequence index, and C_j^i means for a CPU-side workload. In this example, as explained earlier, D_j^i and C_j^i are defined offline.

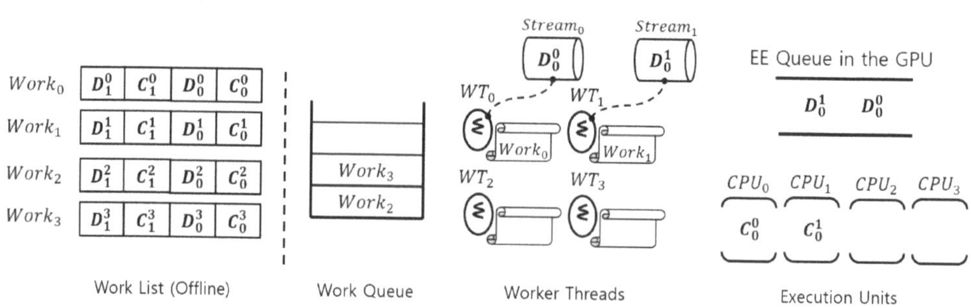

Figure 4. Snapshot of functional blocks in the proposed solution architecture through a walk-through example of SiamRPN++.

In the figure, the execution order of the DNN and CPU workloads inside the work list starts from the right to the left. $Work_0$ and $Work_1$ are already assigned to the worker-threads WT_0 and WT_1, respectively. Thus, C_0^0 and C_0^1, which are the first CPU-side workloads, are running on the two CPUs, and the following DNN workloads D_0^0 and D_1^0 are assigned to their streams. Note that in our design, each worker thread has its own dedicated stream;

$Work_i$ has $Stream_i$. Through the streams, D_0^0 and D_1^0 are enqueued to the EE queue in the GPU, and depending on the SM (streaming multiprocessor) availability inside the GPU, D_0^0 and D_1^0 can be executed simultaneously. Since $Work_2$ and $Work_3$ are still in the work queue, these two works are not assigned to any of the two remaining worker threads WT_2 and WT_3. Once WT_2 pulls the queue from $Work_2$, C_0^2 of WT_2 takes CPU_2 and then starts to assign D_0^2 to $Stream_2$. Even on embedded edge devices with only one GPU, this capability allows multiple trackers to perform their tracking tasks in parallel.

4.5. Placement of Works Constituting CSWinTT

As shown in Table 2, the computational cost that the encoder block dominates in CSWinTT is substantial. On the basis of this profile data, before proceeding further, we closely analyze the encoder blocks with special focus on MHA blocks, which are commonly included in transformer-based DNN models. Figure 5 details multi-head attention MHA_E performed inside the encoder block. The output of the backbone is converted into Q (queries), K (keys), and V (values) tensors through embedding. Each tensor consists of vectors as many as the number of heads and is represented as $Q = Concat(Q_0, Q_1, \cdots, Q_7), K = Concat(K_0, K_1, \cdots, K_7)$ and $V = Concat(V_0, V_1, \cdots, V_7)$, respectively. One of the heads $head_i$ takes Q_i, K_i, V_i as input and outputs i^{th} attention value matrix AVM_i through $Attention(Q_i, K_i, V_i)$ mechanism. Finally, the outputs of each transformer head $(AVM_0, AVM_1, \cdots, AVM_7)$ are concatenated together. To sum up, the final result of MHA_E is obtained as below [7]:

$$MHA_E(Q, K, V) = Concat(AVM_0, AVM_1, \cdots, AVM_7) \quad (2)$$
$$where \quad AVM_i = Attention(Q_i, K_i, V_i) \quad (3)$$
$$= softmax(\frac{Q_i K_i^T}{\sqrt{d_k}}) V_i \quad (4)$$

where d_k means the dimension of key.

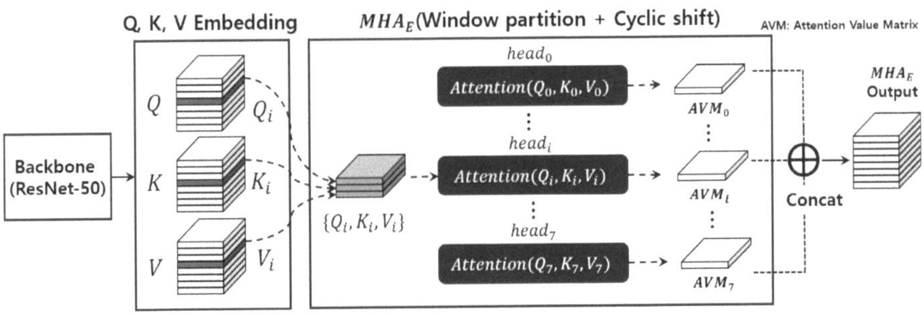

Figure 5. Workflow of multi-head attention in the encoder.

The point to note here is that each head has independent input, and thus all heads in MHA_E can be executed in parallel and independently with each other by model parallelism [24]. Furthermore, as explained earlier, MAH_E includes window partitioning and cyclic shifting, and has a high computational complexity that occupies about 58.7% of the total execution time. Accordingly, we can expect a significant reduction of CSWinTT execution time by processing all the heads in MHA_E in parallel inside the GPU.

For easy understanding, we explain how model parallelism applied to MHA_E described above actually works through a figure. In Figure 6, we illustrate the operational flow of the proposed solution architecture with a walk-through example in a sequence of five works. Different from the case of SiamRPN++, based on the computation analysis of CSWinTT, a work can be the entire CSWinTT tracker itself or one of the transformer heads in MHA of the encoder. In the figure, $head_i$ means the ith transformer head and MHA_E performs multi-head attention in the encoder.

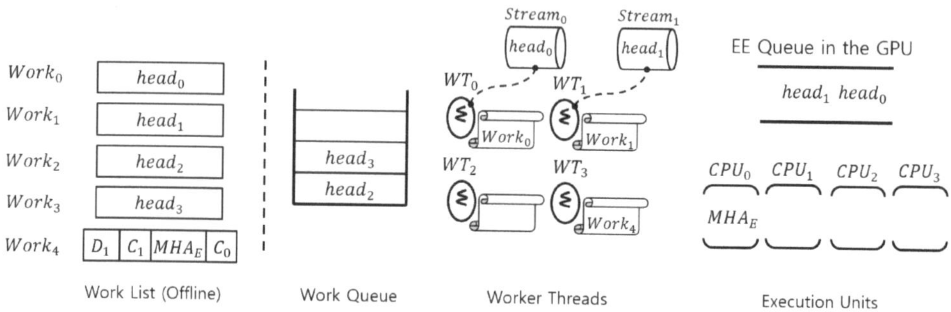

Figure 6. Snapshot of functional blocks in the proposed solution architecture through a walk-through example of CSWinTT.

To demonstrate the effect of model parallelism in the tracker, unlike the case of SiamRPN++, we just take an example of only one CSWinTT tracker. After the CPU-side workload C_0 finishes, WT_3 on CPU_0 performs $Work_4$. Next, MHA_E brings both $Work_0$ and $Work_1$ one after the other from the work queue, assigning transformer heads $head_0$ and $head_1$ to their designated streams. Finally, both $head_0$ and $head_1$ are launched to the EE queue for parallel execution in side of the GPU.

An edge device such as NVIDIA AGX Xavier [25] has eight CPUs and one GPU, and CSWinTT uses eight transformer heads in MHA_E. Therefore, if we run one CSWinTT tracker on that device, Figure 6 is changed to have nine works and nine worker threads. In this case, all transformer heads ($head_0 \sim head_7$) are performed in parallel as much as possible, and then C_1 holds one CPU (possibly CPU_0) to process the obtained result parallelly executed by $head_0 \sim head_7$.

5. Experiments

In this section, we present the experimental verification we have conducted to validate the efficacy of the proposed solution approach. First, we explain implementation method and then elaborate on measurement results along with the relevant analysis.

5.1. Implementation Details

Basically, we take DNN modules built in both SiamRPN++ and CSWinTT trackers from the PyTorch framework [9]. All the threads in the worker-thread pool in Figure 3 are threads of the same process. This multi-threading scheme caters for several benefits in terms of scheduling management; all the functions assembled in all the worker threads are controlled under one single address space, providing the same synchronization primitives and truthful data sharing [5].

DNN modules generated from PyTorch framework are made up of Python-based code, and each module is executed by the Python interpreter. In such an execution environment, GIL (global interpreter lock) enables only one thread to hold the access permission of the Python interpreter, preventing multiple DNN modules from running on several threads [26]. The higher the number of DNN workloads that launch the kernel to the GPU, the more work-conserving the GPU is, so CPU-side multi-threading is indispensable [21]. To clear up the innate constraints of GIL, we propose a new execution methodology and Figure 7 shows the before and after.

In our DNN execution method, to apply the C++-based execution environment, using *TorchScript*, DNN modules of the both trackers are changed into a *ScriptModule* [27]. After combining the libtorch library and *ScriptModule*, we compile it with a C++ compiler *g++*, and then obtain an executable file. Since using the compiled C++-programmed execution file is not controlled by GIL, multi-threaded programming is possible, and accordingly, multiple DNN workloads from DNN modules can be issued to the GPU at the same time.

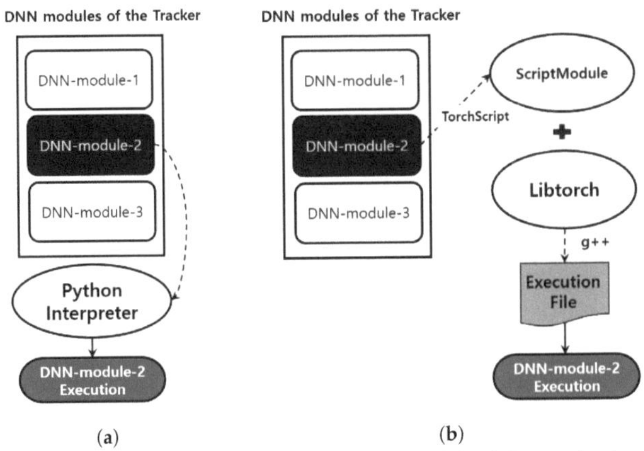

Figure 7. Comparing the execution environment: (**a**) existing mechanism for DNN module execution with Python interpreter and (**b**) proposed execution methodology.

5.2. Experimental Setup

For a more comprehensive verification, we adopt an edge device with limited computing resources and a server computing system, which is the opposite, as the target systems. We take the Jetson AGX Xavier platform [25] as the target edge device, and a GPU server equipped with 4 × NVIDIA RTX A6000 [22] for the target server computing system. The detailed both hardware and software specifications of the target systems are presented in Tables 3 and 4.

Table 3. Specification of the target edge device.

	Classification	Description
HW	CPU	8-core ARM v8.2 Carmel 64-bit CPU, 8 MB L2, 4 MB L3 cache
	GPU	512-core Volta GPU with Tensor cores
	Memory	32 GB 256-Bit LPDDR4x, 137 GB/s
	Storage	32 GB eMMC 5.1
SW	Kernel Ver.	Linux 4.9.140
	SW Package	JetPack 4.2
	CUDA Ver.	CUDA v10.2

Table 4. Specification of the target GPU-server computing system.

	Classification	Description
HW	CPU	16-core, 64 MB L3 cache, 3.9 GHz
	GPU	NVIDIA RTX A6000, 336 Tensor Cores, 10,752 CUDA Cores, 48 GB Memory, 309.7 TFLOPS
	Memory	4 × 64 GB DDR4 PC4
	Storage	1 × SSD 1.92 TBG 2.5″ SATA
SW	Kernel Ver.	Linux 5.15.0
	SW Package	MPI Horovod, NVIDIA GPU Monitoring SW
	CUDA Ver.	CUDA v11.6

As for workloads, TrackingNet [23] with 225,589 images is used. When the trackers are running on the edge device, only 3350 images, randomly selected, out of 225,589 total images of the TrackingNet dataset were used considering the storage space in the target edge device. Whereas for the GPU server, all images of TrackingNet were used.

To demonstrate the usefulness and practicality of the proposed solution approach under various conditions, all the experimental results are from the target edge device as well as from the target GPU server, and we diversified the validation methodologies suitable for each experimental stage. We basically evaluated the proposed solution approach against the original SiamRPN++ and CSWinTT trackers. In each graph, legend *Org.* denotes when the images from TrackingNet are processed by the original SiamRPN++ or CSWinTT trackers without any modifications, whereas *Sol.* is the case under our proposed solution approach.

We checked whether the approach we proposed is less accurate compared to the original trackers and observed how much it was contributing to the improvement of execution speed-up. As for the accuracy measurement, the evaluation metrics, area under the curve (AUC), precision (P), and normalized precision (P_{normal}) were adopted [7,23,28]. Given the ground truth bounding box (BB^{gt}) and the tracked one (BB^{tr}), the success score (i.e., overlap score) is defined as $S = \frac{|BB^{gt} \cap BB^{tr}|}{|BB^{gt} \cup BB^{tr}|}$, where $\cup$ and $\cap$ imply the union and the intersection of BB^{gt} and BB^{tr}, respectively, and $|\cdot|$ means the number of pixels in that area [7,23,28]. The number of frames whose success score S is greater than the given threshold is measured. Using this number, the success plot is obtained to display the proportion of the success frames where the thresholds are ranged from 0 to 1. Ultimately, we can get AUC (area under the curve) from the success plot. The precision is defined as $P = \|C^{tr} - C^{gt}\|_2$, where C^{tr} and C^{gt} denote the centers of the tracker bounding box and the one of ground truth, respectively [7,23,28].

To confirm the speed improvement, in the case of edge devices, the time taken by trackers to process 3350 randomly selected images is compared through graphs, and in the case of GPU servers, the time to process all the images in TrackingNet is measured. In addition, the average frame per second (FPS) result is simultaneously presented on each experimental graph.

5.3. Experimental Results

5.3.1. Inference Accuracy

As shown in Figure 7, DNN modules formed into the trackers are converted into *ScriptModule*. Then *ScriptModule* is compiled together with the libtorch library, creating an execution binary by the proposed execution environment. This may cause an inference accuracy gap compared to the existing interpreter-based method. Thus, to closely examine this, we measure the accuracy in terms of AUC, P, and P_{normal}.

Tables 5 and 6 show the results from running SiamRPN++, and Tables 7 and 8 are the cases for CSWinTT. Overall, comparing the accuracy of the tracker itself, it can be seen that the accuracy of CSwinTT is relatively excellent in both the previous execution environment and the proposed one.

When running SiamRPN++, we can see that the accuracy is slightly improved in three aspects in the proposed execution environment in both edge devices and GPU servers, but they are almost similar.

Table 5. SiamRPN++ running on the target edge device.

	AUC	P	P_{normal}
Org.	77.19	78.1	88.97
Sol.	78.56	80.45	90.32

Table 6. SiamRPN++ running on the target GPU server.

	AUC	P	P_{normal}
Org.	60.92	58.495	70.49
Sol.	63.65	61.9	73.57

The case of CSWinTT showed the opposite result to the SiamRPN++ case. In the case of both edge device and GPU server, the proposed method showed a small accuracy drop. When running on the edge device, AUC decreased by 3.6, P by 6.3, and P_{normal} by 2.576, indicating larger values than in the case of the GPU server.

Table 7. CSWinTT running on the target edge device.

	AUC	P	P_{normal}
Org.	93.32	96.82	97.51
Sol.	93.72	90.52	94.93

Table 8. CSWinTT running on the target GPU server.

	AUC	P	P_{normal}
Org.	90.04	88.63	91.47
Sol.	88.46	87.19	89.78

5.3.2. Inference Speed

Here, we report on the comparison result of the execution speed when two identical trackers are running simultaneously; multiple detection-free trackers are running together. The y-axis of all graphs means the time it takes for the tracker to track all the images used in the experiment. Therefore, the smaller the value, the higher the performance, and of course, the higher the FPS, the higher the execution speed.

Figure 8 displays the result when two identical SiamRPN++ trackers are running together. As can be seen in the figure, when our proposed approach is applied, the FPS increase rate is 32% and 24% on the edge device and GPU server, respectively. As can be seen in Table 1, the percentage of execution time occupied by the backbone and RPN, which are DNN workloads performed on the GPU, is smaller than that of time using the CPU, i.e., CPU dependence is relatively high. Compared with the GPU server, the difference in performance between the CPU and the GPU in the edge device is relatively smaller than that in the GPU server. Therefore, the execution speed improvement in the edge device is about 8% higher.

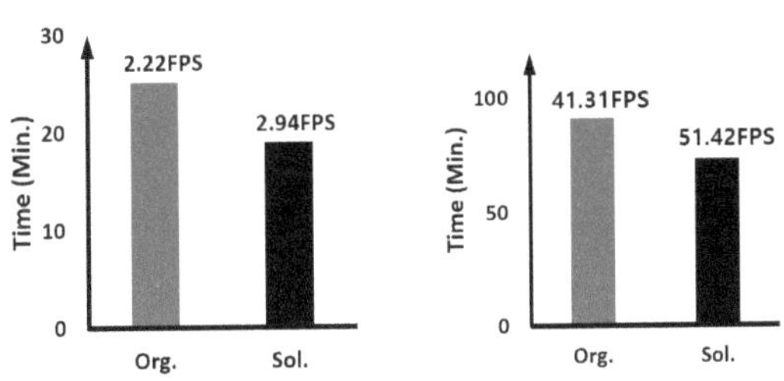

Figure 8. Comparing the execution time and speed when two identical SiamRPN++ trackers are running: (**a**) on the edge device and (**b**) on the GPU server.

Figure 9 shows the result when two identical CSWinTT trackers are running together. As we can see, when the proposed approach is applied, the FPS increase rate is 43% and 55% on the edge device and GPU server, respectively. This is the opposite result from the higher FPS increase rate on the edge device when we experimented with the SiamRPN++ tracker. As shown in Table 2, compared to SiamRPN++, CSWinTT has high computational dependence on the GPU, and MHA accounts for 58.7% of the total computation. Since the GPU server has a GPU with much better parallel processing capability than the embedded GPU of the edge device and the heads of MHA_E use multi-threading and multi-stream, the FPS performance of the GPU server is remarkable.

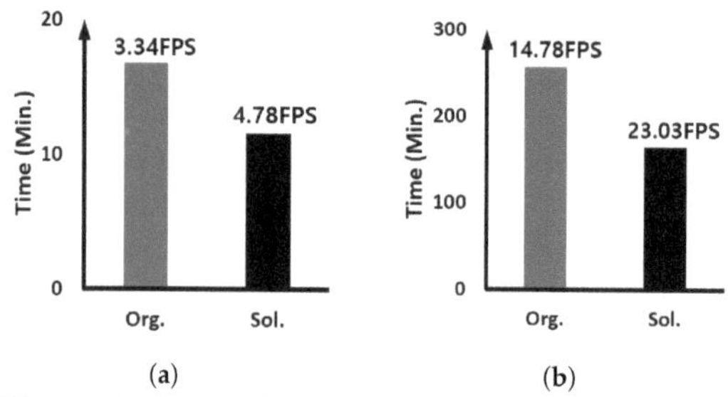

Figure 9. Comparing the execution time and speed when two identical CSWinTT trackers are running: (**a**) on the edge device and (**b**) on the GPU server.

Since MHA_E of CSWinTT itself is parallelized, we verify that the proposed parallel scheduling technique works even when only one CSWinTT model is running. Figure 10 is the result of the experiment. As can be seen through the figure, the degree of FPS performance improvement is more noticeable when only one CSWinTT is running. This means that the case when the GPU embedded in the GPU server we target has 8 head operations of one CSWinTT is more effective compared to when 16 head operations of two CSWinTT trackers are mapped in parallel to SMs inside the GPU through the work queue.

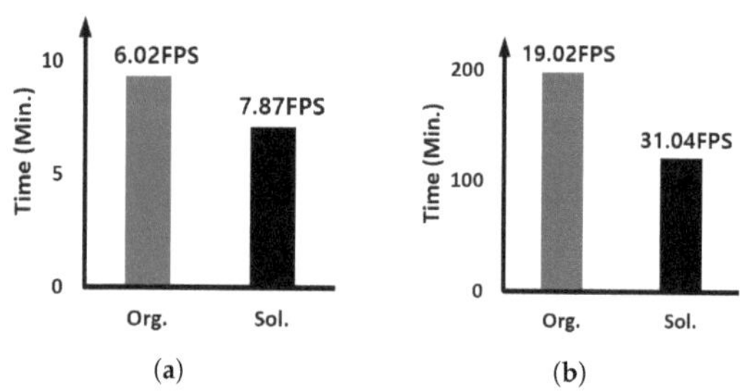

Figure 10. Comparing the execution time and speed when one CSWinTT tracker is running: (**a**) on the edge device and (**b**) on the GPU server.

Next, only the effect when MHA is processed in parallel is verified, and Figures 11 and 12 compare the results with and without the parallel execution of heads in MHA in the case of the edge device and the case for the GPU server, respectively. For the case of the edge device, the FPS increase rate is higher when only one CSWinTT tracker is executed compared to the case with two trackers. However, in case of GPU server, the result was the exact opposite. This result implies that if the internal hardware resource of the GPU is sufficient to perform multiple head operations in parallel, the effect of MHA parallel processing using CUDA stream can be maximized. However, on the edge device using the GPU with the limited hardware resource, the overall effect of multi-threading rather than MHA parallelization is more significant.

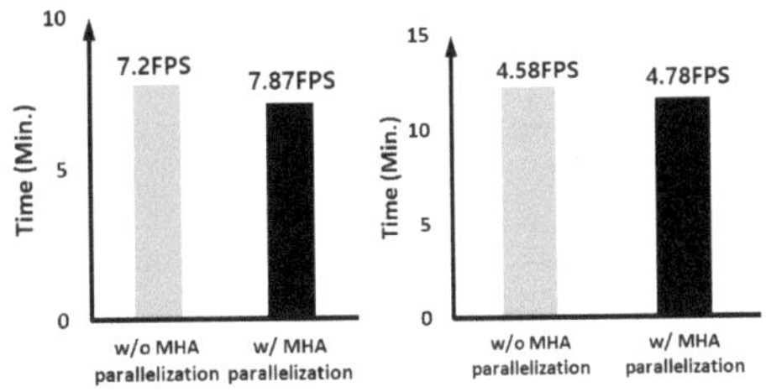

(a) (b)

Figure 11. Comparing the MHA parallelization effect on the edge device: (**a**) 1× CSWinTT and (**b**) 2× CSWinTT.

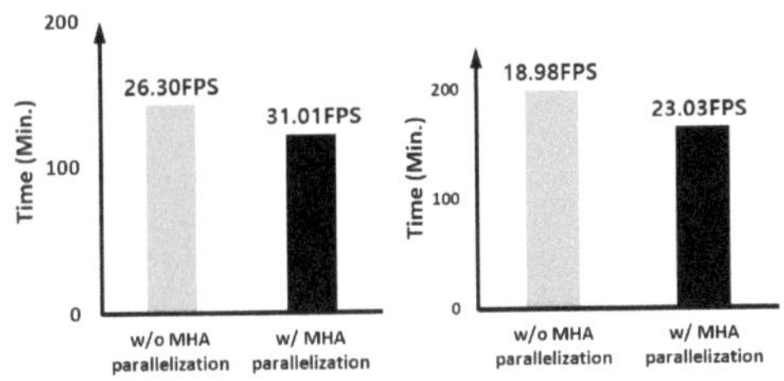

(a) (b)

Figure 12. Comparing the MHA parallelization effect on the GPU server: (**a**) 1× CSWinTT and (**b**) 2× CSWinTT.

6. Conclusions

Object tracking technology is widely used in areas such as crime prevention, facility safety, traffic control, and information collection. Especially, detection-free object tracking technology that can track objects that are not of a predefined class has been highlighted as crucial in these applications. In this paper, we presented a framework that efficiently schedules the workloads inside detection-free trackers to work out the computing-related issues that occur when two or more detection-free-tracking tasks are running simultaneously. To achieve this, first, the computational structures of the Siamese-network-based tracker and the transformer-based tracker, which exhibit excellent tracking performance, are analyzed, and a scheduling unit suitable for each tracker is determined offline. At runtime, multi-threading allows trackers to use multiple CPUs concurrently, delivering multiple DNN workloads included in trackers to the GPU at the same time. By doing so, the GPU is kept work-conserving. As a result of experimental validation, when tracking two user-specified objects, the proposed scheduling framework led to a 55% performance improvement without reducing tracking accuracy.

Author Contributions: Conceptualization, M.K.; Methodology, M.K.; Software, M.K., I.K., J.Y. and H.K.; Validation, I.K., J.Y. and H.K.; Formal analysis, M.K.; Investigation, M.K.; Data curation, I.K., J.Y. and H.K.; Writing—original draft, M.K.; Writing—review & editing, M.K.; Supervision, M.K.; Project administration, M.K.; Funding acquisition, M.K. All authors have read and agreed to the published version of the manuscript.

Funding: This research was financially supported by Hansung University.

Institutional Review Board Statement: Not applicable.

Informed Consent Statement: Not applicable.

Data Availability Statement: Publicly available datasets were analyzed in this study. The data can be found in this link: https://tracking-net.org/ (accessed on 15 September 2022).

Conflicts of Interest: The authors declare no conflicts of interest.

References

1. Liu, L.; Xing, J.; Ai, H.; Ruan, X. Hand posture recognition using finger geometric feature. In Proceedings of the 21st International Conference on Pattern Recognition, Tsukuba, Japan, 11–15 November 2012; pp. 565–568.
2. Tak, S.; Lee, J.D.; Song, J.; Kim, S. Development of AI-Based Vehicle Detection and Tracking System for C-ITS Application. *J. Adv. Transp.* **2021**, *2021*, 4438861. [CrossRef]
3. Xing, J.; Ai, H.; Lao, S. Multiple Human Tracking Based on Multi-view Upper-Body Detection and Discriminative Learning. In Proceedings of the 20th International Conference on Pattern Recognition, Istanbul, Turkey, 23–26 August 2010; pp. 1698–1701.
4. Zhang, G.; Vela, P.A. Good features to Track for Visual SLAM. In Proceedings of the IEEE Conference on Computer Vision and Pattern Recognition, Boston, MA, USA, 7–12 June 2015; pp. 1373–1382.
5. Lim, C.; Kim, M. ODMDEF: On-Device Multi-DNN Execution Framework Utilizing Adaptive Layer-Allocation on General Purpose Cores and Accelerators. *IEEE Access* **2021**, *9*, 85403–85417. [CrossRef]
6. Li, B.; Wu, W.; Wang, Q.; Zhang, F.; Xing, J.; Yan, J. SiamRPN++: Evolution of Siamese Visual Tracking With Very Depp Networks. In Proceedings of the IEEE/CVF Conference on Computer Vision and Pattern Recognition, Long Beach, CA, USA, 15–20 June 2019; pp. 4282–4291.
7. Song, Z.; Yu, J.; Chen, Y.P.C.; Yang, W. Transformer Tracking With Cyclic Shifiting Window Attention. In Proceedings of the IEEE/CVF Conference on Computer Vision and Pattern Recognition, New Orleans, LA, USA, 18–24 June 2022; pp. 8791–8800.
8. TensorFlow. Available online: https://www.tensorflow.org (accessed on 15 February 2023).
9. PyTorch. Available online: https://pytorch.org (accessed on 15 February 2023).
10. Bertinetto, L.; Valmadre, J.; Henriques, J.F.; Vedaldi, A.; Torr, P.H.S. Fully-Convolutional Siamese Networks for Object Tracking. In Proceedings of the Computer Vison—ECCV, Amsterdam, The Netherlands, 8–10 October 2016; pp. 850–865.
11. Li, B.; Yan, J.; Wu, W.; Zhu, Z.; Hu, X. High Performance Visual Tracking With Siamese Region Proposal Network. In Proceedings of the IEEE Conference on Computer Vision and Pattern Recognition, Salt Lake City, UT, USA, 18–23 June 2018; pp. 8971–8980.
12. He, K.; Zhang, X.; Ren, S.; Sun, J. Deep Residual Learning for Image Recognition. In Proceedings of the IEEE Conference on Computer Vision and Pattern Recognition, Las Vegas, NV, USA, 27–30 June 2016; pp. 770–778.
13. Krizhevsky, A.; Sutskever, I.; Hinton, G.E. ImageNet classification with deep convolutional neural networks. *Commun. ACM* **2017**, *60*, 84–90. [CrossRef]
14. Jiang, W.; Trulls, E.; Hosang, J.; Tagliasacchi, A.; Yi, K.M. COTR: Correspondence Transformer for Matching Across Images. In Proceedings of the IEEE/CVF International Conference on Computer Vision, Montreal, BC, Canada, 11–17 October 2021; pp. 6207–6217.
15. Chen, X.; Yan, B.; Zhu, J.; Wang, D.; Yang, X.; Lu, H. Transformer Tracking. In Proceedings of the IEEE/CVF Conference on Computer Vision and Pattern Recognition, Nashville, TN, USA, 20–25 June 2021; pp. 8126–8135.
16. Yan, B.; Peng, H.; Fu, J.; Wang, D.; Lu, H. Learning Saptio-Temporal Transformer for Visual Tracking. In Proceedings of the IEEE/CVF International Conference on Computer Vision, Montreal, BC, Canada, 11–17 October 2021; pp. 10448–10457.
17. Liu, Z.; Lin, Y.; Cao, Y.; Hu, H.; Wei, Y.; Zhang, Z.; Lin, S.; Guo, B. Swin Transformer: Hierarchical Vision Transformer Using Shifted Windows. In Proceedings of the IEEE/CVF International Conference on Computer Vision, Montreal, BC, Canada, 11–17 October 2021; pp. 10021–10022.
18. NVIDIA. CUDA Deep Neural Network(cuDNN) | NVIDIA Developer. Available online: https://developer.nvidia.com/cudnn (accessed on 15 February 2023).
19. NVIDIA. CUDA Toolkit—Free Tools and Training | NVIDIA Developer. Available online: https://developer.nvidia.com/cuda-toolkit (accessed on 15 February 2023).
20. NVIDIA. CUDA Multi-Process Service. Available online: https://docs.nvidia.com/deploy/mps/index.html (accessed on 15 February 2023).
21. Cho, H.; Kim, M. gCFS: Completely fair scheduling on multiple GPUs for improved multi-DNN execution in terms of performance isolation. *J. Supercomput.* **2022**, *79*, 5851–5877. [CrossRef]
22. NVIDIA. NVIDIA RTX 6000 Ada Generation Graphics Card. Available online: https://www.nvidia.com/en-us/design-visualization/rtx-6000/ (accessed on 15 February 2023).
23. Muller, M.; Bibi, A.; Alsubaihi, S.; Ghanem, B. TrackingNet: A Large-Scale Dataset and Benchmark for Object Tracking in the Wild. In Proceedings of the European Conference on Computer Vision, Munich, Germany, 8–14 September 2018; pp. 300–317.
24. Aaron, H.; Deepak, N.; Amar, P.; Vivek, S.; Nikhil, R.; Gregory, R.; Phillip, B. PipeDream: Fast and Efficient Pipeline Parallel DNN Training. *arXiv* **2018**, arXiv:1806.03377.

25. NVIDIA. Jetson AGX Xavier Developer Kit | NVIDIA Developer. Available online: https://developer.nvidia.com/embedded/jetson-agx-xavier-developer-kit (accessed on 15 February 2023).
26. Ajitsaria, A. What Is the Python Global Interpreter Lock (GIL)? Available online: https://realpython.com/python-gil (accessed on 15 February 2023).
27. TorchScript. Available online: https://pytorch.org/docs/master/jit.html (accessed on 15 February 2023).
28. Wu, Y.; Lim, J.; Yang, M.H. Online object tracking: A benchmark. In Proceedings of the IEEE Conference on Computer Vision and Pattern Recognition (CVPR), Portland, OR, USA, 23–28 June 2013; pp. 2411–2418.

sensors

MDPI

Article

A Novel Ranging and IMU-Based Method for Relative Positioning of Two-MAV Formation in GNSS-Denied Environments

Jia Cheng, Peng Ren * and Tingxiang Deng

School of Telecommunications Engineering, Xidian University, Xi'an 710071, China;
21011210141@stu.xidian.edu.cn (J.C.); 21011210317@stu.xidian.edu.cn (T.D.)
* Correspondence: pren@xidian.edu.cn

Abstract: Global Navigation Satellite Systems (GNSS) with weak anti-jamming capability are vulnerable to intentional or unintentional interference, resulting in difficulty providing continuous, reliable, and accurate positioning information in complex environments. Especially in GNSS-denied environments, relying solely on the onboard Inertial Measurement Unit (IMU) of the Micro Aerial Vehicles (MAVs) for positioning is not practical. In this paper, we propose a novel cooperative relative positioning method for MAVs in GNSS-denied scenarios. Specifically, the system model framework is first constructed, and then the Extended Kalman Filter (EKF) algorithm, which is introduced for its ability to handle nonlinear systems, is employed to fuse inter-vehicle ranging and onboard IMU information, achieving joint position estimation of the MAVs. The proposed method mainly addresses the problem of error accumulation in the IMU and exhibits high accuracy and robustness. Additionally, the method is capable of achieving relative positioning without requiring an accurate reference anchor. The system observability conditions are theoretically derived, which means the system positioning accuracy can be guaranteed when the system satisfies the observability conditions. The results further demonstrate the validity of the system observability conditions and investigate the impact of varying ranging errors on the positioning accuracy and stability. The proposed method achieves a positioning accuracy of approximately 0.55 m, which is about 3.89 times higher than that of an existing positioning method.

Keywords: relative positioning; GNSS-denied environments; ranging; IMU; EKF; MAVs

1. Introduction

Micro Aerial Vehicles (MAVs), which refer to small Unmanned Aerial Vehicles (UAVs), have gained popularity in aerial robotics due to their small size, agility, and versatility. It is possible for MAVs to perform a wide range of tasks, including surveillance, inspection, mapping, and environmental monitoring. However, achieving coordinated missions for MAV formations is a challenging task, as it requires precise and reliable navigation and positioning capabilities. Furthermore, such missions involve complex maneuvers and high synchronization between multiple MAVs, and any errors or inaccuracies in navigation or positioning can result in mission failure or even accidents. Therefore, developing robust and efficient navigation and positioning methods is crucial for enabling MAVs to perform coordinated missions effectively and safely.

Urban or forest environments are renowned for their complexity and heterogeneity, which pose significant challenges to wireless communication systems due to issues such as multipath propagation and shadowing [1]. In this context, Hermosilla et al. [2] proposed the use of street-based urban metrics descriptions to quantify various spatial patterns of urban types constructed at different periods, and provided eight different types of urban environments (including urban, residential, and industrial areas) based on their structural characteristics, such as building height distribution or vegetation coverage. These differences may impact wireless signal propagation in different ways. For example, historic urban areas with lower building heights and narrower streets may experience more severe multipath effects, while emerging industrial areas with taller buildings and wider streets may face more severe signal blockage issues. As a result, there are numerous challenges in receiving Global Navigation Satellite System (GNSS) signals in urban or forest environments due to the weak anti-interference ability of GNSS [3]. Despite some related studies [4–6] having reduced the error in GNSS-based positioning systems through multi-information fusion technology, intentional interference may render GNSS unusable in certain specialized fields, such as military applications. Therefore, achieving relative positioning of MAVs in GNSS-denied scenarios has become a potentially fruitful area of research.

The common methods for achieving relative positioning in GNSS-denied environments can be classified into the following three categories: visual positioning [7–17], inertial navigation positioning [18–22], and radio positioning [23,24]. The development status of each category of positioning technology is systematically discussed below, and the advantages and limitations of each technology are analyzed, providing good inspiration for the study of relative positioning methods for MAVs in GNSS-denied environments.

The visual localization methods can be broadly categorized into map-based localization [7–11] and map-free localization [12–17], depending on whether prior visual maps are utilized. (Note that visual refers to visual information, which is typically collected using sensors such as cameras or laser scanners. Visual maps can be understood as maps constructed based on visual information and used for visual localization and navigation.) In the map-based localization, a pre-constructed visual map was used to aid in localization, and the steps involved in map construction and updating, image retrieval, feature point extraction and matching, and precise localization have been studied extensively [7–10]. The main focus of [7] was the construction of 3D maps, while reference [8] addressed map quality issues by removing outliers and tracking lanes. Map matching problems were mainly addressed in [9–11]. In contrast, map-free visual localization methods do not rely on prior visual maps and instead estimate the pose of the object and surrounding environment. This category can be further divided into Visual Simultaneous Localization and Mapping (VSLAM) [12–15] and Structure from Motion (SFM) [16,17]. The VSLAM is designed for real-time processing, making it well-suited for applications such as robotics and autonomous vehicles, while SFM prioritizes accuracy and is more appropriate for offline processing applications such as digital reconstruction of scenes [16]. Specific techniques within these categories have also been put forward. For example, an efficient distributed particle filter (EDPF) was proposed in [12] to address the difficulty of sampling high-dimensional state spaces in range-only SLAM. Reference [13] proposed a weight-optimized particle filter-based algorithm for monocular visual SLAM, which aimed to improve the slow environmental interference repair speed of traditional filtering SLAM algorithms. Three-level parallel optimization was adopted in [14], including the direct method, feature-based method, and pose graph optimization. The method in [15] involves two stages: the first stage implements a local SLAM process based on filtering techniques, while the second stage utilizes optimization-based techniques for constructing and maintaining a consistent global map of the environment, which includes addressing the loop closure problem. To estimate the state of a MAV, the main filtering technique employed is the Extended Kalman Filter (EKF). In terms of optimization techniques, three methods are used: a local bundle adjustment technique to minimize the total reprojection error, the minimization of the Perspective-n-Point (PnP) problem, and graph optimization to correct the global map. In [16], 3D point clouds of close-range images were generated using SFM technology. Homologous features between different images were found to determine the shooting direction and position of each image. Finally, reference [17] used optimization to enhance the convergence rate of SFM by retrieving maximum internal similarity between images. The chosen images and query results were used to reconstruct a three-dimensional scene and estimate relative camera positions with SFM.

The inertial navigation positioning is an autonomous method that provides accurate short-term navigation and positioning without relying on external information or emitting radiation. However, the general positioning system is prone to an error accumulation due to the second-order integration operation [18–22]. Various studies have been conducted to address this issue. For instance, a 3D trajectory planning method based on Particle Swarm Optimization-A star (PSO-A*) algorithm was proposed in [18] to solve the yaw problem of intelligent aircraft. In [19], the nonlinear error model was considered, and internal measurement information was utilized to correct the nonlinear error of inertial navigation. The zero-velocity detection algorithm was adopted to compensate for the accumulated error of pedestrian inertial navigation [20,21]. Additionally, reference [22] employed Convolutional Neural Networks (CNN) to reduce the error of MEMS IMU sensors.

There are various wireless positioning methods based on radio technology, mainly including infrared positioning [23], Ultra Wide Band (UWB) positioning [24], and radio frequency identification (RFID) positioning [25]. In [23], the angle of an object equipped with a positioning device was measured using an infrared laser beam emitted by a lighthouse base station, and the position of the object was calculated. An indoor positioning system based on improved adaptive Kalman filter (IAKF) was proposed in [24] that calculated the distance between the tag and the base stations using the time it took for UWB signals to travel, and then applied a triangulation algorithm to acquire the position of the tag. Similar to [24], reference [25] differs mainly in the method of distance measurement, using RFID technology to calculate the distance. In addition, Ref. [25] mentions that RFID positioning tags can be classified into two types, active and passive, depending on their power requirements. Passive tags mainly involve backscattering communication, and therefore do not require direct power supply, but they have limited communication range compared to active tags. Therefore, RFID technology is mostly used for indoor product tracking. Furthermore, the design, installation, and maintenance of RF navigation systems can be expensive and complex, which may pose challenges for deploying them in certain applications, such as small unmanned aerial vehicles.

In addition, multi-technology fusion is an important means to improve the positioning performance. An important trend in VSLAM is to integrate visual sensor data with other sensor data [26,27]. A method was proposed in [26] to integrate Inertial Measurement Unit (IMU) and dynamic VSLAM, which avoided the static assumption of common SLAM algorithms and solved the problems of fast vehicle motion and insufficient light. This method exhibited higher robustness compared with pure visual dynamic SLAM systems. In [27], a SLAM autonomous positioning algorithm combining magnetometer, IMU, and monocular camera was proposed to address the initialization instability and drift problem in the visual-inertial SLAM (VI-SLAM) algorithm. In wireless positioning, UWB technology is a potentially productive area of research [28–30] in multi-information fusion positioning due to its advantages such as strong penetration ability, low power consumption, small impact of multipath effects, and high positioning accuracy [31–33]. In [28], UWB ranging was used in an indoor positioning scenario, and the position was jointly estimated by fusing the ranging and IMU information through cooperative positioning, which reduced the position drift of Inertial Navigation System (INS). A relative positioning method based on trilateration was proposed for the multi-mobile user mutual positioning scenario [29], where the UWB ranging and IMU information were fused and integrated into a probabilistic framework for cooperative positioning fault recovery. In [30], the pedestrian navigation was realized based on IMU and UWB ranging, and a zero-velocity detection algorithm and single anchor point reference were used. Due to the inherent error drift of IMU, relying solely on IMU for inertial navigation positioning is uncommon. Typically, the fusion of multi-sensor information is required to reduce the positioning errors [26–35].

Authors have two observations on the existing research on positioning methods.

- In the visual-based positioning research, the positioning accuracy is weakened in low-light environments and line-of-sight limitations lead to poor system robustness. These factors pose significant challenges for MAVs to efficiently perform collaborative tasks in diverse environments.
- In the research on IMU-based relative positioning, the multi-sensor fusion is mostly employed to reduce the inertial drift. Solely relying on IMU for dead reckoning cannot achieve long-term high-precision positioning [18–22]. When fusing with radio information, there are restrictions on MAV cooperative formation tasks, such as the existence of zero-velocity detection [31,33] and fixed reference anchor [33]. In addition, there is a situation where two MAVs cannot be positioned [32].

Motivated by the above facts, this paper investigates a relative positioning method of a two-MAV cooperative formation fusing inter-vehicle distance and IMU information in GNSS-denied environments, which does not require the use of other auxiliary devices such as odometers, except for the IMU and the devices used for ranging. The main contributions of this paper are summarized as follows:

1. A novel method for relative positioning in GNSS-denied scenarios is proposed based on ranging and IMU information. The method utilizes the EKF algorithm to jointly estimate the relative positions of MAVs, providing continuous, precise, and reliable information for the formation without being affected by the error accumulation problem of IMU. Additionally, the method is capable of achieving relative positioning for an arbitrary number of nodes without requiring an accurate reference anchor. This innovative approach offers notable advantages over existing methods and has great potential for applications in various fields of aerial robotics.
2. Theoretical derivations of the system observability conditions are presented, along with the specific expressions. The conditions indicate that the system positioning accuracy and reliability can be guaranteed when the flight trajectory of the MAV formation satisfies the observability conditions. Failure to satisfy these conditions may result in decreased positioning accuracy and reliability, as well as a possibility of divergence.
3. Monte Carlo simulations were conducted, where the correctness of the system observability conditions was verified and the effects of different ranging errors on the positioning accuracy and reliability were investigated. Moreover, the positioning error was reduced by approximately 3.89 times compared to an existing positioning method [32].

The rest of this paper is organized as follows. Section 2 describes the system model. Section 3 presents a detailed description of the relative positioning method. Section 4 derives the system observability conditions and provides the specific expressions. Section 5 presents simulation results, followed by Section 6 which concludes this paper.

2. System Model

In this paper, we consider a system model for relative localization of two cooperative MAVs in GNSS-denied environments, as shown in Figure 1. The definitions of the reference coordinate systems are explained as follows. The global coordinate system is the Earth-fixed North-East-Down (NED) coordinate system, denoted as n, and assumed to be an inertial frame. The origin of the body-fixed horizontal coordinate system (denoted as h_i, i = 1, 2) for MAV i (i = 1, 2) is located at its center of gravity, with x-y plane and z-axis paralleling those of n, respectively, meaning that h_i is obtained by rotating the n around its z-axis by a yaw angle φ.

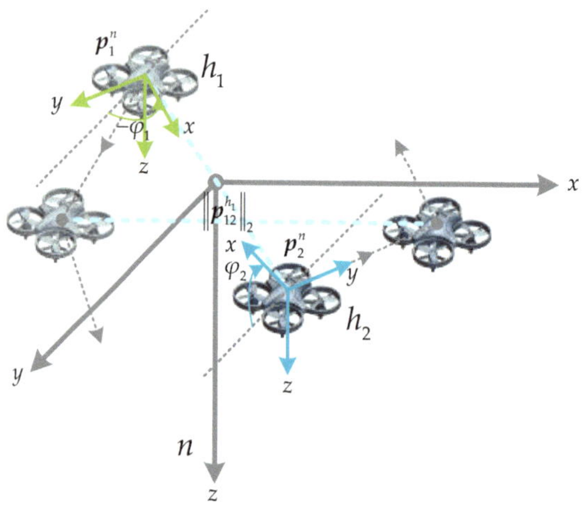

Figure 1. Description of the system model (Gray: NED coordinate system (n); Olive: body-fixed horizontal coordinate system (h_1); Light blue: body-fixed horizontal coordinate system (h_2)).

It is worth noting that this paper does not use the typical body-fixed coordinate system (denoted as b_i, i = 1, 2) for the MAV i, which is represented by Euler angles with respect to n. According to the 321-rotation sequence, the corresponding Euler angles are yaw angle φ, pitch angle θ, and roll angle γ, respectively. The reason for using h_i instead of b_i is to simplify the kinematic relationships and minimize the impact of unnecessary factors, such as near-hovering states with small roll and pitch angles.

In the system model, MAV i can measure its own state variables, which include acceleration, angular velocity, and velocity in h_i. Furthermore, by means of wireless communication, MAV i can also obtain velocity and distance information from the other one in the system framework, where two MAVs utilize inter-vehicle distance information to enhance the accuracy and robustness of relative positioning in the system framework.

The relative motion of two MAVs is described in h. Let $\boldsymbol{p}_i^n$ denote the position vector of MAV i in n, where the subscript i refers to the MAV number and the superscript n indicates the vector is projected onto n. The projection of the relative position of MAV k (k = 1, 2, but $k \neq i$) with respect to MAV i in h_i can be expressed as:

$$\boldsymbol{p}_{ik}^{h_i} = \boldsymbol{C}_n^{h_i}(\boldsymbol{p}_k^n - \boldsymbol{p}_i^n) \tag{1}$$

where $\boldsymbol{C}_n^{h_i}$ is the coordinate transformation matrix from n to h_i (see Equation (2)). It is only dependent on the yaw angle φ_i of MAV i, since x-y plane and z-axis of h_i are parallel to those of n, respectively.

$$\boldsymbol{C}_n^{h_i} = \begin{bmatrix} \cos\varphi_i & -\sin\varphi_i & 0 \\ \sin\varphi_i & \cos\varphi_i & 0 \\ 0 & 0 & 1 \end{bmatrix} \tag{2}$$

It can be inferred from Equations (1) and (2) that the z-component of the relative position $\boldsymbol{p}_{ik}^{h_i}$ corresponds to the difference in height (which can be measured by a barometric altimeter) between MAV i and MAV k in n. Therefore, the system model can be simplified from three-dimensional space to a two-dimensional plane in the horizontal direction.

Some of the parameters and symbols used in this section are shown in Table 1.

Table 1. Descriptions of parameters or symbols in Section 2.

Parameters or Symbols	Descriptions
n	Earth-fixed North-East-Down (NED) coordinate system
h	Body-fixed horizontal coordinate system
b	Body-fixed coordinate system
φ	Yaw angle
θ	Pitch angle
γ	Roll angle
$\lVert\cdot\rVert_2$	2-norm
$p_i^n \in \mathbb{R}^3$	Position vector of MAV i in n
$p_{ik}^{h_i} \in \mathbb{R}^3$	Projection of the relative position of MAV k with respect to MAV i in h_i
$C_n^{h_i} \in \mathbb{R}^{3\times 3}$	Coordinate transformation matrix from n to h_i

3. A Ranging and IMU-Based Relative Positioning Method

We propose a relative positioning method that integrates the inter-vehicle distance with on-board IMU information and employs the EKF algorithm for optimal position estimation. The acquired data is transformed into the corresponding framework to establish the state differential equation. Based on the measurement information, the state of the MAVs is updated. The system observability analysis is conducted before and after the EKF algorithm's prediction and update process to ensure the system positioning accuracy. Specifically, in Section 3.1, the state differential equation between two MAVs is simply derived, and the observation model and the overall system process are given in Sections 3.2 and 3.3, respectively.

3.1. State Differential Model

The nonlinear state vector system for the localization model is defined as:

$$\begin{cases} \dot{\boldsymbol{x}} = \boldsymbol{f}(\boldsymbol{x}, \boldsymbol{u}) \\ \boldsymbol{y} = \boldsymbol{h}(\boldsymbol{x}) \end{cases} \tag{3}$$

where the vectors $\boldsymbol{x} \in \mathbb{R}^n$, $\boldsymbol{u} \in \mathbb{R}^m$, $\boldsymbol{y} \in \mathbb{R}^l$ are the state vector, input vector, and output vector of the system, respectively. $\boldsymbol{f}(\boldsymbol{x}, \boldsymbol{u})$ is the system state differential vector function containing the vector parameters $\boldsymbol{x}$ and $\boldsymbol{u}$, and $\boldsymbol{h}(\boldsymbol{x})$ is the observation equation related to the state vector $\boldsymbol{x}$.

According to the discussion in Section 2, and considering the relative positioning model of two MAVs, let $\boldsymbol{p} = \boldsymbol{R}(\boldsymbol{p}_2 - \boldsymbol{p}_1) \in \mathbb{R}^2$, which represents the projection of the relative position of MAV 2 with respect to MAV 1 in h_1, where $\boldsymbol{R}$ is the two-dimensional case of Equation (2). Specifically, it is equal to:

$$\boldsymbol{R} = \begin{bmatrix} \cos(\Delta\varphi) & -\sin(\Delta\varphi) \\ \sin(\Delta\varphi) & \cos(\Delta\varphi) \end{bmatrix} \tag{4}$$

where $\Delta\varphi = \varphi_2 - \varphi_1$ is the yaw angle difference between MAV 1 and MAV 2.

Moreover, the other variables in the nonlinear system are defined below. The yaw rate difference is denoted by $\Delta\dot{\varphi} = r_2 - r_1$ (where r_i is the yaw rate of MAV i, i = 1, 2). The projections of velocity and acceleration of MAV i in h_i are denoted as $\boldsymbol{v}_i \in \mathbb{R}^2$ and $\boldsymbol{a}_i \in \mathbb{R}^2$, respectively.

It should be noted that the yaw rate r_i and acceleration $\boldsymbol{a}_i$ mentioned earlier refer to the horizontal plane components in h_i and cannot be equated simply with the actual measured values $\boldsymbol{\omega}_i \in \mathbb{R}^3$ and $\boldsymbol{s}_i \in \mathbb{R}^3$ of the gyroscope and accelerometer. In particular, the latter must undergo a conversion process, which can be expressed as:

$$r_i = \frac{\sin\gamma_i}{\cos\theta_i}\cdot\omega_{iy} + \frac{\cos\gamma_i}{\cos\theta_i}\cdot\omega_{iz},\ i = 1,\ 2 \tag{5}$$

$$a_i = \begin{bmatrix} \cos\theta_i & \sin\gamma_i \sin\theta_i & \cos\gamma_i \sin\theta_i \\ 0 & \cos\gamma_i & -\sin\gamma_i \end{bmatrix} \cdot s_i,\ i = 1,\ 2 \tag{6}$$

where ω_{iy} and ω_{iz} represent the y-axis component (true pitch rate) and z-axis component (true yaw rate) of ω_i, respectively. γ_i and θ_i are the roll angle and pitch angle of MAV i, respectively. Proofs of Equations (5) and (6) are provided in Appendix A for reference.

Let the state and input vector be $x = \left[p^T, \Delta\varphi, v_1^T, v_2^T\right]^T \in \mathbb{R}^7$ and $u = \left[r_1, r_2, a_1^T, a_2^T\right]^T \in \mathbb{R}^6$, respectively. The state differential equation of the system is simply derived as follows.

The derivative of the relative position p with respect to time t is (Note that in Equation (7), the coordinate transformation matrix R does not act on v_1. The reason is that, in the scenario of mutual positioning between two MAVs, when calculating the relative position of MAV 2 with respect to MAV 1 in h_1, there is no need to transform the velocity of MAV 1 in its own coordinate system h_1 by left-multiplying R):

$$\begin{aligned} \frac{dp}{dt} &= \frac{dR(p_2 - p_1)}{dt} \\ &= \frac{dR}{dt}(p_2 - p_1) + R\frac{d(p_2 - p_1)}{dt} \\ &= -r_i^{\times} R(p_2 - p_1) + Rv_2 - v_1 \\ &= -r_1^{\times} p + Rv_2 - v_1 \end{aligned} \tag{7}$$

where the antisymmetric matrix $r_i^{\times}$ of cross product in two-dimensional case is equal to:

$$r_i^{\times} = \begin{bmatrix} 0 & -r_i \\ r_i & 0 \end{bmatrix},\ i = 1,\ 2 \tag{8}$$

v_i is obtained by coordinate transformation of $v_i^n \in \mathbb{R}^2$ in n, which is equal to $v_i = Rv_i^n$. Taking the derivative of the velocity v_i with respect to time t:

$$\begin{aligned} \frac{dv_i}{dt} &= \frac{dRv_i^n}{dt} \\ &= \frac{dR}{dt}v_i^n + R\frac{dv_i^n}{dt} \\ &= -r_i^{\times} Rv_i^n + a_i \\ &= -r_i^{\times} v_i + a_i \end{aligned} \tag{9}$$

According to Equations (7) and (9), and the definition of the yaw rate difference $\Delta\dot{\varphi}$, the state differential equation in Equation (3) can be written as:

$$\dot{x} = f(x, u) = \begin{bmatrix} -r_1^{\times} p + Rv_2 - v_1 \\ r_2 - r_1 \\ -r_1^{\times} v_1 + a_1 \\ -r_2^{\times} v_2 + a_2 \end{bmatrix} \in \mathbb{R}^7 \tag{10}$$

3.2. Observation Model

The observations involved in the positioning model are the distance information $\|p\|_2$ between two MAVs and the velocity v_i of MAV i (where i = 1, 2). Converting the distance information to $\|p\|_2^2/2$ can simplify the system observability theoretical analysis in Section 4 without affecting the analysis results [36]. However, the distance observation is still selected as $\|p\|_2$ in the experimental verification. The corresponding observation model is expressed as:

$$y = h(x) = \begin{bmatrix} \|p\|_2^2/2 \\ v_1 \\ v_2 \end{bmatrix} \in \mathbb{R}^5 \tag{11}$$

3.3. Overall System Process

The system process as a whole is depicted in Figure 2. In the system framework illustrated in Figure 1, the MAVs acquire their motion states through onboard sensors, encompassing acceleration, angular velocity, and velocity, which are transformed into h_1 via transformation Blocks T1, T2, and T3, respectively.

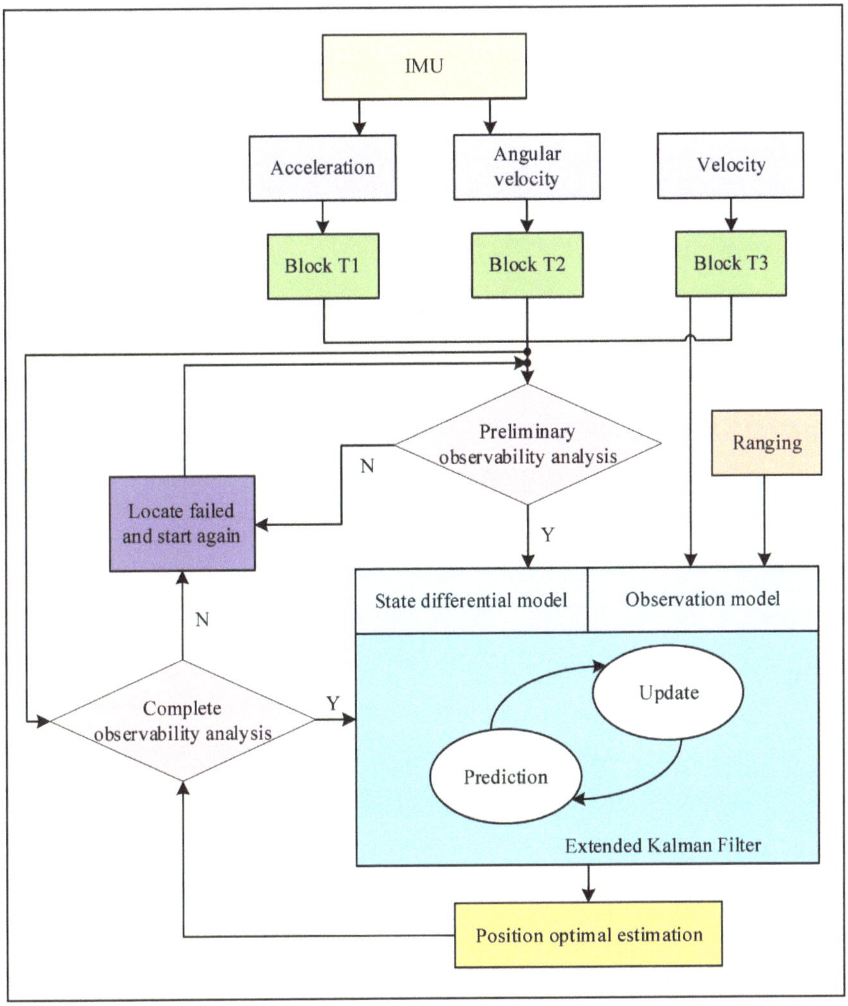

Figure 2. The overall system process (including information acquisition and transformation, preliminary observability analysis, EKF algorithm, and complete system observability analysis).

Subsequently, a preliminary observability analysis is conducted to determine whether the observability conditions presented in Equations (30)–(32) of Section 4 are satisfied. Provided that the conditions are met, the positioning accuracy is adequately guaranteed. Otherwise, the system fails to localize and awaits the next sampling interval to retry localization.

One important aspect of the system process is that the state differential equation is constructed based on the MAVs' states. In addition, the real-time distance measurement between two MAVs and the target MAV's velocity are obtained through radio communication, with the measurement equation constructed accordingly. The EKF algorithm is then used to estimate the optimal relative position of the target MAV.

A complete system observability analysis is performed based on the optimal position estimation and the MAVs' states, as shown in Equation (29). The result is expected to be similar to that of the preliminary observability analysis, with the only difference being the expression of the conditions.

Some of the parameters and symbols used in this section are shown in Table 2.

Table 2. Descriptions of parameters or symbols in Section 3.

Parameters or Symbols	Descriptions
$x \in \mathbb{R}^7$	State vector
$u \in \mathbb{R}^6$	Input vector
$y \in \mathbb{R}^5$	Output vector
$\dot{x} = f(\cdot)$	State differential vector function
$h(\cdot)$	Observation function
$p \in \mathbb{R}^2$	Projection of the relative position of MAV 2 with respect to MAV 1 in h_1
$R \in \mathbb{R}^{2\times 2}$	Two-dimensional case of $C_n^{h_i}$
$\Delta\varphi$	Yaw angle difference between MAV 1 and MAV 2
$\Delta\dot{\varphi}$	Yaw rate difference between MAV 1 and MAV 2
r_i	Yaw rate of MAV i
$v_i \in \mathbb{R}^2$	Projection of velocity of MAV i in h_i
$a_i \in \mathbb{R}^2$	Projection of acceleration in h_i
$\omega_i \in \mathbb{R}^3$	Gyroscope measurement
$s_i \in \mathbb{R}^3$	Accelerometer measurement
$r_i^\times \in \mathbb{R}^{2\times 2}$	Antisymmetric matrix of cross product
$v_i^n \in \mathbb{R}^2$	Velocity of MAV i in n

4. System Observability Analysis

Considering the nonlinear state vector system shown in Equation (3), the system observability is analyzed by means of Lie derivative. The multiple Lie derivatives are defined as follows:

$$L_f^0 h = h \tag{12}$$

$$L_f^i h = L_f\left(L_f^{i-1} h\right) = J\left(L_f^{i-1} h\right) \cdot f,\ i \in \mathbb{N}^* \tag{13}$$

where $J\left(L_f^i h\right)$ represents the Jacobi matrix of $L_f^i h$. Furthermore, the observability matrix of the nonlinear system is:

$$H = \begin{bmatrix} J\left(L_f^0 h\right) \\ J\left(L_f^1 h\right) \\ \vdots \\ J\left(L_f^i h\right) \end{bmatrix},\ i \in \mathbb{N}^* \tag{14}$$

When the observable matrix H has full rank, the nonlinear system is considered locally weakly observable [37].

The first term of the observability matrix H is equal to:

$$J\left(L_f^0 h\right) = J(h) = \begin{bmatrix} p^T & 0 & 0_{1\times 2} & 0_{1\times 2} \\ 0_{2\times 2} & 0_{2\times 1} & E_2 & 0_{2\times 2} \\ 0_{2\times 2} & 0_{2\times 1} & 0_{2\times 2} & E_2 \end{bmatrix} = \begin{bmatrix} p^T & 0 & 0_{1\times 4} \\ 0_{4\times 2} & 0_{4\times 1} & E_4 \end{bmatrix} \tag{15}$$

Since the cross term of the last four rows and columns in the first term of the observability matrix $\boldsymbol{H}$ is the unit matrix, the rank of $\boldsymbol{H}$ cannot be increased by its corresponding observation functions. Therefore, only the distance observation corresponding to the first row in Equation (11), denoted as $h_1(\boldsymbol{x}) = \|\boldsymbol{p}\|_2^2/2$, needs to be considered.

The first-order Lie derivative corresponding to the distance observation $h_1(\boldsymbol{x})$ is equal to:

$$L_f^1 h_1 = J\left(L_f^0 h_1\right) \cdot f = \boldsymbol{p}^T\left(-\boldsymbol{r}_1^{\times} \boldsymbol{p} + \boldsymbol{R}\boldsymbol{v}_2 - \boldsymbol{v}_1\right) \tag{16}$$

Calculating the Jacobi matrix of the first-order Lie derivative, the second term of the observability matrix $\boldsymbol{H}$ can be obtained:

$$J\left(L_f^1 h_1\right) = \begin{bmatrix} (\boldsymbol{R}\boldsymbol{v}_2 - \boldsymbol{v}_1) \\ \boldsymbol{p}^T \frac{\partial \boldsymbol{R}}{\partial \Delta\varphi} \boldsymbol{v}_2 \\ -\boldsymbol{p} \\ \boldsymbol{R}^T \boldsymbol{p} \end{bmatrix}^T \tag{17}$$

Given that the full rank of the observability matrix $\boldsymbol{H}$ is seven and the first term $J\left(L_f^0 \boldsymbol{h}\right)$ has rank five, calculation of the second-order Lie derivative is necessary. The second-order Lie derivative is given by:

$$\begin{aligned} L_f^2 h_1 &= J\left(L_f^1 h_1\right) \cdot f \\ &= (\boldsymbol{R}\boldsymbol{v}_2 - \boldsymbol{v}_1)^T\left(-\boldsymbol{r}_1^{\times} \boldsymbol{p} + \boldsymbol{R}\boldsymbol{v}_2 - \boldsymbol{v}_1\right) + \boldsymbol{p}^T \frac{\partial \boldsymbol{R}}{\partial \Delta\varphi} \boldsymbol{v}_2 (r_2 - r_1) \\ &\quad -\boldsymbol{p}^T\left(-\boldsymbol{r}_1^{\times} \boldsymbol{v}_1 + \boldsymbol{a}_1\right) + \boldsymbol{p}^T \boldsymbol{R}\left(-\boldsymbol{r}_2^{\times} \boldsymbol{v}_2 + \boldsymbol{a}_2\right) \\ &= \boldsymbol{v}_2^T \boldsymbol{v}_2 - 2\boldsymbol{v}_1^T \boldsymbol{R}\boldsymbol{v}_2 + \boldsymbol{v}_1^T \boldsymbol{v}_1 + \boldsymbol{p}^T \frac{\partial \boldsymbol{R}}{\partial \Delta\varphi} \boldsymbol{v}_2 r_2 - \boldsymbol{p}^T \boldsymbol{a}_1 + \boldsymbol{p}^T \boldsymbol{R}\left(-\boldsymbol{r}_2^{\times} \boldsymbol{v}_2 + \boldsymbol{a}_2\right) \end{aligned} \tag{18}$$

where the simplification is achieved using Equations (19) and (20). The result shows that the yaw rate r_1 of MAV 1 is completely cancelled out.

$$\boldsymbol{R}^T \boldsymbol{R} = \boldsymbol{E}_2 \tag{19}$$

$$\boldsymbol{R}^T \boldsymbol{r}_1^{\times} = -\frac{\partial \boldsymbol{R}^T}{\partial \Delta\varphi} r_1 \tag{20}$$

The third term of the observability matrix $\boldsymbol{H}$ is the Jacobi matrix of the second-order Lie derivative (see Equation (18)), which can be expressed as:

$$J\left(L_f^2 h_1\right) = \begin{bmatrix} \frac{\partial L_f^2 h_1}{\partial \boldsymbol{p}} & \frac{\partial L_f^2 h_1}{\partial \Delta\varphi} & \frac{\partial L_f^2 h_1}{\partial \boldsymbol{v}_1} & \frac{\partial L_f^2 h_1}{\partial \boldsymbol{v}_2} \end{bmatrix} \tag{21}$$

where the specific expressions of each term are derived simply as follows:

$$\begin{aligned} \frac{\partial L_f^2 h_1}{\partial \boldsymbol{p}} &= \boldsymbol{v}_2^T\left(\frac{\partial \boldsymbol{R}^T}{\partial \Delta\varphi} r_2 - \boldsymbol{r}_2^{\times T} \boldsymbol{R}^T\right) - \boldsymbol{a}_1^T + \boldsymbol{a}_2^T \boldsymbol{R}^T \\ &= -\boldsymbol{a}_1^T + \boldsymbol{a}_2^T \boldsymbol{R}^T \end{aligned} \tag{22}$$

$$\begin{aligned} \frac{\partial L_f^2 h_1}{\partial \Delta\varphi} &= -2\boldsymbol{v}_2^T \frac{\partial \boldsymbol{R}^T}{\partial \Delta\varphi} \boldsymbol{v}_1 + \boldsymbol{v}_2^T\left(\frac{\partial^2 \boldsymbol{R}^T}{\partial \Delta\varphi^2} r_2 - \boldsymbol{r}_2^{\times T} \frac{\partial \boldsymbol{R}^T}{\partial \Delta\varphi}\right)\boldsymbol{p} + \boldsymbol{a}_2^T \frac{\partial \boldsymbol{R}^T}{\partial \Delta\varphi} \boldsymbol{p} \\ &= -2\boldsymbol{v}_2^T \frac{\partial \boldsymbol{R}^T}{\partial \Delta\varphi} \boldsymbol{v}_1 + \boldsymbol{a}_2^T \frac{\partial \boldsymbol{R}^T}{\partial \Delta\varphi} \boldsymbol{p} \end{aligned} \tag{23}$$

$$\frac{\partial L_f^2 h_1}{\partial \boldsymbol{v}_1} = -2\boldsymbol{v}_2^T \boldsymbol{R}^T + 2\boldsymbol{v}_1^T \tag{24}$$

$$\begin{aligned} \frac{\partial L_f^2 h_1}{\partial \boldsymbol{v}_2} &= 2\boldsymbol{v}_2^T - 2\boldsymbol{v}_1^T \boldsymbol{R} + \boldsymbol{p}^T\left(\frac{\partial \boldsymbol{R}}{\partial \Delta\varphi} r_2 - \boldsymbol{R}\boldsymbol{r}_2^{\times}\right) \\ &= 2\boldsymbol{v}_2^T - 2\boldsymbol{v}_1^T \boldsymbol{R} \end{aligned} \tag{25}$$

It can be seen from the above simplified results that the yaw rate r_2 of MAV 2 is completely offset. In the event that the combination matrix $\boldsymbol{A}$ consisting of the Jacobi matrixes of the zero-order, first-order, and second-order Lie derivatives corresponding to the distance observation $h_1(\boldsymbol{x})$ has full rank, the rank of observability matrix $\boldsymbol{H}$ is guaranteed to have full rank, indicating that the nonlinear system is observable.

$$\boldsymbol{A} = \begin{bmatrix} \boldsymbol{p}^T & 0 \\ \boldsymbol{v}_2^T\boldsymbol{R}^T - \boldsymbol{v}_1^T & \boldsymbol{p}^T\frac{\partial \boldsymbol{R}}{\partial \Delta\varphi}\boldsymbol{v}_2 \\ -\boldsymbol{a}_1^T + \boldsymbol{a}_2^T\boldsymbol{R}^T & -2\boldsymbol{v}_1^T\frac{\partial \boldsymbol{R}}{\partial \Delta\varphi}\boldsymbol{v}_2 + \boldsymbol{p}^T\frac{\partial \boldsymbol{R}}{\partial \Delta\varphi}\boldsymbol{a}_2 \end{bmatrix} \in \mathbb{R}^{3\times 3} \tag{26}$$

In light of the above discussion, the observable system needs to satisfy the following condition:

$$|\boldsymbol{A}| \neq 0 \tag{27}$$

Multiplying each element in the third column of determinant $|\boldsymbol{A}|$ with the corresponding algebraic cofactor, adding and expanding to calculate, we can obtain the following expression:

$$\begin{aligned} |\boldsymbol{A}| &= -\boldsymbol{p}^T\frac{\partial \boldsymbol{R}}{\partial \Delta\varphi}\boldsymbol{v}_2 \cdot \left(-\boldsymbol{a}_1^T + \boldsymbol{a}_2^T\boldsymbol{R}^T\right)\boldsymbol{R}_{\Delta\varphi=\pi/2}\boldsymbol{p} \\ &\quad + \left(-2\boldsymbol{v}_1^T\frac{\partial \boldsymbol{R}}{\partial \Delta\varphi}\boldsymbol{v}_2 + \boldsymbol{p}^T\frac{\partial \boldsymbol{R}}{\partial \Delta\varphi}\boldsymbol{a}_2\right) \cdot \left(\boldsymbol{v}_2^T\boldsymbol{R}^T - \boldsymbol{v}_1^T\right)\boldsymbol{R}_{\Delta\varphi=\pi/2}\boldsymbol{p} \\ &= \left(\boldsymbol{p}^T\frac{\partial \boldsymbol{R}}{\partial \Delta\varphi}\left(\boldsymbol{v}_2\boldsymbol{a}_1^T - \boldsymbol{a}_2\boldsymbol{v}_1^T\right) - 2\boldsymbol{v}_1^T\frac{\partial \boldsymbol{R}}{\partial \Delta\varphi}\left(\boldsymbol{v}_2\boldsymbol{v}_2^T\boldsymbol{R}^T - \boldsymbol{v}_2\boldsymbol{v}_1^T\right)\right)\boldsymbol{R}_{\Delta\varphi=\pi/2}\boldsymbol{p} \end{aligned} \tag{28}$$

Considering the properties of matrix $\boldsymbol{R}_{\Delta\varphi=\pi/2}$, when $\boldsymbol{p} \neq \boldsymbol{0}$ and Equation (28) satisfies the following inequality (where m is an arbitrary constant), Equation (27) holds, and the nonlinear system is observable.

$$\boldsymbol{p}^T\frac{\partial \boldsymbol{R}}{\partial \Delta\varphi}\left(\boldsymbol{v}_2\boldsymbol{a}_1^T - \boldsymbol{a}_2\boldsymbol{v}_1^T\right) - 2\boldsymbol{v}_1^T\frac{\partial \boldsymbol{R}}{\partial \Delta\varphi}\left(\boldsymbol{v}_2\boldsymbol{v}_2^T\boldsymbol{R}^T - \boldsymbol{v}_2\boldsymbol{v}_1^T\right) \neq m\boldsymbol{p}^T \tag{29}$$

While the inequality (29) is not intuitive in revealing the motion constraints of MAVs, it can be used to extract some more obvious conditions (as seen in Equations (30)–(32)), which greatly aid our comprehension of Equation (29).

$$\boldsymbol{p} \neq \boldsymbol{0} \tag{30}$$

$$\boldsymbol{v}_i \neq \boldsymbol{0} \text{ or } \boldsymbol{a}_i \neq \boldsymbol{0},\ i = 1,\ 2 \tag{31}$$

$$\boldsymbol{v}_1 \neq n\boldsymbol{R}\boldsymbol{v}_2 \text{ or } (\boldsymbol{a}_1 \neq \boldsymbol{0} \text{ or } \boldsymbol{a}_2 \neq \boldsymbol{0}) \tag{32}$$

Equation (30) serves as a prerequisite for Equation (29), which means that the relative position between two MAVs cannot be equal to zero. Equation (31) indicates that the MAVs cannot remain stationary. Both the conditions are obvious. Equation (32) shows that two MAVs cannot fly in parallel unless at least one of the MAVs has a non-zero acceleration, where n is an arbitrary constant.

According to the system observability analysis in this section, we derived the motion conditions that must be satisfied by MAVs (see Equations (29)–(32)), which means if the relative motion between two MAVs in the system violates the observability conditions, the positioning accuracy and reliability of MAVs cannot be guaranteed. In such a case, by actively intervening in the relative motion of MAVs to satisfy the observability conditions, MAVs in the system can still achieve mutual positioning.

Some of the parameters and symbols used in this section are shown in Table 3.

Table 3. Descriptions of parameters or symbols in Section 4.

Parameters or Symbols	Descriptions
$L_f^i h$	The i-th order Lie derivative of $\boldsymbol{h}$ with respect to f
$J(\cdot)$	Jacobi matrix
$\boldsymbol{H}$	Observability matrix
$\boldsymbol{E}_i$	The i-th order identity matrix
∂	Partial derivative
$A \in \mathbb{R}^{3\times 3}$	Combination matrix
$\lvert\cdot\rvert$	Determinant

5. Results

In this section, we employ computer simulations to demonstrate the performance of the proposed relative positioning method. Section 5 is structured as follows: Section 5.1 describes the experimental setup, followed by Section 5.2 which verifies the system observability conditions. In Section 5.3, we investigate the impact of different ranging errors on the positioning accuracy and stability. Finally, an error comparison experiment with an existing method is conducted in Section 5.4.

5.1. Experimental Setup

Assuming two MAVs, A and B, in the system framework depicted in Figure 1, we use the proposed relative positioning method to calculate the position of MAV B relative to MAV A. The specific parameter settings for the simulation experiment are listed in Table 4, where deg $= \pi/180$ and $mg = 9.8 \times 10^{-3}$.

Table 4. Parameter settings.

Parameters/(Unit)	Values
Filtering step/(s)	0.1
Total simulation duration/(s)	1000
Constant gyroscope drift/(rad/s)	0.05 deg/3600
Constant accelerometer bias/$(\mathrm{m/s^2})$	0.1 mg
Barometric altimeter error/(m)	1
Ranging error/(m)	2
Velocity error/(m/s)	0.1
Noise matrix	$diag(0, 0, 10^{-8}, 10^{-6}, 10^{-6}, 10^{-6}, 10^{-6})$
Initial error	$[2, 2, 2\mathrm{deg}, 0.1, 0.1, 0.1, 0.1]^T$

Explanations for certain simulation parameters in Table 4 are provided below. The gyroscope and accelerometer parameters are typical electrical parameters of commercial IMU inertial sensors (such as MPU6050) commonly available on the market [38]. The error of the barometric altimeter was referenced from [39], where an ultrahigh resolution pressure sensor based on percolative metal nanoparticle arrays was designed. The sensor has an ultra-high resolution of 0.5 Pa and high sensitivity of 0.13 kPa^{-1}, and the pressure range and sensitivity can be adjusted by changing the thickness of the PET membrane, which extends the working pressure range to 40 kPa. In actual altitude tests, the altitude measurement sensitivity was calculated as $-0.00025\ \mathrm{m}^{-1}$ according to the slope of the response curve. RMS noise analysis shows that the accuracy of the sensor can be as low as 1 m. As a high-precision barometric sensor, the sensor can be used for high-resolution barometric altimeters (Note that it is possible to have different pressure at the same height but at different horizontal locations. However, in this work, we mainly focus on the relative positioning of two MAVs, where two MAVs are generally not distant from each other. Thus, it is reasonable to assume that the pressure difference of two MAVs only comes from the altitude level) [39]. The ranging error in this paper is referenced from [40], which proposed an improved through-the-wall (TTW) NLOS ranging method using UWB technology to achieve ranging errors of 0–2 m in NLOS environments. In addition, the noise used in the simulation is close to actual noise, and an adaptive adjustment method is employed for the noise matrix of the system to better adapt to the changes in noise characteristics and maintain good filtering performance.

In accordance with the system observability analysis presented in Section 4, the flight trajectory of the two MAVs needs to satisfy Equation (29) to guarantee the system positioning accuracy and reliability. A trajectory satisfying the requirements of Equation (29), called Trajectory 1, can be defined as follows: the straight-line AB rotates around the geometric center O in the horizontal plane, while O undergoes sinusoidal motion. The MAV height is measured using a barometric altimeter. The motion parameters are listed in Table 5. Setting the rotational angular velocity of AB and sinusoidal angular velocity of O to zero in Table 5, results in a straight-line flight of the MAVs in parallel, which is denoted as Trajectory 2.

Table 5. Motion parameter settings.

Parameters/(Unit)	Values
Rotational angular velocity of $AB/(\mathrm{rad/s})$	0.03
Sinusoidal angular velocity of $O/(\mathrm{rad/s})$	0.03
Linear velocity of $O/(\mathrm{m/s})$	20
Initial angle of MAV $A/(\mathrm{rad})$	$\pi/3$
Initial angle of MAV $B/(\mathrm{rad})$	π

5.2. Verification of the System Observability Conditions

Section 5.2 aims to validate the system observability conditions derived in Section 4 and evaluate the effectiveness of the proposed relative positioning method.

In Section 5.1 of the experimental setup, two trajectories for the MAVs are defined: Trajectory 1 and Trajectory 2. Trajectory 1 satisfies the system observability condition shown in Equation (29), while Trajectory 2 fails to comply with it. More specifically, Trajectory 2 violates the observability sub-condition shown in Equation (32).

Figure 3 shows the filtered error results for the two trajectories, while the corresponding relative motion trajectories are presented in Figure 4. The average filtering errors for x and y axes of Trajectory 1 are approximately 1.22 m and 0.57 m, respectively, with the errors gradually converging to zero over time. In contrast, Trajectory 2 has significantly larger average filtering errors of 6.23 m and 3.86 m for x and y axes, respectively, exceeding those of Trajectory 1 by factors of 5.1 and 6.77. Furthermore, the errors of Trajectory 2 tend to diverge over time. These results indicate that system observability conditions are critical for ensuring the higher positioning accuracy and reliability, whereas failing to meet it, the system accuracy and reliability cannot be guaranteed.

For the purpose of avoiding experimental randomness, Monte Carlo simulation experiments were conducted to investigate the error statistical characteristics of Trajectory 1 and Trajectory 2. Figure 5 presents the results of 100 Monte Carlo simulation experiments for error statistics. The upper and lower dashed lines of the error curves represent the $+3\sigma$ and -3σ boundaries, respectively, while the solid lines represent the error mean. The boundary values represent the positioning accuracy, and the boundary range represents the positioning stability. The specific data for Figure 5 is listed in Table 6. It can be concluded from Figure 5 and Table 6 that the positioning accuracy and stability of Trajectory 2 are worse than those of Trajectory 1, and the positioning error of Trajectory 2 exhibits a diverging trend.

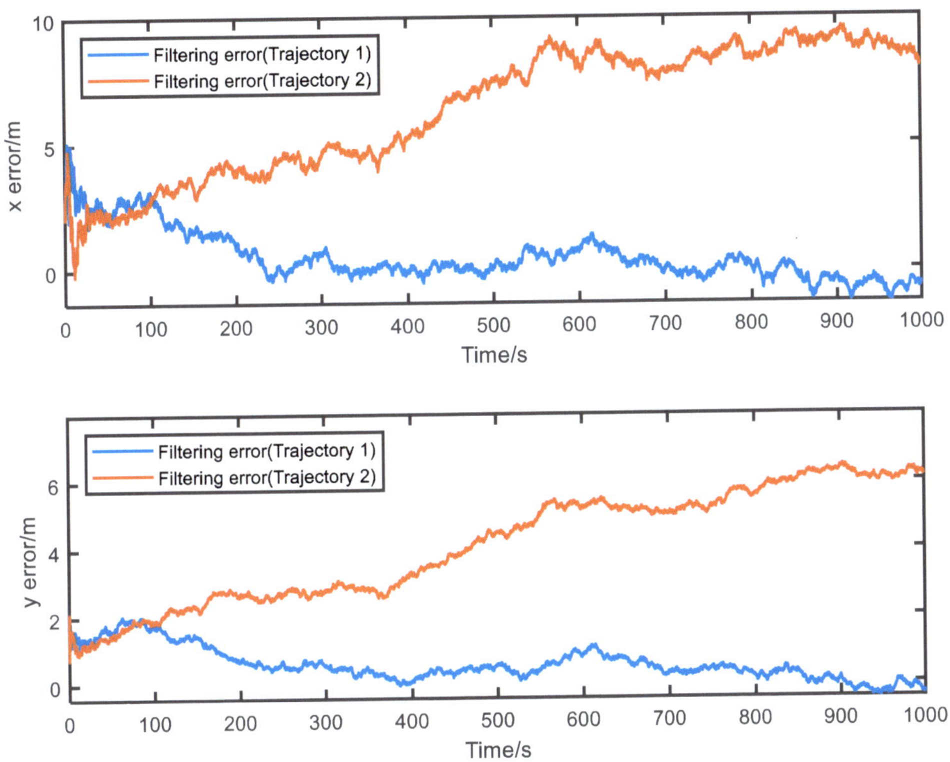

Figure 3. Filtering errors for the two trajectories.

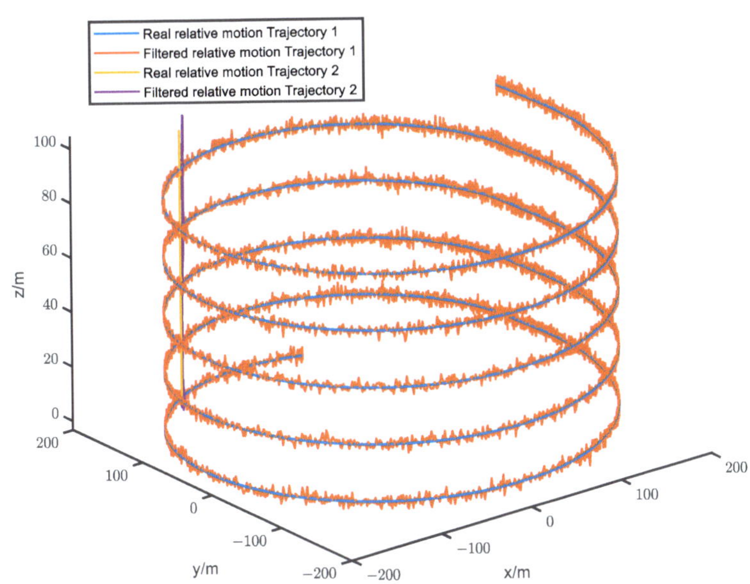

Figure 4. The relative motion trajectories of MAV B relative to A.

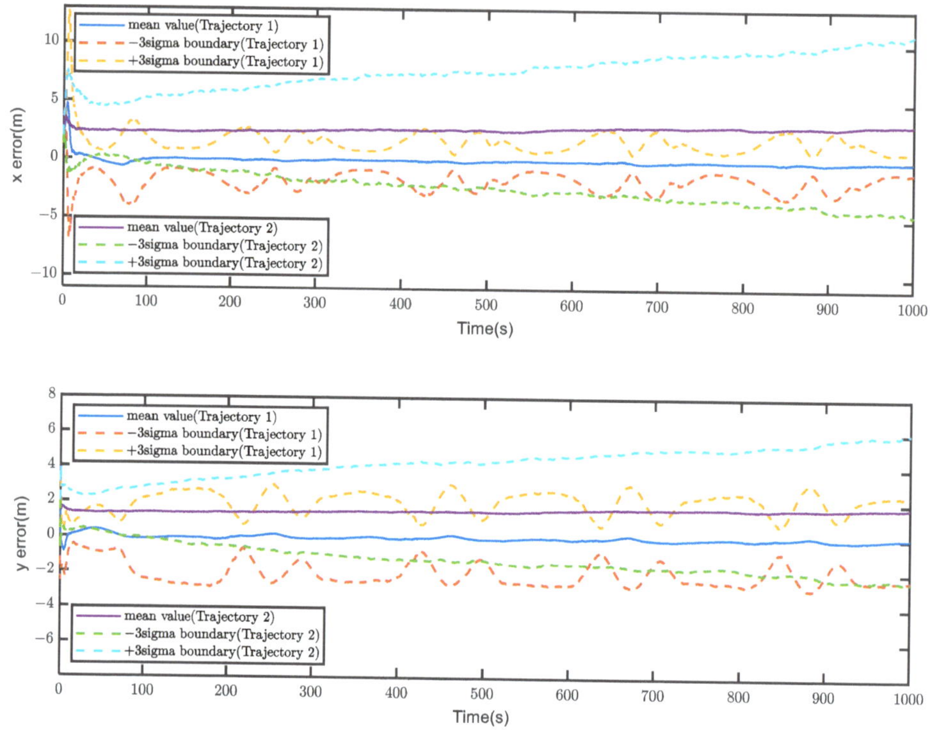

Figure 5. 100 Monte Carlo simulations, which are conducted to verify the system observability conditions.

Table 6. Mean statistical data from 100 Monte Carlo simulation tests.

	Parameters	+3σ Boundary	Mean	−3σ Boundary
Trajectory 1	x	1.62 m	−0.04 m	−1.69 m
	y	2.07 m	0.02 m	−2.03 m
Trajectory 2	x	8.47 m	2.44 m	−2.59 m
	y	4.19 m	1.39 m	−3.41 m
$\frac{\text{Trajectory 2}}{\text{Trajectory 1}}$	x	5.23	/	1.53
	y	2.02	/	1.18

The simulation experiments in Section 5.2 demonstrate that the system has smaller positioning errors when it satisfies the observability conditions, versus larger errors with a divergent trend when it fails to meet the conditions. The result further confirms the validity of the system observability theoretical analysis in Section 4.

5.3. The Influence of Different Ranging Errors on the Positioning Accuracy and Stability

In this section, we investigate the impact of different ranging errors on the positioning accuracy and stability of the proposed relative localization method, which integrates ranging and IMU information. The ranging errors for Trajectory 1 are set to four levels: 1 m, 2 m, 4 m, and 8 m. The results of a single experiment's positioning errors are shown in Figure 6. Since Trajectory 1 satisfies the system observability conditions, the corresponding positioning accuracy is within 2 m after approximately 100 s, despite the different ranging errors, and the system remains stable.

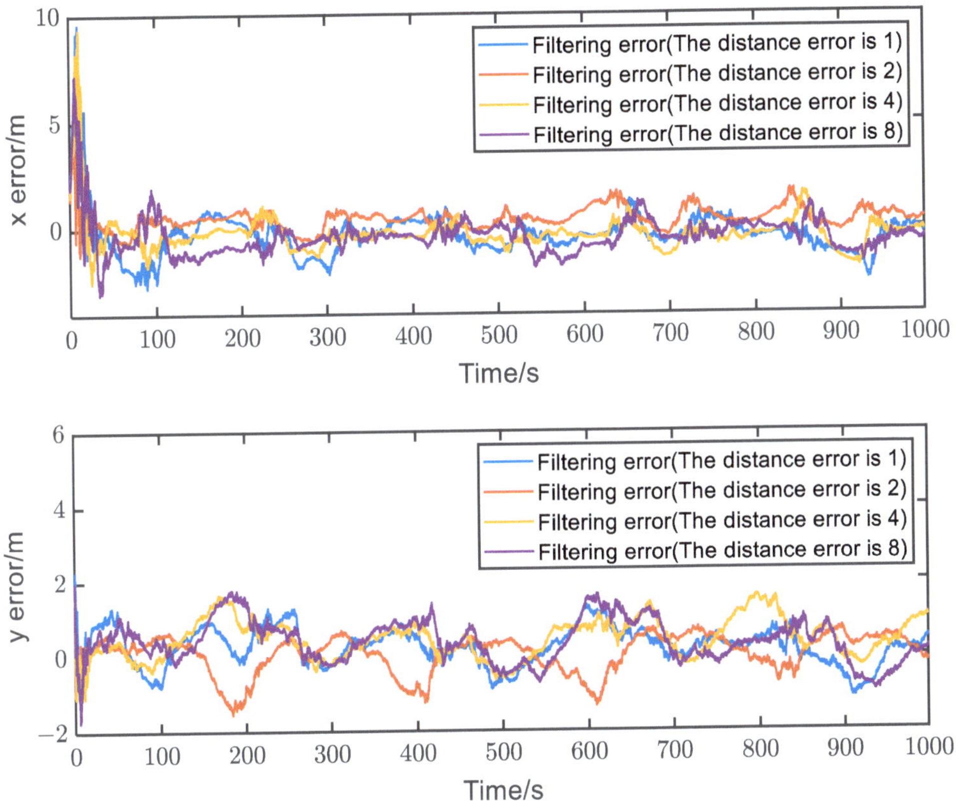

Figure 6. The positioning errors corresponding to different ranging errors, where four ranging error levels of 1 m, 2 m, 4 m, and 8 m are set.

Subsequently, 100 Monte Carlo simulations are conducted, and the simulation results are presented in Figure 7. It can be visually observed that the larger the ranging error, the worse the system's positioning accuracy and stability. However, increasing ranging errors did not cause the positioning accuracy and stability of the proposed system to deteriorate significantly, meaning the system did not diverge, but only increased the error. Specifically, the average boundary value increased from about 2 m to around 3 m as ranging errors increased from 1 m to 8 m. These experimental results indicate that different ranging errors have a limited impact on the system's positioning accuracy and stability.

5.4. Error Comparison with an Existing Positioning Method

In Section 5.4, we compare the performance of the proposed method with an existing method introduced in [32].

To demonstrate the rationality of the parameter settings in the comparative experiment, we offer the following explanations: The UWB ranging error in [32] was set to 2 m, and the IMU used was the 9-DOF MPU-9150, with assumed displacement errors of 0.2 m and orientation errors of 0.1 rad. The rationality of the parameters has been previously confirmed in [32]. Therefore, we adopt the ranging error of 2 m from [32]. In addition, the IMU data used in this paper is from MPU-6050, whose data parameters are essentially the same as those of MPU-9150. Based on these considerations, we conclude that our parameter settings are appropriate for the purposes of our simulation.

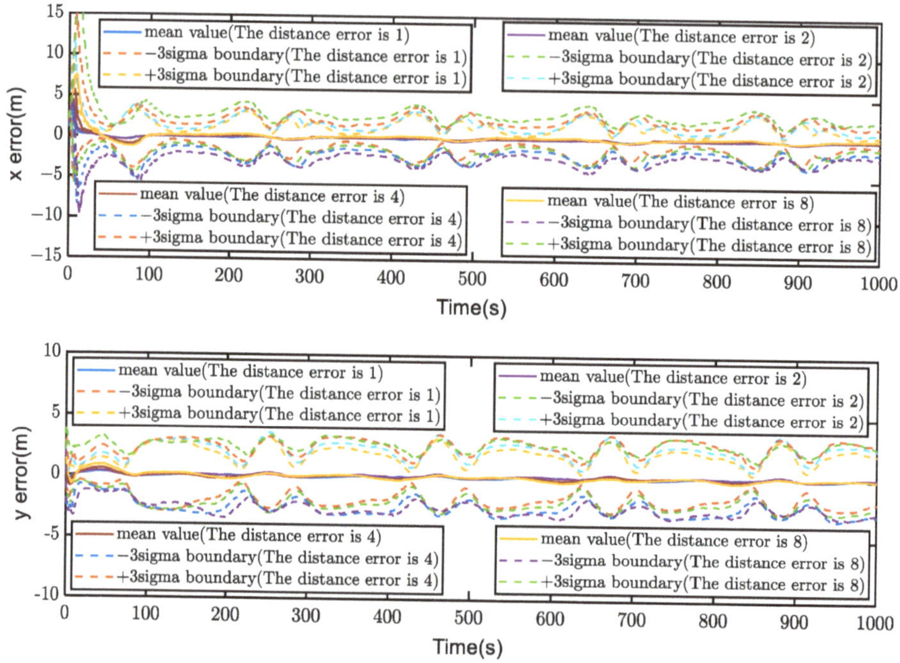

Figure 7. 100 Monte Carlo simulations, which are conducted to investigate the impact of different ranging errors on the positioning accuracy and stability.

The experimental results of the two methods are shown in Figure 8. It can be observed that the average positioning error of the proposed method in this paper is approximately 0.55 m, while that of the method in [32] is about 2.14 m, which is approximately 3.89 times the error of the proposed method. In contrast, the relative positioning method proposed in this paper, which fuses ranging and IMU information, achieves better positioning accuracy.

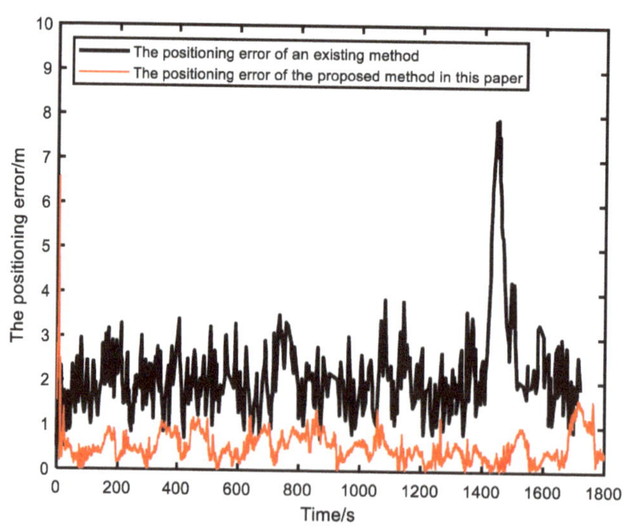

Figure 8. A comparison of the positioning errors between the method proposed in this paper and that in [32] is presented.

At the initial startup phase of the positioning system, EKF may have uncertainty about the initial state, that is, the initial estimate of the system state may not be accurate enough, which may lead to large errors. As time goes on, since the system has started to operate, through measurement updates and filtering, the state estimate of the system will gradually become more accurate, and the error will gradually decrease until it stabilizes.

6. Conclusions

In this paper, we have presented a novel cooperative relative positioning method for MAVs in GNSS-denied scenarios. Within the framework of the system model, we have employed the EKF algorithm to fuse ranging and IMU information, addressing the problem of IMU error drift and enabling high-precision and robust MAV positioning. In addition, our proposed method has the following several distinctive features: it eliminates the requirement for precise reference anchors, expands the application scope of the system, and solves the problem of limited number of nodes in MAV formation. Furthermore, we have theoretically derived the system observability conditions and provided specific expressions that guarantee the system positioning accuracy when the system satisfies the observability conditions. We have verified the correctness of the observability theoretical analysis through simulations and investigated the influence of different ranging errors on the positioning accuracy and stability. Finally, we have demonstrated the superiority of the proposed method through simulation comparisons with an existing method.

We have verified the proposed method through simulation experiments. To ensure that the experiments are as close to reality as possible, the system parameters considered in this paper are taken from the official datasheet of the sensor and previous research [32,38–40]. However, there is still a certain gap between simulation and reality, mainly in the following aspects:

1. The presence of obstacles, wind conditions, and other environmental factors may affect the performance of MAVs in practical applications.
2. Simulated noise cannot fully reflect the noise experienced by MAVs in actual applications.
3. There may be a certain difference in air pressure at the same altitude but different horizontal positions due to the influence of factors such as atmospheric pressure, temperature, and humidity, which may result in some spatial variations. In addition, the scale of pressure gradient varies across different regions. As a result, the barometric altimeter readings may be affected and may introduce some measurement bias.
4. In practical applications, the airflow generated by the movement of MAVs may have an impact on the flight stability and control of MAVs.

These factors are difficult to model accurately in simulation experiments. Therefore, actual verification will direct our research in the future.

Author Contributions: Conceptualization, J.C.; methodology, J.C.; software, J.C.; validation, J.C., P.R. and T.D.; data curation, J.C.; writing—original draft preparation, J.C.; writing—review and editing, J.C. and P.R.; supervision, P.R. All authors have read and agreed to the published version of the manuscript.

Funding: This work was supported in part by the National Natural Science Foundation of China (Grant No. 61971320, 62001361) and in part by the Key Industrial Innovation Chain Project in Industrial Domain (Grant No. 2023-ZDLGY-50).

Institutional Review Board Statement: Not applicable.

Informed Consent Statement: Not applicable.

Data Availability Statement: The data presented in this study are available on request from the corresponding author. The data are not publicly available due to the regulations of our laboratory.

Acknowledgments: The authors would like to thank the anonymous reviewers, and Long Yang for their insightful comments and suggestions on the research that led to this paper.

Conflicts of Interest: The authors declare no conflict of interest.

Appendix A.

Appendix A.1. Proof of Equation (5)

The relationship between the Earth-fixed North-East-Down coordinate system n and the typical body-fixed coordinate system b is shown in Figure A1. The following unit vectors are defined:

$$\boldsymbol{n}_1 \triangleq \begin{bmatrix} 1 & 0 & 0 \end{bmatrix}^T, \boldsymbol{n}_2 \triangleq \begin{bmatrix} 0 & 1 & 0 \end{bmatrix}^T, \boldsymbol{n}_3 \triangleq \begin{bmatrix} 0 & 0 & 1 \end{bmatrix}^T \quad \text{(A1)}$$

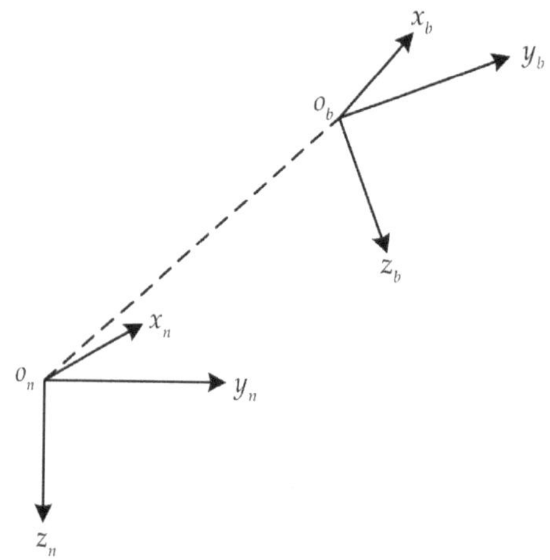

Figure A1. The relationship between the Earth-fixed North-East-Down (NED) coordinate system n and the body-fixed coordinate system b.

The unit vectors along $o_n x_n$, $o_n y_n$, and $o_n z_n$ axes in n are represented by $\boldsymbol{n}_1$, $\boldsymbol{n}_2$, and $\boldsymbol{n}_3$, respectively. The following relationships are satisfied by the unit vectors along $o_b x_b$, $o_b y_b$, and $o_b z_b$ axes in b:

$${}^b\boldsymbol{b}_1 = \boldsymbol{n}_1, {}^b\boldsymbol{b}_2 = \boldsymbol{n}_2, {}^b\boldsymbol{b}_3 = \boldsymbol{n}_3 \quad \text{(A2)}$$

Furthermore, the unit vectors along $o_b x_b$, $o_b y_b$, and $o_b z_b$ axes in n are represented by ${}^n\boldsymbol{b}_1$, ${}^n\boldsymbol{b}_2$, and ${}^n\boldsymbol{b}_3$, respectively.

The schematic diagram of the rotational relationship from the Earth-fixed North-East-Down coordinate system n to the body-fixed coordinate system b is shown in Figure A2. Assuming $on_1n_2n_3$ is a right-handed Cartesian coordinate system, the following three rotations are performed: First, the $on_1n_2n_3$ system is rotated about the positive on_3-axis by an angle φ, resulting in the $ok_1k_2k_3$ system, which shares the same oz axis as the $on_1n_2n_3$ system. Next, the $ok_1k_2k_3$ system is rotated about the positive ok_2-axis by an angle θ, resulting in the $oe_1e_2e_3$ system, which shares the same oy axis as the $ok_1k_2k_3$ system. Finally, the $oe_1e_2e_3$ system is rotated about the positive oe_1-axis by an angle γ, resulting in the $ob_1b_2b_3$ system, which shares the same ox axis as the $oe_1e_2e_3$ system. The rotation sequence and the sign of each rotation angle can be denoted as '(+3)(+2)(+1)', or simply '321' by omitting the plus sign, where the numbers 1, 2, and 3 represent the rotations about ox, oy, and oz axes, respectively, and the plus sign in the parentheses indicates the positive direction of rotation according to the right-hand rule.

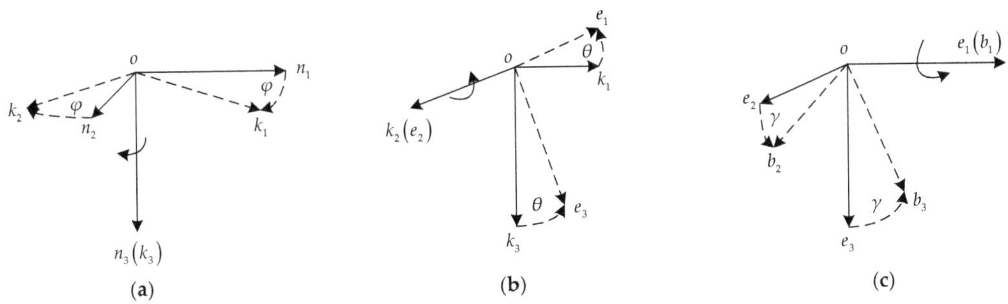

Figure A2. Define Euler angles by the 321-rotation sequence. (**a**) Yaw angle φ; (**b**) Pitch angle θ; (**c**) Roll angle γ.

Assuming that the body angular velocity is ${}^{b}\boldsymbol{\omega} = \begin{bmatrix}\omega_{xb} & \omega_{yb} & \omega_{zb}\end{bmatrix}^{T}$, the relationship between the attitude change rate and the body angular velocity is given by the following equation [41]:

$$ {}^{b}\boldsymbol{\omega} = \dot{\varphi}\cdot{}^{b}\boldsymbol{k}_3 + \dot{\theta}\cdot{}^{b}\boldsymbol{e}_2 + \dot{\gamma}\cdot{}^{b}\boldsymbol{b}_1 \tag{A3} $$

where,

$$ {}^{b}\boldsymbol{b}_1 = \begin{bmatrix}1 & 0 & 0\end{bmatrix}^{T} \tag{A4} $$

$$ \begin{aligned} {}^{b}\boldsymbol{e}_2 &= \boldsymbol{R}_e^b\cdot{}^{e}\boldsymbol{e}_2 = \boldsymbol{R}_e^b\cdot\boldsymbol{n}_2 = \boldsymbol{R}_x(\gamma)\cdot\boldsymbol{n}_2 \\ &= \begin{bmatrix}1 & 0 & 0\\ 0 & \cos\gamma & \sin\gamma\\ 0 & -\sin\gamma & \cos\gamma\end{bmatrix}\begin{bmatrix}0\\1\\0\end{bmatrix} \\ &= \begin{bmatrix}0 & \cos\gamma & -\sin\gamma\end{bmatrix}^{T} \end{aligned} \tag{A5} $$

$$ \begin{aligned} {}^{b}\boldsymbol{k}_3 &= \boldsymbol{R}_k^b\cdot{}^{k}\boldsymbol{k}_2 = \boldsymbol{R}_k^b\cdot\boldsymbol{n}_3 = \boldsymbol{R}_x(\gamma)\boldsymbol{R}_y(\theta)\cdot\boldsymbol{n}_3 \\ &= \begin{bmatrix}1 & 0 & 0\\ 0 & \cos\gamma & \sin\gamma\\ 0 & -\sin\gamma & \cos\gamma\end{bmatrix}\begin{bmatrix}\cos\theta & 0 & -\sin\theta\\ 0 & 1 & 0\\ \sin\theta & 0 & \cos\theta\end{bmatrix}\begin{bmatrix}0\\0\\1\end{bmatrix} \\ &= \begin{bmatrix}-\sin\theta & \sin\gamma\cos\theta & \cos\gamma\cos\theta\end{bmatrix}^{T} \end{aligned} \tag{A6} $$

According to Equations (A3)–(A6), the following equation is obtained:

$$ \begin{bmatrix}\omega_{xb}\\ \omega_{yb}\\ \omega_{zb}\end{bmatrix} = \begin{bmatrix}1 & 0 & -\sin\theta\\ 0 & \cos\gamma & \sin\gamma\cos\theta\\ 0 & -\sin\gamma & \cos\gamma\cos\theta\end{bmatrix}\begin{bmatrix}\dot{\gamma}\\ \dot{\theta}\\ \dot{\varphi}\end{bmatrix} \tag{A7} $$

Then we can obtain the following expression by applying the inverse transformation to Equation (A7):

$$ \begin{aligned} \begin{bmatrix}\dot{\gamma}\\ \dot{\theta}\\ \dot{\varphi}\end{bmatrix} &= \begin{bmatrix}1 & 0 & -\sin\theta\\ 0 & \cos\gamma & \sin\gamma\cos\theta\\ 0 & -\sin\gamma & \cos\gamma\cos\theta\end{bmatrix}^{-1}\begin{bmatrix}\omega_{xb}\\ \omega_{yb}\\ \omega_{zb}\end{bmatrix} \\ &= \begin{bmatrix}1 & \sin\gamma\tan\theta & \cos\gamma\tan\theta\\ 0 & \cos\gamma & -\sin\gamma\\ 0 & \frac{\sin\gamma}{\cos\theta} & \frac{\cos\gamma}{\cos\theta}\end{bmatrix}\begin{bmatrix}\omega_{xb}\\ \omega_{yb}\\ \omega_{zb}\end{bmatrix} \end{aligned} \tag{A8} $$

where,

$$ \dot{\varphi} = \frac{\sin\gamma}{\cos\theta}\omega_{yb} + \frac{\cos\gamma}{\cos\theta}\omega_{zb} \tag{A9} $$

Since the yaw angles in both coordinate systems (h and b) are identical, the yaw rate in the horizontal body-fixed coordinate system h is the same as that given by Equation (A9). Therefore, Equation (5) is confirmed.

Appendix A.2. Proof of Equation (6)

According to the relationship between the direction cosine matrix and the equivalent rotation vector, the transformation matrix from b to n in Figure A2 can be obtained as follows:

$$\begin{aligned} C_b^n &= \begin{bmatrix} \cos\varphi & -\sin\varphi & 0 \\ \sin\varphi & \cos\varphi & 0 \\ 0 & 0 & 1 \end{bmatrix} \begin{bmatrix} \cos\theta & 0 & \sin\theta \\ 0 & 1 & 0 \\ -\sin\theta & 0 & \cos\theta \end{bmatrix} \begin{bmatrix} 1 & 0 & 0 \\ 0 & \cos\gamma & -\sin\gamma \\ 0 & \sin\gamma & \cos\gamma \end{bmatrix} \\ &= \begin{bmatrix} \cos\varphi\cos\theta & -\sin\varphi\cos\gamma + \cos\varphi\sin\theta\sin\gamma & \sin\varphi\sin\gamma + \cos\varphi\sin\theta\cos\gamma \\ \sin\varphi\cos\theta & \cos\varphi\cos\gamma + \sin\varphi\sin\theta\sin\gamma & -\cos\varphi\sin\gamma + \sin\varphi\sin\theta\cos\gamma \\ -\sin\theta & \cos\theta\sin\gamma & \cos\theta\cos\gamma \end{bmatrix} \end{aligned} \tag{A10}$$

In this paper, we use the fixed-body horizontal coordinate system h, whose yaw angle is the same as that of the typical fixed-body coordinate system b, and x-y plane is parallel to that of the NED coordinate system n. Therefore, the transformation matrix from b to h is equivalent to C_b^n with $\varphi = 0$, which is given by:

$$C_b^h = \begin{bmatrix} \cos\theta & \sin\theta\sin\gamma & \sin\theta\cos\gamma \\ 0 & \cos\gamma & -\sin\gamma \\ -\sin\theta & \cos\theta\sin\gamma & \cos\theta\cos\gamma \end{bmatrix} \tag{A11}$$

Since this paper considers the two-dimensional case, according to Equation (A11), the transformation matrix of Equation (6) can be obtained.

References

1. Suzuki, T.; Matsuo, K.; Amano, Y. Rotating GNSS Antennas: Simultaneous LOS and NLOS Multipath Mitigation. *GPS Solut.* **2020**, *24*, 86. [CrossRef]
2. Hermosilla, T.; Palomar-Vázquez, J.; Balaguer-Beser, Á.; Balsa-Barreiro, J.; Ruiz, L.A. Using street based metrics to characterize urban typologies. *Comput. Environ. Urban Syst.* **2014**, *44*, 68–79. [CrossRef]
3. Nicola, M.; Falco, G.; Morales Ferre, R.; Lohan, E.-S.; de la Fuente, A.; Falletti, E. Collaborative Solutions for Interference Management in GNSS-Based Aircraft Navigation. *Sensors* **2020**, *20*, 4085. [CrossRef] [PubMed]
4. José, B.B. Aplicación de Sistemas GNSS y SIG a Infraestructuras de Transporte: Estudio Sobre la Conducción Naturalista. Ph.D. Thesis, Universidade da Coruña, La Coruña, Spain, 2014.
5. Wang, S.; Dong, X.; Liu, G.; Gao, M.; Xiao, G.; Zhao, W.; Lv, D. GNSS RTK/UWB/DBA Fusion Positioning Method and Its Performance Evaluation. *Remote Sens.* **2022**, *14*, 5928. [CrossRef]
6. He, S.; Hu, S.; Guo, Q.; Jiang, W. Relative Positioning Method for UAVs Based on Multi-Source Information Fusion. *Math. Probl. Eng.* **2022**, *2022*, 3755861.
7. Zhang, J.; Ren, M.; Wang, P.; Meng, J.; Mu, Y. Indoor Localization Based on VIO System and Three-Dimensional Map Matching. *Sensors* **2020**, *20*, 2790. [CrossRef] [PubMed]
8. Zhu, F.; Shen, Y.; Wang, Y.; Jia, J.; Zhang, X. Fusing GNSS/INS/Vision With A Priori Feature Map for High-Precision and Continuous Navigation. *IEEE Sens. J.* **2021**, *21*, 23370–23381. [CrossRef]
9. Naus, K.; Waz, M. Precision in Determining Ship Position using the Method of Comparing an Omnidirectional Map to a Visual Shoreline Image. *J. Navig.* **2016**, *69*, 391–413. [CrossRef]
10. Krüger, R.; Simeonov, G.; Beck, F.; Ertl, T. Visual Interactive Map Matching. *IEEE Trans. Vis. Comput. Graph.* **2018**, *24*, 1881–1892. [CrossRef]
11. Yoo, D.-H.; Shan, G.; Roh, B.-H. A Vision-based Indoor Positioning Systems utilizing Computer Aided Design Drawing. In Proceedings of the 28th Annual International Conference on Mobile Computing and Networking (ACM MobiCom '22), Sydney, NSW, Australia, 17–21 October 2022; ACM: New York, NY, USA, 2022. [CrossRef]
12. Xiong, J.; Cheong, J.W.; Ding, Y.; Xiong, Z.; Dempster, A.G. Efficient Distributed Particle Filter for Robust Range-Only SLAM. *IEEE Internet Things J.* **2022**, *9*, 21932–21945. [CrossRef]
13. Wang, Y.; Wang, X. Research on SLAM Road Sign Observation Based on Particle Filter. *Comput. Intell. Neurosci.* **2022**, *2022*, 4478978. [CrossRef] [PubMed]
14. Lu, S.; Zhi, Y.; Zhang, S.; He, R.; Bao, Z. Semi-Direct Monocular SLAM With Three Levels of Parallel Optimizations. *IEEE Access* **2021**, *9*, 86801–86810. [CrossRef]

15. Munguia, R.; Trujillo, J.-C.; Guerra, E.; Grau, A. A Hybrid Visual-Based SLAM Architecture: Local Filter-Based SLAM with KeyFrame-Based Global Mapping. *Sensors* **2022**, *22*, 210. [CrossRef] [PubMed]
16. Balsa-Barreiro, J.; Fritsch, D. Generation of visually aesthetic and detailed 3D models of historical cities by using laser scanning and digital photogrammetry. *Digit. Appl. Archaeol. Cult. Herit.* **2018**, *8*, 57–64. [CrossRef]
17. Salarian, M.; Iliev, N.; Çetin, A.E.; Ansari, R. Improved Image-Based Localization Using SFM and Modified Coordinate System Transfer. *IEEE Trans. Multimed.* **2018**, *20*, 3298–3310. [CrossRef]
18. Xing, H.; Zhao, Y.; Zhang, Y.; Chen, Y. 3d trajectory planning of positioning error correction based on pso-a* algorithm. *Comput. Mater. Contin.* **2020**, *65*, 2295–2308. [CrossRef]
19. Chen, D.; Neusypin, K.; Selezneva, M.; Mu, Z. New Algorithms for Autonomous Inertial Navigation Systems Correction with Precession Angle Sensors in Aircrafts. *Sensors* **2019**, *19*, 5016. [CrossRef]
20. Cho, S.Y.; Lee, J.H.; Park, C.G. A Zero-Velocity Detection Algorithm Robust to Various Gait Types for Pedestrian Inertial Navigation. *IEEE Sens. J.* **2022**, *22*, 4916–4931. [CrossRef]
21. Wang, Q.; Liu, K.; Sun, Z.; Cai, M.; Cheng, M. Research on the Heading Calibration for Foot-Mounted Inertial Pedestrian-Positioning System Based on Accelerometer Attitude. *Electronics* **2019**, *8*, 1405. [CrossRef]
22. Chen, H.; Taha, T.M.; Chodavarapu, V.P. Towards Improved Inertial Navigation by Reducing Errors Using Deep Learning Methodology. *Appl. Sci.* **2022**, *12*, 3645. [CrossRef]
23. Maheepala, M.; Joordens, M.A.; Kouzani, A.Z. A Low-Power Connected 3-D Indoor Positioning Device. *IEEE Internet Things J.* **2022**, *9*, 9002–9011. [CrossRef]
24. Lin, S.-H.; Chang Chien, H.-H.; Wang, W.-W.; Lin, K.-H.; Li, G.-J. An Efficient IAKF Approach for Indoor Positioning Drift Correction. *Sensors* **2022**, *22*, 5697. [CrossRef]
25. Park, J.; Kim, Y.-J.; Lee, B.K. Passive Radio-Frequency Identification Tag-Based Indoor Localization in Multi-Stacking Racks for Warehousing. *Appl. Sci.* **2020**, *10*, 3623. [CrossRef]
26. Peng, Z.; Cheng, S.; Li, X.; Li, K.; Cai, M.; You, L. Dynamic Visual SLAM Integrated with IMU for Unmanned Scenarios. In Proceedings of the 2022 IEEE 25th International Conference on Intelligent Transportation Systems (ITSC), Macau, China, 8–12 October 2022; pp. 4247–4253. [CrossRef]
27. Yu, F.; Yu, H.; Wei, Y. Research on Robot Positioning Technology Based on Multi Sensor. In Proceedings of the 2019 International Conference on Computer Network, Electronic and Automation (ICCNEA), Xi'an, China, 27–29 September 2019; pp. 480–485. [CrossRef]
28. Olsson, F.; Rantakokko, J.; Nygårds, J. Cooperative localization using a foot-mounted inertial navigation system and ultrawide-band ranging. In Proceedings of the 2014 International Conference on Indoor Positioning and Indoor Navigation (IPIN), Busan, Republic of Korea, 27–30 October 2014; pp. 122–131. [CrossRef]
29. Liu, R.; Yuen, C.; Do, T.-N.; Jiao, D.; Liu, X.; Tan, U.-X. Cooperative relative positioning of mobile users by fusing IMU inertial and UWB ranging information. In Proceedings of the 2017 IEEE International Conference on Robotics and Automation (ICRA), Singapore, 29 May–3 June 2017; pp. 5623–5629. [CrossRef]
30. Long, K.; Shen, C.; Tian, C.; Zhang, K.; Bhatti, U.A.; Kong, D.F.N.; Feng, S.; Cheng, H. Single UWB Anchor Aided PDR Heading and Step Length Correcting Indoor Localization System. *IEEE Access* **2021**, *9*, 11511–11522. [CrossRef]
31. Zhao, X.; Yan, G.; Chang, T.; Liang, C.H.; Wang, Z.; Fu, H. Antenna design for ultra-wideband through wall radar. In Proceedings of the 2017 3rd IEEE International Conference on Computer and Communications (ICCC), Chengdu, China, 13–16 December 2017; pp. 1121–1124. [CrossRef]
32. Xu, H.; Zhu, Y.; Wang, G. On the anti-multipath performance of UWB signals in indoor environments. In Proceedings of the ICMMT 4th International Conference on Microwave and Millimeter Wave Technology, Beijing, China, 18–21 August 2004; pp. 163–166. [CrossRef]
33. Lou, X.; Zhao, Y. High-Accuracy Positioning Algorithm Based on UWB. In Proceedings of the 2019 International Conference on Artificial Intelligence and Advanced Manufacturing (AIAM), Dublin, Ireland, 17–19 October 2019; pp. 71–75. [CrossRef]
34. Yuan, K.; Wang, H.; Zhang, H. Robot Position Realization Based on Multi-sensor Information Fusion Algorithm. In Proceedings of the 2011 Fourth International Symposium on Computational Intelligence and Design, Hangzhou, China, 28–30 October 2011; pp. 294–297. [CrossRef]
35. Zheng, W.; Wang, J.; Wang, Z. Multi-sensor fusion based real-time hovering for a quadrotor without GPS in assigned position. In Proceedings of the 2016 Chinese Control and Decision Conference (CCDC), Yinchuan, China, 28–30 May 2016; pp. 3605–3610. [CrossRef]
36. Zhou, X.S.; Roumeliotis, S.I. Robot-to-Robot Relative Pose Estimation from Range Measurements. *IEEE Trans. Robot.* **2008**, *24*, 1379–1393. [CrossRef]
37. Velimir, J. Abstract control systems: Controllability and observability. *SIAM J. Control* **1970**, *8*, 424–439.
38. Cismas, A.; Ioana, M.; Vlad, C.; Casu, G. Crash Detection Using IMU Sensors. In Proceedings of the 2017 21st International Conference on Control Systems and Computer Science (CSCS), Bucharest, Romania, 29–31 May 2017; pp. 672–676. [CrossRef]
39. Chen, M.; Luo, W.; Xu, Z.; Zhang, X.; Xie, B.; Wang, G.; Han, M. An ultrahigh resolution pressure sensor based on percolative metal nanoparticle arrays. *Nat. Commun.* **2019**, *10*, 4024. [CrossRef] [PubMed]
40. Silva, B.; Hancke, G.P. Ranging Error Mitigation for Through-the-Wall Non-Line-of-Sight Conditions. *IEEE Trans. Ind. Inform.* **2020**, *16*, 6903–6911. [CrossRef]

41. Ducard, G.J.J. *Fault-Tolerant Flight Control and Guidance Systems: Practical Methods for Small Unmanned Aerial Vehicles*; Springer Science & Business Media: Berlin/Heidelberg, Germany, 2009.

sensors

MDPI

Article

IMU/UWB Fusion Method Using a Complementary Filter and a Kalman Filter for Hybrid Upper Limb Motion Estimation

Yutong Shi [1], Yongbo Zhang [1,2,*], Zhonghan Li [1], Shangwu Yuan [1] and Shihao Zhu [1]

1 School of Aeronautic Science and Engineering, Beihang University, Beijing 100191, China; 17375384@buaa.edu.cn (Y.S.); zhonghan@buaa.edu.cn (Z.L.); yuanshangwu@buaa.edu.cn (S.Y.); 2205221@buaa.edu.cn (S.Z.)

2 Aircraft and Propulsion Laboratory, Ningbo Institute of Technology, Beihang University, Ningbo 315100, China

* Correspondence: zhangyongbo@buaa.edu.cn

Abstract: Motion capture systems have enormously benefited the research into human–computer interaction in the aerospace field. Given the high cost and susceptibility to lighting conditions of optical motion capture systems, as well as considering the drift in IMU sensors, this paper utilizes a fusion approach with low-cost wearable sensors for hybrid upper limb motion tracking. We propose a novel algorithm that combines the fourth-order Runge–Kutta (RK4) Madgwick complementary orientation filter and the Kalman filter for motion estimation through the data fusion of an inertial measurement unit (IMU) and an ultrawideband (UWB). The Madgwick RK4 orientation filter is used to compensate gyroscope drift through the optimal fusion of a magnetic, angular rate, and gravity (MARG) system, without requiring knowledge of noise distribution for implementation. Then, considering the error distribution provided by the UWB system, we employ a Kalman filter to estimate and fuse the UWB measurements to further reduce the drift error. Adopting the cube distribution of four anchors, the drift-free position obtained by the UWB localization Kalman filter is used to fuse the position calculated by IMU. The proposed algorithm has been tested by various movements and has demonstrated an average decrease in the RMSE of 1.2 cm from the IMU method to IMU/UWB fusion method. The experimental results represent the high feasibility and stability of our proposed algorithm for accurately tracking the movements of human upper limbs.

Keywords: motion estimation; inertial measurement unit (IMU); ultrawideband (UWB); Madgwick orientation filter; Kalman filter

1. Introduction

Motion capture (MoCap) systems provide technical support for the operation of space robots, which has enormously benefited the research into human–computer Interaction in the aerospace field. By capturing human upper limb movement and mapping it to space robot motion, space operators can remotely control space robot arms and allow them to perform complex and elaborate actions, such as grasping and handling. Thus, we need to accurately estimate the movement of human upper limbs [1–3]. Optical Mocap systems employ multiple cameras to capture the positions of reflective markers and have been widely used in human motion tracking. Although these systems share precise real-time tracking and strong anti-interference abilities, they are expensive and suffer from light and occlusion problems [4,5]. In order to flexibly track human movements, it is necessary to employ ubiquitous wearable sensors.

With the intensive study and rapid advancement of micro-electro-mechanical systems, wearable inertial measurement units (IMUs) have been extensively adopted for 3-D human motion tracking due to their low cost, portability, and high independence [6]. In many of the proposed IMU Mocap methods, accelerometers and gyroscopes are fused to estimate the body segment orientation [7,8], including the knee joint angle [9,10], arm adduction angle [11], and shoulder and elbow joint angles [12,13]. However, it has been illustrated that there is obvious drift error of the attitude angle calculated by the gyroscope, which contributes to the poor estimation accuracy and stability. Considering that the accelerometers can only compensate for the tilt angles relative to the direction of gravity, it is necessary to introduce magnetometers to provide compensation for the yaw angles, which is known as nine-axis IMU or the magnetic, angular rate, and gravity (MARG) system [14]. Data fusion of three inertial sensors can be achieved by adopting either a complementary filter [14–17] or a classic Kalman filter [18,19]. The Kalman filter has been widely used as an orientation filter for the majority of the proposed methods. Wu et al. [20–23] employed a particle filter and an unscented Kalman filter for information fusion and developed a micro-sensor Mocap system to achieve real-time tracking. Roetenberg et al. [24] designed a Kalman filter and improved the orientation estimation through the compensation of magnetic disturbances. Considering that the Kalman filter requires knowledge of noise distribution and is prone to computational load and parameter tuning issues, many researchers tend to use the complementary filter. Fourati et al. [25] proposed a complementary observer based on the quaternion method for motion tracking. However, only adopting inertial sensors fusion for motion estimation will still cause drift, so it is expected that a stable and high-precision technology will be introduced to fuse with IMU.

Considering the flexibility and high sampling rate of IMU, ultrawideband (UWB) technology is attractive for data fusion due to its low cost, portability, and low energy consumption [26,27], but it requires a clear line-of-sight (LOS) channel [28]. Compared with traditional UWB positioning methods consisting of maximum likelihood estimation (MLE), linearized least square estimation (LLSE), and weighted centroid estimation (WCE), the extended Kalman filter (EKF) performs with less computational time and less complexity [29]. Applying UWB and IMU fusion for human motion tracking can not only compensate for the low sampling rate of the UWB system, but reduce the drift of IMU as well. Depending on whether the fusion is based on raw time or location, data fusion methods are divided into tightly coupled [30] and loosely coupled [31,32] categories. Kok et al. [33] provided tightly coupled IMU and UWB fusion using an optimization method that showed good performance in terms of handling outliers, but suffered from clock skew challenges. Therefore, in order to facilitate the calculation, the loosely coupled method is often employed. Zihajehzadeh et al. [32,34,35] proposed a Kalman-filter-based IMU and UWB fusion method without a magnetometer and accurately captured lower body motion under magnetic disturbances. However, when extending the rotation matrices to the upper limb, their algorithm may suffer from the singularity problem. Zhang et al. [36] adopted a Mahony filter and quaternion for foot attitude estimation via IMU and UWB fusion, but the use of acceleration double integration to obtain the position would lead to huge cumulative errors.

In this paper, the human upper limb is taken as our research subject, and a novel IMU/UWB data fusion method is proposed for 3-D motion estimation by applying the Runge–Kutta Madgwick filter, the UWB localization Kalman filter, and the IMU/UWB Kalman filter. On the basis of the established kinematics model of the upper limbs, the quaternion method was employed to calculate the attitude angle to avoid the gimbal lock problem. Our proposed algorithm comprises the following two novel aspects: (1) we combined the Madgwick RK4 complementary orientation filter and Kalman filter for motion tracking, the Madgwick RK4 filter was employed to reduce gyroscope drift without leading to noise distribution, and the Kalman filter was implemented for UWB localization and fusion with known error distribution from the UWB system; and (2) the drift-free position obtained by the UWB localization system was used to fuse the position calculated by IMU for upper limb motion estimation, which enormously reduced the drift caused by the double integration of acceleration. The good experimental results represent the high feasibility and stability of our proposed algorithm.

The rest of the paper is structured as follows. In Section 2, the theoretical fusion method for 3-D upper limb tracking is described. The experimental setup and protocol are demonstrated in Section 3. Then, the experimental results are shown and analyzed in Section 4. Finally, conclusions are provided in Section 5.

2. Theoretical Method

In this section, the proposed information fusion algorithm for human upper limb motion estimation is described. As shown in Figure 1, we first simplified the human upper limb as a kinematics model with three joints (shoulder, elbow, and wrist) and two segments (upper arm and forearm) [21]. The shoulder joint was set as a fixed point with three degrees of freedom (DoFs) and two DoFs for the elbow and wrist [37]. Taking the right arm as an example, two IMU sensors were arranged on the lateral side above the wrist and the elbow, respectively, as well as two UWB tags. The navigation coordinate frame (n) was set as the reference system, which was consistent with the coordinate system of UWB and MoCap. The body frame (b) was attached to each body segment where the sensors were located. It aligned with the n-frame initially. Neglecting the installation errors, the sensor frame (s) was attached to each IMU and aligned with the b-frame to track the movements of human upper limbs.

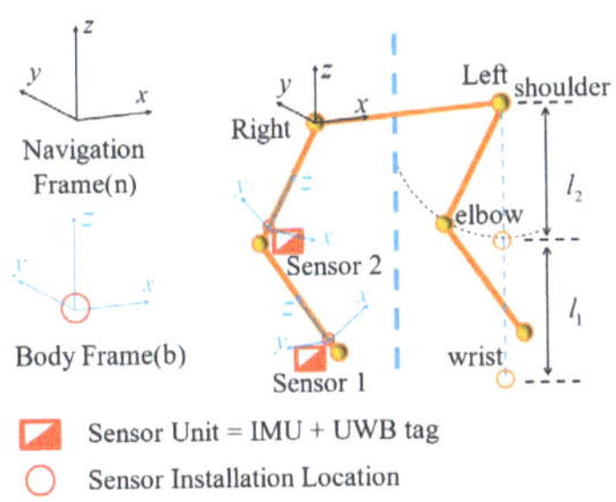

Figure 1. Upper limb kinematics model and sensor unit arrangement.

The overall framework of the proposed algorithm consists of three parts, illustrated in Figure 2, including (1) Quaternion RK4 Based Madgwick Orientation Complementary Filter, (2) UWB localization Kalman filter, and (3) IMU/UWB Fusion Kalman filter. For the purpose of estimating the 3-D spatial trajectory of the movements of human upper limbs, we first employed the quaternion method to calculate the attitude angle using gyroscope measurements. However, the integration of gyroscope measurement errors contributed to an accumulating error in the quaternion algorithm [14]. Therefore, we considered adopting the Madgwick orientation filter to improve the motion estimation accuracy through the optimal fusion of the accelerometer, gyroscope, and magnetometer of the MARG system. Then, in order to further reduce the drift error, we proposed that the UWB localization system be combined with the IMU sensors. An extended Kalman filter was utilized to fuse IMU and UWB in order to perform stable and high-precision motion estimation of the human upper limbs.

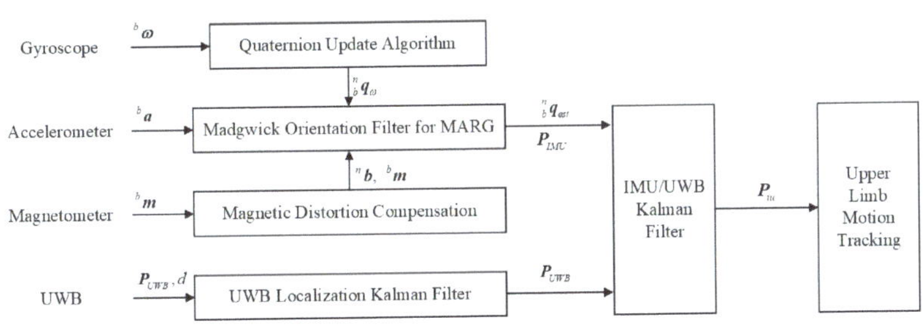

Figure 2. Framework of the proposed upper limb motion tracking algorithm.

2.1. Quaternion RK4 Based Madgwick Orientation Complementary Filter

2.1.1. Fourth-Order Runge–Kutta-Based Quaternion Update Algorithm

Rotation transformation can be described as a vector rotating around a specified rotation axis with a certain angle in a coordinate system. The unit quaternion is defined as

$$\boldsymbol{q} = q_1 + \boldsymbol{i}q_2 + \boldsymbol{j}q_3 + \boldsymbol{k}q_4 \tag{1}$$

For the same vector ${}^n\boldsymbol{r}$ and ${}^b\boldsymbol{r}$, defined in the n-frame and b-frame, respectively, ${}^n\boldsymbol{r}'$ and ${}^b\boldsymbol{r}'$ are their extended forms, containing a 0 inserted as the real part [14]. ${}^n_b\boldsymbol{q}$ describes the rotation of the n-frame relative to the b-frame, so ${}^n\boldsymbol{r}'$ can be expressed as

$$\begin{aligned} {}^n\boldsymbol{r}' &= {}^n_b\boldsymbol{q} \otimes {}^b\boldsymbol{r}' \otimes {}^n_b\boldsymbol{q}^* = {}^n_b\boldsymbol{q} \otimes \left[M'({}^n_b\boldsymbol{q}^*){}^b\boldsymbol{r}'\right] = M({}^n_b\boldsymbol{q})M'({}^n_b\boldsymbol{q}^*){}^b\boldsymbol{r}' \\ &= \begin{bmatrix} q_1 & -q_2 & -q_3 & -q_4 \\ q_2 & q_1 & -q_4 & q_3 \\ q_3 & q_4 & q_1 & -q_2 \\ q_4 & -q_3 & q_2 & q_1 \end{bmatrix} \begin{bmatrix} q_1 & q_2 & q_3 & q_4 \\ -q_2 & q_1 & -q_4 & q_3 \\ -q_3 & q_4 & q_1 & -q_2 \\ -q_4 & -q_3 & q_2 & q_1 \end{bmatrix} \begin{bmatrix} 0 \\ x \\ y \\ z \end{bmatrix} \end{aligned} \tag{2}$$

$${}^n_b\boldsymbol{q} \otimes {}^b\boldsymbol{r}' = \begin{bmatrix} q_1 & -q_2 & -q_3 & -q_4 \\ q_2 & q_1 & -q_4 & q_3 \\ q_3 & q_4 & q_1 & -q_2 \\ q_4 & -q_3 & q_2 & q_1 \end{bmatrix} \begin{bmatrix} 0 \\ x \\ y \\ z \end{bmatrix} \tag{3}$$

where $\boldsymbol{q}^*$ is the conjugate of $\boldsymbol{q}$; the symbol $\otimes$ denotes the quaternion product [21], as defined in (3); $M({}^n_b\boldsymbol{q})$ is the left multiplication matrix of the quaternion ${}^n_b\boldsymbol{q}$; and $M'({}^n_b\boldsymbol{q}^*)$ is the right multiplication matrix of the conjugate quaternion ${}^n_b\boldsymbol{q}^*$. Furthermore, the rotation can be represented by a rotation matrix.

$${}^n\boldsymbol{r} = {}^n_b\boldsymbol{C}{}^b\boldsymbol{r} \tag{4}$$

where ${}^n_b\boldsymbol{C}$ represents the rotation of the n-frame relative to the b-frame.

$${}^n_b C = \begin{bmatrix} (q_1{}^2 + q_2{}^2 - q_3{}^2 - q_4{}^2) & 2(q_2q_3 - q_1q_4) & 2(q_2q_4 + q_1q_3) \\ 2(q_2q_3 + q_1q_4) & (q_1{}^2 - q_2{}^2 + q_3{}^2 - q_4{}^2) & 2(q_3q_4 - q_1q_2) \\ 2(q_2q_4 - q_1q_3) & 2(q_3q_4 + q_1q_2) & (q_1{}^2 - q_2{}^2 - q_3{}^2 + q_4{}^2) \end{bmatrix} \tag{5}$$

The quaternion update algorithm uses the angular velocity increment in the sample period measured by the IMU sensors to calculate the quaternion at each time in order to update the human motion data. The quaternion derivative is given by

$$\begin{aligned} {}^n_b\dot{\boldsymbol{q}} &= \tfrac{1}{2}{}^n_b\boldsymbol{q} \otimes {}^b\boldsymbol{\omega} = \tfrac{1}{2}M({}^n_b\boldsymbol{q}){}^b\boldsymbol{\omega} = \tfrac{1}{2}M'\left({}^b\boldsymbol{\omega}\right){}^n_b\boldsymbol{q} \\ &= \frac{1}{2}\begin{bmatrix} q_1 & -q_2 & -q_3 & -q_4 \\ q_2 & q_1 & -q_4 & q_3 \\ q_3 & q_4 & q_1 & -q_2 \\ q_4 & -q_3 & q_2 & q_1 \end{bmatrix} \begin{bmatrix} 0 \\ {}_b\omega_x \\ {}_b\omega_y \\ {}_b\omega_z \end{bmatrix} = \frac{1}{2}\begin{bmatrix} 0 & -{}_b\omega_x & -{}_b\omega_y & -{}_b\omega_z \\ {}_b\omega_x & 0 & {}_b\omega_z & -{}_b\omega_y \\ {}_b\omega_y & -{}_b\omega_z & 0 & {}_b\omega_x \\ {}_b\omega_z & {}_b\omega_y & -{}_b\omega_x & 0 \end{bmatrix} \begin{bmatrix} q_1 \\ q_2 \\ q_3 \\ q_4 \end{bmatrix} \end{aligned} \tag{6}$$

where ${}^b\boldsymbol{\omega} = \begin{bmatrix} 0 & {}^b\omega_x & {}_b\omega_y & {}_b\omega_z \end{bmatrix}^T$ is the angular velocity measured by the gyroscope.

Given the initial value and the rotational angular velocity, the orientation at each time ${}^n_b\boldsymbol{q}_{\omega,t}$ can be obtained by numerically integrating the quaternion derivative ${}^n_b\dot{\boldsymbol{q}}_{\omega,t}$.

$${}^n_b\boldsymbol{q}_{\omega,t} = {}^n_b\hat{\boldsymbol{q}}_{est,t-1} + {}^n_b\dot{\boldsymbol{q}}_{\omega,t}\Delta t = {}^n_b\hat{\boldsymbol{q}}_{est,t-1} + \frac{1}{2}\Delta t{}^n_b\hat{\boldsymbol{q}}_{est,t-1} \otimes {}^b\boldsymbol{\omega}_t \tag{7}$$

where Δt is the sampling period, ${}^n_b\hat{\boldsymbol{q}}_{est,t-1}$ is the estimation of the quaternion at the previous time, and ${}_b\boldsymbol{\omega}_t$ is the angular velocity at time t.

Although increasing the order of the above algorithm can improve the computational accuracy, the complexity is also increasing. The Runge–Kutta method is a high-precision single-step algorithm; the most classic and widely used is the fourth-order Runge–Kutta algorithm. It can perform iterative operations on definite solution problems with known initial values and equations without solving differential equations, which enormously reduces the computational complexity. In this paper, the calculation of human upper limb motion can be regarded as the initial value problem of a differential equation. The calculation formulas are as follows.

$$
\begin{aligned}
\Delta_b^n \dot{q}_1 &= \tfrac{1}{2} {}_b^n\hat{q}_{est,t-1} \otimes {}^b\omega_{t-1} \\
\Delta_b^n \dot{q}_2 &= \tfrac{1}{2}\left({}_b^n\hat{q}_{est,t-1} + \frac{\Delta_b^n \dot{q}_1}{2}\right) \otimes {}^b\omega_{t-1/2} \\
\Delta_b^n \dot{q}_3 &= \tfrac{1}{2}\left({}_b^n\hat{q}_{est,t-1} + \frac{\Delta_b^n \dot{q}_2}{2}\right) \otimes {}^b\omega_{t-1/2} \\
\Delta_b^n \dot{q}_4 &= \tfrac{1}{2}\left({}_b^n\hat{q}_{est,t-1} + \Delta_b^n \dot{q}_3\right) \otimes {}^b\omega_t \\
{}_b^n q_{\omega,t} &= {}_b^n\hat{q}_{est,t-1} + \tfrac{1}{6}\Delta t\left(\Delta_b^n \dot{q}_1 + 2\Delta_b^n \dot{q}_2 + 2\Delta_b^n \dot{q}_3 + \Delta_b^n \dot{q}_4\right)
\end{aligned} \tag{8}
$$

where ${}^b\omega_{t-1/2}$ is the angular velocity at the intermediate time between t − 1 and t.

The quaternion fourth-order Runge–Kutta algorithm is designed to interpolate in the integration interval, and the slope is iteratively optimized at each step of the calculation to obtain an updated value.

2.1.2. Madgwick RK4 Orientation Complementary Filter for MARG

By formulating an objective function, this filter fuses the data of the tri-axis accelerometer, gyroscope, and magnetometer of the MARG system, then iteratively optimizes it to calculate the orientation ${}_b^n\hat{q}$.

Let the predefined reference vector in the n-frame be ${}^n\hat{d}$ and the sensor measurement in the b-frame be ${}^b\hat{s}$. The objective function is defined as

$$
\begin{aligned}
&\min_{{}_b^n\hat{q} \in \mathbb{R}^4} f\left({}_b^n\hat{q}, {}^n\hat{d}, {}^b\hat{s}\right) \\
&f\left({}_b^n\hat{q}, {}^n\hat{d}, {}^b\hat{s}\right) = {}_b^n\hat{q}^* \otimes {}^n\hat{d} \otimes {}_b^n\hat{q} - {}^b\hat{s} \\
&{}^n\hat{d} = \begin{bmatrix} 0 & d_x & d_y & d_z \end{bmatrix} \\
&{}^b\hat{s} = \begin{bmatrix} 0 & s_x & s_y & s_z \end{bmatrix}
\end{aligned} \tag{9}
$$

By recording the initial quaternion ${}_b^n\hat{q}_0$ and step size μ, the estimation of ${}_b^n\hat{q}_{n+1}$ is obtained for n iterations, adopting the general gradient descent algorithm illustrated in (10). The symbol ∇ represents the gradient of the objective function, which can be computed by the objective function and its Jacobian, as shown in (11).

$$
{}_b^n q_{k+1} = {}_b^n\hat{q}_k - \mu \frac{\nabla f\left({}_b^n\hat{q}_k, {}^n\hat{d}, {}^b\hat{s}\right)}{\left\| \nabla f\left({}_b^n\hat{q}_k, {}^n\hat{d}, {}^b\hat{s}\right) \right\|}, \ k = 0, 1, 2 \cdots n \tag{10}
$$

$$
\nabla f\left({}_b^n\hat{q}_k, {}^n\hat{d}, {}^b\hat{s}\right) = J^T\left({}_b^n\hat{q}_k, {}^n\hat{d}\right) f\left({}_b^n\hat{q}_k, {}^n\hat{d}, {}^b\hat{s}\right) \tag{11}
$$

The specific algorithm expression of the accelerometer is described in (12). Normalized gravity ${}^n\hat{g}$ and normalized accelerometer measurements ${}^b\hat{a}$ substituted ${}^n\hat{d}$ and ${}^b\hat{s}$, respectively.

$$
\begin{aligned}
&{}^n\hat{g} = \begin{bmatrix} 0 & 0 & 0 & 1 \end{bmatrix} \\
&{}^b\hat{a} = \begin{bmatrix} 0 & a_x & a_y & a_z \end{bmatrix} \\
&f_g\left({}_b^n\hat{q}, {}^b\hat{a}\right) = \begin{bmatrix} 2(q_2 q_4 - q_1 q_3) - a_x \\ 2(q_1 q_2 + q_3 q_4) - a_y \\ 2(0.5 - q_2^2 - q_3^2) - a_z \end{bmatrix} \\
&J_g\left({}_b^n\hat{q}\right) = \begin{bmatrix} -2q_3 & 2q_4 & -2q_1 & 2q_2 \\ 2q_2 & 2q_1 & 2q_4 & 2q_3 \\ 0 & -4q_2 & -4q_3 & 0 \end{bmatrix}
\end{aligned} \tag{12}
$$

The accelerometer compensated the tilt angle of motion estimation obtained by the gyroscope measurement, but the heading angle was not compensated, as this requires the introduction of a magnetometer. The magnetometer can be used to measure the strength and direction of the magnetic field and to determine the orientation of a device.

As for the magnetometer, the geomagnetic field ${}^{n}\hat{\boldsymbol{b}}$, as the substitution of ${}^{n}\hat{\boldsymbol{d}}$, can be decomposed into a horizontal axis and a vertical axis theoretically. Additionally, the normalized magnetometer measurement ${}^{b}\hat{\boldsymbol{m}}$ is the substitution of ${}^{b}\hat{\boldsymbol{s}}$. The specific form can be written as

$$\begin{aligned}
{}^{n}\hat{\boldsymbol{b}} &= \begin{bmatrix} 0 & 0 & b_y & b_z \end{bmatrix} \\
{}^{b}\hat{\boldsymbol{m}} &= \begin{bmatrix} 0 & m_x & m_y & m_z \end{bmatrix} \\
f_b\left({}^{n}_{b}\hat{\boldsymbol{q}}, {}^{n}\hat{\boldsymbol{b}}, {}^{b}\hat{\boldsymbol{m}}\right) &= \begin{bmatrix} 2b_y(q_1q_4 + q_2q_3) + 2b_z(q_2q_4 - q_1q_3) - m_x \\ 2b_y(0.5 - q_2^2 - q_4^2) + 2b_z(q_1q_2 + q_3q_4) - m_y \\ 2b_y(q_3q_4 - q_1q_2) + 2b_z(0.5 - q_2^2 - q_3^2) - m_z \end{bmatrix} \\
J_b\left({}^{n}_{b}\hat{\boldsymbol{q}}, {}^{n}\hat{\boldsymbol{b}}\right) &= \begin{bmatrix} 2b_yq_4 - 2b_zq_3 & 2b_yq_3 + 2b_zq_4 & 2b_yq_2 - 2b_zq_1 & 2b_yq_1 + 2b_zq_2 \\ 2b_zq_2 & -4b_yq_2 + 2b_zq_1 & 2b_zq_4 & -4b_yq_4 + 2b_zq_3 \\ -2b_yq_2 & -2b_yq_1 - 4b_zq_2 & 2b_yq_4 - 4b_zq_3 & 2b_yq_3 \end{bmatrix}
\end{aligned} \tag{13}$$

However, the magnetometer may be disturbed by the bias of hard iron and soft iron, causing errors in the measurement direction of the earth's magnetic field, so magnetic distortion compensation needs to be employed.

$${}^{n}\hat{\boldsymbol{h}}_t = \begin{bmatrix} 0 & h_x & h_y & h_z \end{bmatrix} = {}^{n}_{b}\hat{\boldsymbol{q}}_{est,t-1} \otimes {}^{b}\hat{\boldsymbol{m}}_t \otimes {}^{n}_{b}\hat{\boldsymbol{q}}^{*}_{est,t-1} \tag{14}$$

$${}^{n}\hat{\boldsymbol{b}}_t = \begin{bmatrix} 0 & 0 & \sqrt{h_x^2 + h_y^2} & h_z \end{bmatrix} \tag{15}$$

where ${}^{n}\hat{\boldsymbol{h}}_t$ is the normalized magnetometer measurement in the n-frame and ${}^{n}\hat{\boldsymbol{b}}_t$ is the compensated geomagnetic field at time t.

Combining the specific algorithms of the accelerometer and magnetometer, the formulas were developed as follows.

$$\begin{aligned}
f_{g,b}\left({}^{n}_{b}\hat{\boldsymbol{q}}, {}^{b}\hat{\boldsymbol{a}}, {}^{n}\hat{\boldsymbol{b}}, {}^{b}\hat{\boldsymbol{m}}\right) &= \begin{bmatrix} f_g\left({}^{n}_{b}\hat{\boldsymbol{q}}, {}^{b}\hat{\boldsymbol{a}}\right) \\ f_b\left({}^{n}_{b}\hat{\boldsymbol{q}}, {}^{n}\hat{\boldsymbol{b}}, {}^{b}\hat{\boldsymbol{m}}\right) \end{bmatrix} \\
J_{g,b}\left({}^{n}_{b}\hat{\boldsymbol{q}}, {}^{n}\hat{\boldsymbol{b}}\right) &= \begin{bmatrix} J_g^T\left({}^{n}_{b}\hat{\boldsymbol{q}}\right) \\ J_b^T\left({}^{n}_{b}\hat{\boldsymbol{q}}, {}^{n}\hat{\boldsymbol{b}}\right) \end{bmatrix}
\end{aligned} \tag{16}$$

By substituting (16) into (11), the gradient of the combined objective function can be written as

$$\nabla f = J_{g,b}^T\left({}^{n}_{b}\hat{\boldsymbol{q}}_{est,t-1}, {}^{n}\hat{\boldsymbol{b}}\right) f_{g,b}\left({}^{n}_{b}\hat{\boldsymbol{q}}_{est,t-1}, {}^{b}\hat{\boldsymbol{a}}, {}^{n}\hat{\boldsymbol{b}}, {}^{b}\hat{\boldsymbol{m}}\right) \tag{17}$$

And the estimated quaternion ${}^{n}_{b}\boldsymbol{q}_{\nabla,t}$ calculated at time t can be given by

$${}^{n}_{b}\boldsymbol{q}_{\nabla,t} = {}^{n}_{b}\hat{\boldsymbol{q}}_{est,t-1} - \mu_t \frac{\nabla f}{\|\nabla f\|} \tag{18}$$

Employing the Madgwick orientation filter, an estimated rotation ${}^{n}_{b}\boldsymbol{q}_{est,t}$ was obtained by fusing ${}^{n}_{b}\boldsymbol{q}_{\omega,t}$ and ${}^{n}_{b}\boldsymbol{q}_{\nabla,t}$.

$${}^{n}_{b}\boldsymbol{q}_{est,t} = \gamma_t {}^{n}_{b}\boldsymbol{q}_{\nabla,t} + (1 - \gamma_t) {}^{n}_{b}\boldsymbol{q}_{\omega,t},\ 0 \le \gamma_t \le 1 \tag{19}$$

where γ_t represents weight, ensuring that the weighted divergence of ${}^{n}_{b}\boldsymbol{q}_{\omega,t}$ equals the weighted convergence of ${}^{n}_{b}\boldsymbol{q}_{\nabla,t}$.

According to [14], it is known that when noise augmentation of μ_t is assumed to be extremely high, (19) can be rewritten as

$$ {}^n_b q_{est,t} = {}^n_b \hat{q}_{est,t-1} + {}^n_b \dot{q}_{est,t}\Delta t = {}^n_b \hat{q}_{est,t-1} + \Delta t\left({}^n_b \dot{q}_{\omega,t} - \beta\frac{\nabla f}{\|\nabla f\|}\right) \tag{20} $$

where β is the divergence rate of ${}^n_b q_{\omega,t}$.

By substituting (8) into (20), it can be written in the fourth-order Runge–Kuta form as

$$ {}^n_b q_{est,t} = {}^n_b \hat{q}_{est,t-1} + \Delta t\left(\frac{1}{6}(\Delta^n_b \dot{q}_1 + 2\Delta^n_b \dot{q}_2 + 2\Delta^n_b \dot{q}_3 + \Delta^n_b \dot{q}_4) - \beta\frac{\nabla f}{\|\nabla f\|}\right) \tag{21} $$

The process of the quaternion RK4-based Madgwick orientation filter algorithm is summarized as follows in Algorithm 1.

Algorithm 1: Process of quaternion RK4-based Madgwick orientation complementary filter.

Initialization: ${}^n_b q_0$
Input: ${}^b\omega_t, \Delta t, {}^b\hat{a}_t, {}^b\hat{m}_t, \beta$
Output: ${}^n_b q_{est,t}$
1: ${}^n_b \dot{q}_{\omega,t} \leftarrow 0.5\, {}^n_b \hat{q}_{est,t-1} \otimes {}^b\omega_t$
2: ${}^n_b q_{\omega,t} \leftarrow {}^n_b \hat{q}_{est,t-1} + {}^n_b \dot{q}_{\omega,t}\Delta t$
3: $\Delta^n_b \dot{q}_1, \Delta^n_b \dot{q}_2, \Delta^n_b \dot{q}_3, \Delta^n_b \dot{q}_4 \leftarrow {}^n_b \hat{q}_{est,t-1}, {}^b\omega_{t-1}, {}^b\omega_{t-1/2}, {}^b\omega_t$
4: ${}^n_b q_{\omega,t} \leftarrow {}^n_b \hat{q}_{est,t-1} + \Delta t(\Delta^n_b \dot{q}_1 + 2\Delta^n_b \dot{q}_2 + 2\Delta^n_b \dot{q}_3 + \Delta^n_b \dot{q}_4)/6$
5: ${}^n\hat{h}_t \leftarrow {}^n_b \hat{q}_{est,t-1} \otimes {}^b\hat{m}_t \otimes {}^n_b \hat{q}^*_{est,t-1}$; ${}^n\hat{b}_t \leftarrow {}^n\hat{h}_t$
6: $f_{g,b}\left({}^n_b \hat{q}, {}^b\hat{a}, {}^n\hat{b}, {}^b\hat{m}\right) \leftarrow \left[f_g\left({}^n_b \hat{q}, {}^b\hat{a}\right) \quad f_b\left({}^n_b \hat{q}, {}^n\hat{b}, {}^b\hat{m}\right)\right]^T$
7: $J_{g,b}\left({}^n_b \hat{q}, {}^n\hat{b}\right) \leftarrow \left[J^T_g\left({}^n_b \hat{q}\right) \quad J^T_b\left({}^n_b \hat{q}, {}^n\hat{b}\right)\right]^T$
8: $\nabla f \leftarrow J^T_{g,b}\left({}^n_b \hat{q}_{est,t-1}, {}^n\hat{b}\right) f_{g,b}\left({}^n_b \hat{q}_{est,t-1}, {}^b\hat{a}, {}^n\hat{b}, {}^b\hat{m}\right)$
9: ${}^n_b q_{est,t} \leftarrow {}^n_b q_{\omega,t} - \Delta t\beta\frac{\nabla f}{\|\nabla f\|}$
Return: ${}^n_b q_{est,t}$

2.2. UWB Localization Kalman Filter

The UWB positioning system contained a tag-anchor wireless communication channel, as shown in Figure 3. The calculation of the distances between each tag and anchor exploiting double-sided two-way ranging (DS-TWR) method and position estimation employed distances [28,29]. In addition to classical UWB positioning algorithms such as MLE, LLSE, and WCE, we used EKF for position estimation due to its lower complexity and shorter computational time [29].

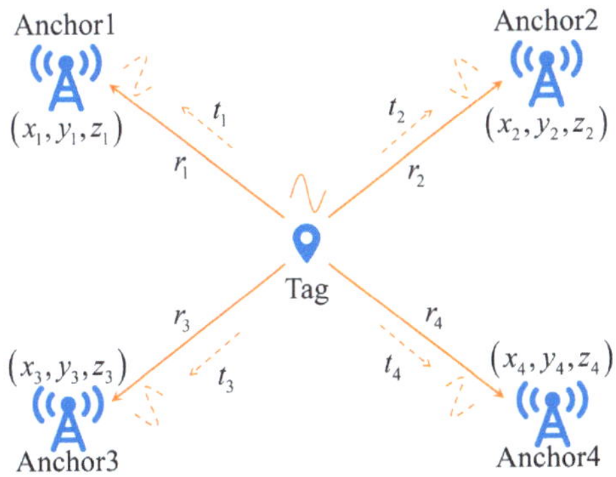

Figure 3. UWB positioning system.

It was defined that the UWB localization system was aligned with the n-frame. The system model adopted for UWB EKF algorithm can be expressed as

$$\begin{aligned} \boldsymbol{x}_{u,t} &= \boldsymbol{x}_{u,t-1} + \boldsymbol{w}_{u,t-1} \\ \boldsymbol{y}_{u,t} &= h_{u,t}(\boldsymbol{x}_{u,t}) + \boldsymbol{v}_{u,t} \end{aligned} \tag{22}$$

where the subscript u represents the UWB localization Kalman filter, and the state vector $\boldsymbol{x}_{u,t} = \boldsymbol{P}_{UWB,t} = \begin{bmatrix} x_t & y_t & z_t \end{bmatrix}^T$ is a 3×1 vector where $\boldsymbol{P}_{UWB,t}$ represents the 3-D position of UWB tags in the n-frame. The measurement vector $\boldsymbol{y}_{u,t} = \begin{bmatrix} d_{1,t} & d_{2,t} & d_{3,t} & d_{4,t} \end{bmatrix}^T$ denotes the distances between each anchor and tag. The expansion form of $h_{u,t}(\boldsymbol{x}_{u,t})$ is expressed in (23). The process noise $\boldsymbol{w}_{u,t-1}$ and measurement noise $\boldsymbol{v}_{u,t}$ are zero mean additional Gaussian white noise with covariance matrices of $\boldsymbol{Q}_{u,t-1}$ and $\boldsymbol{R}_{u,t}$ respectively. These matrices were obtained from the UWB system, and distinguished the ranging accuracy of different anchors.

$$h_{u,t}(\boldsymbol{x}_{u,t}) = \begin{bmatrix} \sqrt{(x_t - x_1)^2 + (y_t - y_1)^2 + (z_t - z_1)^2} \\ \sqrt{(x_t - x_2)^2 + (y_t - y_2)^2 + (z_t - z_2)^2} \\ \sqrt{(x_t - x_3)^2 + (y_t - y_3)^2 + (z_t - z_3)^2} \\ \sqrt{(x_t - x_4)^2 + (y_t - y_4)^2 + (z_t - z_4)^2} \end{bmatrix} \tag{23}$$

where $x_i, y_i, z_i, i = 1, 2, 3, 4$ represent the position of the i-th anchor.

Thus, the Jacobian matrix $\boldsymbol{H}_{u,t}$ could be calculated as shown in (24). And for the sake of brevity, $D_{i,t}$, $i = 1, 2, 3, 4$ is the corresponding row of $h_{u,t}(\boldsymbol{x}_{u,t})$.

$$\boldsymbol{H}_{u,t} = \left.\frac{\partial h_{u,t}(\boldsymbol{x}_{u,t})}{\partial \boldsymbol{x}_{u,t}}\right|_{\boldsymbol{x}_{u,t}} = \begin{bmatrix} \frac{(x_t - x_1)}{D_{1,t}} & \frac{(y_t - y_1)}{D_{1,t}} & \frac{(z_t - z_1)}{D_{1,t}} \\ \frac{(x_t - x_2)}{D_{2,t}} & \frac{(y_t - y_2)}{D_{2,t}} & \frac{(z_t - z_2)}{D_{2,t}} \\ \frac{(x_t - x_3)}{D_{3,t}} & \frac{(y_t - y_3)}{D_{3,t}} & \frac{(z_t - z_3)}{D_{3,t}} \\ \frac{(x_t - x_4)}{D_{4,t}} & \frac{(y_t - y_4)}{D_{4,t}} & \frac{(z_t - z_4)}{D_{4,t}} \end{bmatrix} \tag{24}$$

Assuming that $\boldsymbol{P}_{u,0}$ is the initial state estimation covariance and is known as the prior, as well as the initial, state vector $\boldsymbol{x}_{u,0}$, the posterior estimations $\boldsymbol{P}_{u,t}$ and $\boldsymbol{x}_{u,t}$ were recursively obtained by employing the prediction and update functions. It was defined that $\boldsymbol{P}_{u,t|t-1}$ and $\boldsymbol{x}_{u,t|t-1}$ were the prediction forms illustrated in (25) and could be calculated using the posterior estimations $\boldsymbol{P}_{u,t-1}$ and $\boldsymbol{x}_{u,t-1}$ at time $t-1$.

$$\begin{aligned} \boldsymbol{P}_{u,t|t-1} &= \boldsymbol{P}_{u,t-1} + \boldsymbol{Q}_{u,t-1} \\ \boldsymbol{x}_{u,t|t-1} &= \boldsymbol{x}_{u,t-1} \end{aligned} \tag{25}$$

Then, the prediction and the measurement vector $\boldsymbol{y}_{u,t}$ were used to update the prior estimations, and the posterior estimations at time t could be expressed as

$$\begin{aligned} \boldsymbol{k}_{u,t} &= \boldsymbol{P}_{u,t|t-1}\boldsymbol{H}_{u,t}^T\left[\boldsymbol{H}_{u,t}\boldsymbol{P}_{u,t|t-1}\boldsymbol{H}_{u,t}^T + \boldsymbol{R}_{u,t}\right]^{-1} \\ \boldsymbol{x}_{u,t} &= \boldsymbol{x}_{u,t|t-1} + \boldsymbol{k}_{u,t}\left[\boldsymbol{y}_{u,t} - h_{u,t}\left(\boldsymbol{x}_{u,t|t-1}\right)\right] \\ \boldsymbol{P}_{u,t} &= (\boldsymbol{I}_{3\times 3} - \boldsymbol{k}_{u,t}\boldsymbol{H}_{u,t})\boldsymbol{P}_{u,t|t-1} \end{aligned} \tag{26}$$

where $\boldsymbol{x}_{u,t}$ is the 3-D position of the UWB tag, and will be used as the measurement vector in the IMU/UWB Kalman filter.

2.3. IMU/UWB Fusion Kalman Filter

The innovative aspect of this IMU/UWB Kalman filter is that it uses the drift-free position calculated by the UWB system to compensate for the orientation and position estimated by the IMU system. The information obtained by the IMU system was employed as the state vector, and the position calculated by the UWB system was applied as the measurement vector. The system model can be expressed as

$$\begin{aligned} x_{iu,t} &= f_{iu,t-1}\left(x_{iu,t-1}, u_{iu,t-1}, w_{iu,t-1}\right) \\ y_{iu,t} &= h_{iu,t}\left(x_{iu,t}, v_{iu,t}\right) \end{aligned} \tag{27}$$

where the subscript iu represents the IMU/UWB Kalman filter. $x_{iu,t}$ is the state vector, $u_{iu,t-1}$ is the input vector, $y_{iu,t}$ is the measurement vector, and $w_{iu,t-1}$ and $v_{iu,t}$ are the process noise and measurement noise, respectively.

The state vector $x_{iu,t} = \left[P_{IMU,t} \quad {}^n_b q_{est,t}\right]^T$ is a 7×1 vector where $P_{IMU,t}$ represents the 3-D position of human upper limb and ${}^n_b q_{est,t}$ is the orientation obtained from the Madgwick complementary filter. The state model consists of two parts, and the part ${}^n_b q_{est,t}$ can be expressed as

$$\begin{aligned} {}^n_b q_{est,t} &= {}^n_b q_{est,t-1} + \Delta t\, {}^n_b \dot{q}_{est,t} \\ &= {}^n_b q_{est,t-1} + \Delta t\left({}^n_b \dot{q}_{\omega,t} - \beta \frac{\nabla f}{\|\nabla f\|}\right) \\ &= {}^n_b q_{est,t-1} + \Delta t\left[\frac{1}{2}{}^n_b q_{est,t-1} \otimes \left({}^b\omega_t + {}^b\omega_{b,t}\right) - \beta \frac{\nabla f}{\|\nabla f\|}\right] \end{aligned} \tag{28}$$

where ${}^b\omega_{b,t}$ is the bias of the gyroscope and $\frac{\nabla f}{\|\nabla f\|}$ is the input vector.

Supposing that $P_{IMU,0}$ is the initial position of the IMU sensors, then $P_{IMU,t}$ can be recursively calculated from a geometric perspective using the rotation quaternion ${}^n_b q_{est,t}$ or the rotation matrix ${}^n_b C_t$ in (5), updated by ${}^n_b q_{est,t}$. The position part of the state model is given by (29) and (30).

$$\begin{aligned} P^u_{IMU,1} &= \left[{}^n_b C_{u,1}\left(P^u_{IMU,0}\right)^T\right]^T \\ P^u_{IMU,2} &= \left[{}^n_b C_{u,2}\left(P^u_{IMU,0}\right)^T\right]^T = \left[{}^n_b C_{u,2}\,{}^n_b C^T_{u,1}\left(P^u_{IMU,1}\right)^T\right]^T \\ P^u_{IMU,3} &= \left[{}^n_b C_{u,3}\left(P^u_{IMU,0}\right)^T\right]^T = \left[{}^n_b C_{u,3}\,{}^n_b C^T_{u,2}\left(P^u_{IMU,2}\right)^T\right]^T \\ &\cdots \\ P^u_{IMU,t} &= \left[{}^n_b C_{u,t}\,{}^n_b C^T_{u,t-1}\left(P^u_{IMU,t-1}\right)^T\right]^T \end{aligned} \tag{29}$$

where $P^u_{IMU,t}$ and ${}^n_b C_{u,t}$ represent the position and rotation matrix of the upper arm. The position of the forearm relies on the upper arm, and is given by

$$P^f_{IMU,t} = P^u_{IMU,t} + \left[{}^n_b C_{f,t}\,{}^n_b C^T_{f,t-1}\left(P^f_{IMU,t-1} - P^u_{IMU,t-1}\right)^T\right]^T \tag{30}$$

where $P^f_{IMU,t}$ and ${}^n_b C_{f,t}$ represent the position and rotation matrix of the forearm. When using the Kalman filter for the forearm, the input vector should contain the upper arm position.

The Jacobian matrices $F_{iu,t-1}$ and $L_{iu,t-1}$ were obtained.

$$F_{iu,t-1} = \left.\frac{\partial f_{iu,t-1}}{\partial x_{iu,t-1}}\right|_{x_{iu,t-1}} = \begin{bmatrix} {}^n_b C_t\,{}^n_b C^T_{t-1} & 0_{3\times4} \\ 0_{4\times3} & I_{4\times4} + \frac{1}{2}\Delta t M'\left({}^b\omega_t + {}^b\omega_{b,t}\right) \end{bmatrix} \tag{31}$$

$$L_{iu,t-1} = \left.\frac{\partial f_{iu,t-1}}{\partial w_{iu,t-1}}\right|_{x_{iu,t-1}} = \begin{bmatrix} 0_{3\times4} \\ \frac{1}{2}\Delta t M\left({}^n_b q_{est,t-1}\right) \end{bmatrix} \tag{32}$$

The measurement vector $y_{iu,t} = [P_{UWB,t}]^T$ is a 3×1 vector obtained from the UWB localization Kalman filter, representing the 3-D position of UWB tags in the n-frame. The measurement equation can be expressed as

$$y_{iu,t} = H_{iu,t} x_{iu,t} + v_{iu,t} \tag{33}$$

where $H_{iu,t}$ is the Jacobian matrix, expressed as

$$H_{iu,t} = \begin{bmatrix} I_{3\times3} & 0_{3\times4} \end{bmatrix} \tag{34}$$

The covariance matrices $Q_{iu,t-1}$ and $R_{iu,t}$ of the process noise $w_{iu,t-1}$ and the measurement noise $v_{iu,t}$ were calculated as

$$\begin{aligned} Q_{iu,t-1} &= E\left[w_{iu,t-1} w_{iu,t-1}^T\right] = \sum_G = \sigma_G^2 I_{4\times4} \\ R_{iu,t} &= E\left[v_{iu,t} v_{iu,t}^T\right] = \sum_{P,UWB} = \sigma_{P,UWB}^2 I_{3\times3} \end{aligned} \tag{35}$$

It was assumed that $P_{iu,0}$ and $x_{iu,0}$ were the initial state estimation covariance and state vector, respectively, and the prediction forms $P_{iu,t|t-1}$ and $x_{iu,t|t-1}$ could be recursively calculated using the following equations.

$$\begin{aligned} P_{iu,t|t-1} &= F_{iu,t-1} P_{iu,t-1} F_{iu,t-1}^T + L_{iu,t-1} Q_{iu,t-1} L_{iu,t-1}^T \\ x_{iu,t|t-1} &= f_{iu,t-1}\left(x_{iu,t-1}, u_{iu,t-1}, 0\right) \end{aligned} \tag{36}$$

Then, the prior estimations were updated, and the posterior estimations $P_{iu,t}$ and $x_{iu,t}$ were written as

$$\begin{aligned} k_{iu,t} &= P_{iu,t|t-1} H_{iu,t}^T \left[H_{iu,t} P_{iu,t|t-1} H_{iu,t}^T + R_{iu,t}\right]^{-1} \\ x_{iu,t} &= x_{iu,t|t-1} + k_{iu,t}\left[y_{iu,t} - h_{iu,t}\left(x_{iu,t|t-1}, 0\right)\right] \\ P_{iu,t} &= (I_{7\times7} - k_{iu,t} H_{iu,t}) P_{iu,t|t-1} \end{aligned} \tag{37}$$

where $x_{iu,t}$ is the fused 3-D position in the IMU/UWB Kalman filter, which was used for 3-D motion reconstruction of human upper limbs. A detailed flowchart of the UWB localization and the IMU/UWB Kalman filters is shown in Figure 4. The parameters were set as follows. The initial state estimation covariance $P_{u,0}$ was set to $0.1 \times I_{3\times3}$. All elements of $P_{iu,0}$ were set to 0.1. $x_{u,0}$ and $x_{iu,0}$ were obtained from the UWB system and the optical MoCap system for each trial of movement, respectively.

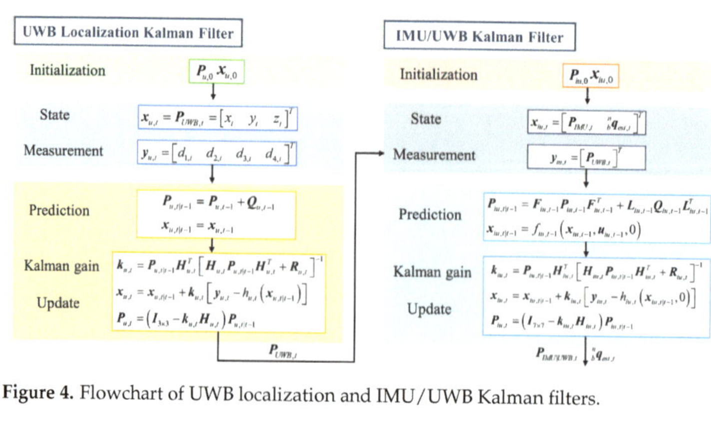

Figure 4. Flowchart of UWB localization and IMU/UWB Kalman filters.

3. Experimental Demonstration

3.1. Experimental Setup

To estimate the movement of human upper limbs, wearable 9-axis MPU9250 inertial sensor units (each sensor unit included a tri-axis accelerometer, a tri-axis gyroscope, and a tri-axis magnetometer) were adopted in our experiments. To avoid the extrusive effect of muscles of the human upper limbs, these two IMU sensors were placed on the lateral side, above the wrist and the elbow, respectively [11,21]. Each IMU sensor was connected to the computer via a serial port module for data transmission. The sample rate of the IMU sensors was set to 100 Hz, and the installation direction is shown in Figure 1.

Additionally, a DW1000 UWB real-time localization system manufactured by Haoru Technology, Dalian, China, was employed to track the 3-D position with an update rate of about 10 Hz. The UWB communication technology was implemented between the anchors and tags through the DS-TWR method based on time-domain transmission of radio signals, and the anchors were also connected to the computer via a serial port module. It provided an accuracy of 10 cm for the X/Y axis and 30 cm for the Z axis. Two UWB tags were fixed on the bracelets with the IMU sensors to avoid relative movement, as shown in Figure 5. To ensure the positioning stability of UWB, four anchors were arranged in our laboratory in a cube distribution.

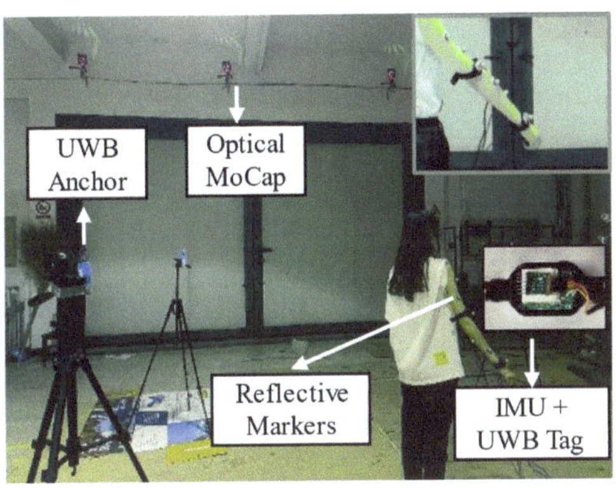

Figure 5. Setup of IMU sensors, UWB system, and optical MoCap system.

The Nokov optical MoCap system captures human motion through sixteen cameras evenly arranged in the laboratory, as illustrated in Figure 5, and is used as a reference system for algorithm verification and further comparison [34]. For each segment of the human upper limb, at least four reflective markers were attached to the skin surface and sensors, constituting an envelope rather than a cluster [9,11,38,39]. The optical MoCap system achieved sub-millimeter positioning accuracy and higher reliability.

3.2. Experimental Protocol

Our experiments aimed to capture several common movements of the human upper limbs, including simple whole-arm movements and combined upper-arm and forearm movements. These common movements comprised motion 1: shoulder flexion/extension and abduction/adduction (the whole arm moving along a circular trajectory); motion 2: shoulder abduction/adduction, internal/external rotation, and elbow flexion/extension; and motion 3: shoulder flexion/extension, internal/external rotation, and elbow flexion/extension, as shown in Figure 6. In each trial, one type of movement was performed periodically, with a duration of about one minute.

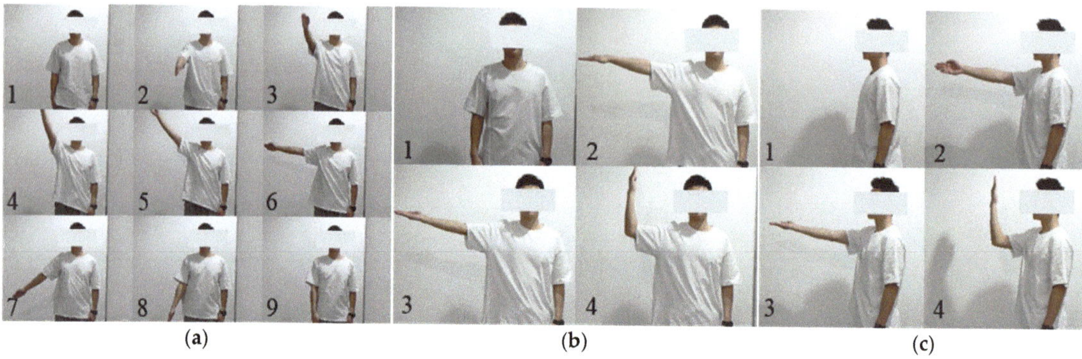

Figure 6. Tested movements. (**a**–**c**) represent motions 1–3, respectively.

As for the UWB system, four anchors arranged in conventional cube distribution were employed, as shown in Figure 7. The above movements were performed by one subject under each distribution throughout the experiment. The subject was a 24-year-old healthy female 165 cm in height and 40 kg in weight. It is important to mention that the optical MoCap system tracked the movement of the human upper limbs for each trial, which could be used to initialize the IMU sensors and obtain the positions of UWB anchors. Before each trial, the IMU sensors were calibrated, and a few seconds of stillness then enhanced the stability of the proposed algorithm.

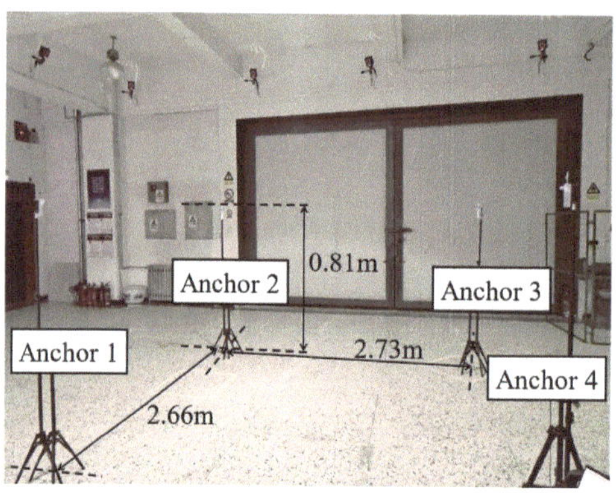

Figure 7. Cube distribution of four anchors of the UWB system.

4. Experimental Results and Discussion

4.1. Performance of Madgwick RK4 Orientation Filter

The performance of the Madgwick orientation filter in terms of attitude angle and 3-D position estimation is shown in this section. Figure 8 shows the Euler angles of the wrist for motion 1, the whole arm moving along a circular trajectory, which is beneficial for us in verifying the tri-axis rotation (roll, pitch, and yaw) of the upper limbs. On the basis of the attitude angle obtained by the quaternion fourth-order Runge–Kutta algorithm and the original second-order update algorithm illustrated in (7), the Madgwick orientation filter was combined, as shown in (21) and (20), respectively, and the tri-axis positions of sensor unit 1 during the movement were calculated as shown in Figure 9. The solid lines represent the positions calculated using various methods from the IMU system, whereas the black dotted lines are the reference positions obtained by the optical Mocap system. As can be seen, the proposed methods were able to track the movements of the upper limbs in spite of the errors at the inflection points.

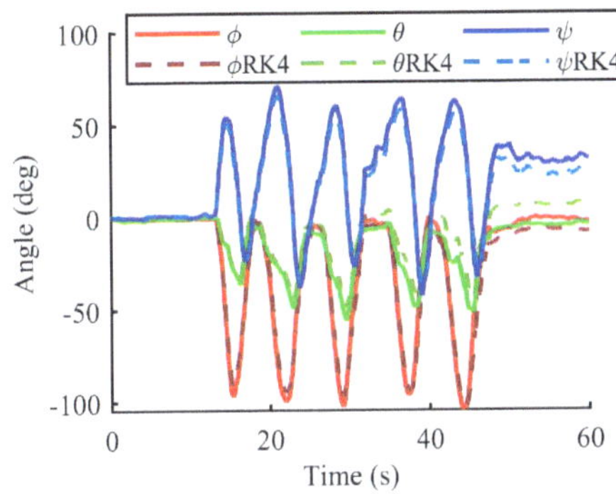

Figure 8. Euler angles of the forearm for motion 1 using the IMU system.

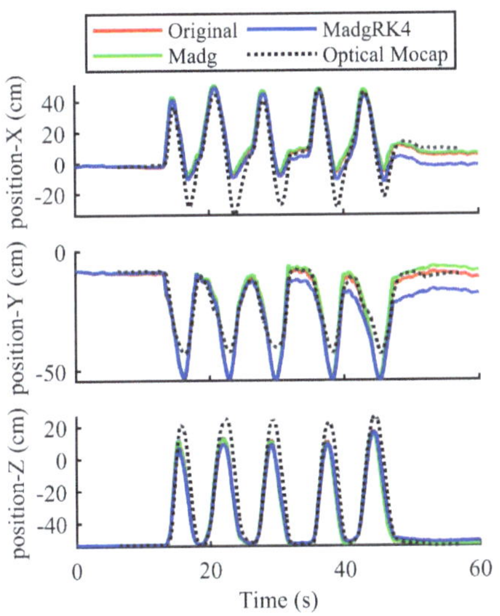

Figure 9. Positions of sensor unit 1 for motion 1.

The position accuracies of the three kinds of movements are summarized in Table 1. It shows the 3-D position accuracy of forearm motion tracking by the fourth-order Runge–Kutta Madgwick method and the original second-order update algorithm compared with the optical Mocap system. According to Table 1, the root mean square error (RMSE) values of tri-axis motion tracking of Madgwick RK4 method varied from 6.17 cm to 12.76 cm, which shows the feasibility of precise motion tracking of the human upper limbs. In order to further reduce the error, it is necessary to utilize the UWB measurements.

Table 1. The tri-axis position accuracy of forearm motion tracking by the Madgwick RK4 method and the original second-order update algorithm compared with the optical Mocap system.

	Position RMSE—Original Second -Order (cm)			Position RMSE—MadgwickRK4 (cm)		
	X	Y	Z	X	Y	Z
Motion 1	9.8361	4.9402	10.1491	8.6124	6.1745	9.3199
Motion 2	7.4554	7.5898	10.7254	6.6717	7.6876	10.5322
Motion 3	13.2050	11.1597	11.2611	12.7563	10.3026	11.9849

4.2. Performance of UWB Localization Kalman Filter

As for the UWB localization Kalman filter, we utilized multiple sets of movements of the human upper limb to verify the accuracy of the UWB positioning system. It is important to mention that the installation direction of the IMU was always the same as that in Figure 1 during the experiment. And the subject stood at the edge of the UWB area to ensure that there was no occlusion between the tags and anchors during the movement of the upper limbs. Before the experiment, we first kept the tag still and verified the static positioning accuracy of UWB. Table 2 shows the static positioning variance of the four-anchor UWB system. The mean values of the tri-axis variances fell within 0.6 cm^2, so it shared the good positioning robustness of the UWB system.

Table 2. Static positioning variance of the four-anchor UWB system.

	Static Positioning Variance (cm^2)		
	X	Y	Z
1	0.2970	0.4106	0.4080
2	0.2803	0.2255	0.7632
3	0.1668	0.3004	0.2671
4	0.1990	0.1649	0.2265
5	0.3813	0.4273	0.9591
mean	0.2649	0.3057	0.5248

Figure 10 shows a comparison of the position of upper limb motion tracking for three kinds of movements by the IMU, UWB, and optical Mocap systems. Motion 1, represented by Figure 10a, consisted of movement along a circular trajectory, which is consistent with of the motion in Figure 9. And motion 2 was a combined motion, represented by Figure 10b, comprising shoulder abduction/adduction, internal/external rotation, and elbow flexion/extension. Motion 3 was another combined motion, represented by Figure 10c and similar to motion 2, consisting of shoulder flexion/extension, internal/external rotation, and elbow flexion/extension. As can be seen in these figures, it is obvious that compared with the position calculated by the IMU sensors, the position calculated by the UWB system was closer to the reference value of the optical Mocap system at certain steps, especially the inflection points. Therefore, as for the measurement, the UWB system can be fused with IMU for upper limb motion estimation to reduce the error caused by the drift of the IMU sensors.

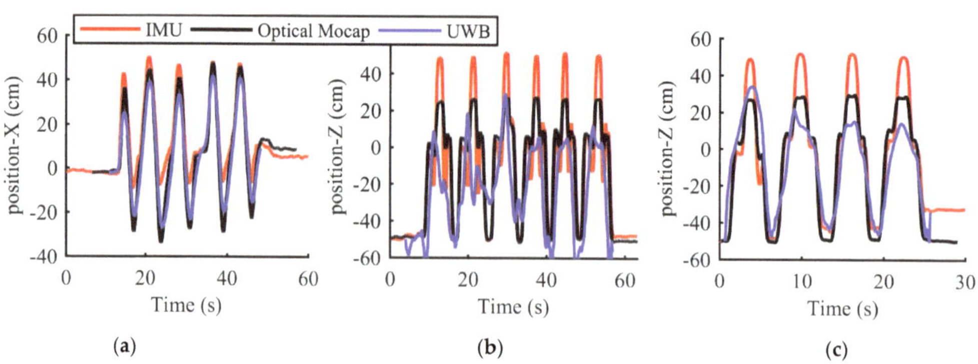

Figure 10. Comparison of the sensor unit 1 position of upper limb motion tracking for three kinds of movements by the IMU, UWB, and optical Mocap systems: (**a**–**c**) represent motion 1, motion 2, and motion 3, respectively.

4.3. Performance of IMU/UWB Kalman Filter

The performance of the IMU/UWB Kalman filter in 3-D upper limb motion tracking is shown in this section. As mentioned previously, by adopting the cube distribution of four anchors, the measurement of UWB was fused with IMU for drift error reduction and better positioning accuracy in motion estimation. The position's RMSE values are illustrated in Table 3, where motion 1, motion 2, and motion 3 are represented in Figure 10a–c, respectively. This shows a comparison of the 3-D position accuracy of various upper limb motion tracking methods, i.e., the IMU method, UWB method, and IMU/UWB fusion method with the optical Mocap system. Although the position accuracy of the UWB system may have sometimes fallen lower than that of the IMU, the adopted Kalman filter adaptively adjusted the weights between them based on the covariances. It can be observed that the tri-axis position RMSE calculated by the IMU/UWB Kalman filter was consistently less than that without UWB fusion. In our experiment, the proposed method was been extensively tested by various movements. For simple movements like motion 1, consisting of a circular trajectory, the tri-axis position RMSE values were all less than 10 cm. Compared with the IMU method, the relative errors calculated by the IMU/UWB fusion method were reduced by 40%, 3.6%, and 25.5% in the X-axis, Y-axis, and Z-axis, respectively. Complex combined movements such as motion 2 and 3 always consisted of multiple simple movements, such as shoulder flexion/extension, abduction/adduction, internal/external rotation, and elbow flexion/extension. Despite the fact that the tri-axis errors with combined motions are usually more significant than those with simple movements, the IMU/UWB fusion method still demonstrated better position accuracy than the IMU method, with a maximum RMSE of 12.2 cm. And our proposed methodology achieved an average decrease in the RMSE of 1.2 cm from the IMU method to the IMU/UWB fusion method. In comparison, by transferring the angle RMSE into position RMSE, the position RMSE values illustrated in Table 3 fell within the accuracy range of forearm motion without the constraints represented in [21].

Table 3. Comparison of the 3-D position accuracy of various upper limb motion tracking methods, i.e., the IMU method, UWB method, and IMU/UWB fusion method with the optical Mocap system.

	Position RMSE—IMU (cm)			Position RMSE—UWB (cm)			Position RMSE—IMU/UWB (cm)		
	X	Y	Z	X	Y	Z	X	Y	Z
Motion 1	9.1448	8.6423	8.5407	5.7678	4.3311	12.8657	5.4827	8.3291	6.3664
Motion 2	5.1256	8.5176	10.7520	11.5741	5.6851	15.9024	4.9606	7.8206	9.7357
Motion 3	12.7599	10.2743	11.9922	16.0117	12.5021	13.7803	12.2170	9.0845	11.1169

Figure 11 shows a comparison of the 3-D spatial trajectory of upper limb motion reconstructed by the proposed algorithm and the real movement for motion 3. As can be seen in this figure, the proposed method was able to accurately track the movement of the human upper limb, and showed high feasibility and stability.

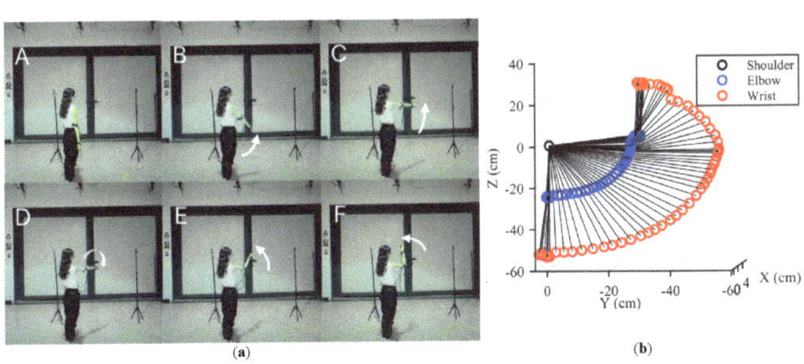

Figure 11. 3-D spatial trajectory of sensor units 1 and 2 for motion 3. (**a**) Sequences of the real movement. (**b**) Motion reconstructed by the proposed algorithm.

4.4. Power Consumption and Cost

According to the datasheet of the MPU9250 which was adopted in our experiment, the power consumption of the sensor depends on the selected operating mode and sensor configuration. In general, the device consumes between 8.3 μA and 3.9 mA. When the voltage is 3.3 V, the power consumption is as follows: in low-power mode, it ranges from 27.4 μW to 66.7 μW; in normal mode, with both the accelerometer and the gyroscope enabled, it varies from 31.5 μW to 764.7 μW; and in normal mode with all three sensors (accelerometer, gyroscope, and magnetometer) enabled, the power consumption ranges from 37.3 μW to 926.7 μW. The average price for a single MPU9250 chip is between USD 1 and 5.

As for the UWB system, according to the DW1000 datasheet, the typical power consumption values for the device in different modes are as follows: in idle mode, 24.75 mW; in receive mode, 56.84 mW; in transmit mode (at 6.8 Mbps data rate), around 270.76 mW; and in sleep mode, 20 nW. Typically, the cost of a single DW1000 module ranges from around USD 10 to 25.

However, the power consumption of an optical MoCap system can range from a few hundred watts to several kilowatts. This is because the cameras employed in this system require significant processing power and strict light conditions. And the cost of a basic system with a few cameras and basic software is around USD 1000. Therefore, compared with the optical MoCap system, the IMU/UWB system which we adopted exhibits extremely lower power consumption and is more economical and applicable.

5. Conclusions

In this paper, we proposed a novel method for hybrid upper limb motion tracking and 3-D positioning by fusing the IMU and UWB systems. We first simplified the human upper limb as a kinematics model and employed the quaternion method to calculate the attitude angle of each segment. To compensate for the accumulated error of gyroscope measurement, the fourth-order Runge–Kutta Madgwick orientation filter was adopted to improve the accuracy of 3-D motion tracking through the optimal fusion of the accelerometer, gyroscope, and magnetometer of the MARG system. The RMSE values varied from 6.17 cm to 12.76 cm, which shows feasibility for the precise motion tracking of human upper limbs. The error was mainly caused by drift of the IMU sensors. And it was inevitable that there would be a slight misalignment of coordinate frames at the initial moment.

In order to further reduce the drift error, we combined the UWB localization system with the IMU sensors. The static positioning variance of the four-anchor UWB system was tested, and the mean values of the tri-axis variances were within 0.6 cm^2, so it shared good positioning robustness. By employing the UWB localization Kalman filter, the accuracy of UWB was verified by multiple sets of movements of the human upper limbs.

Adopting the four-anchor UWB system, we employed various movements to test the IMU/UWB Kalman filter. The experimental results represent that our proposed fusion algorithm achieved an average decrease in the RMSE of 1.2 cm from the IMU method to the IMU/UWB fusion method. With high feasibility and stability, our proposed algorithm was able to accurately track the movements of the human upper limbs.

In future work, we aim to study the effects of various distributions of multiple anchors on UWB localization accuracy, as well as to extend the proposed algorithm to the whole-body motion estimation.

Author Contributions: Conceptualization, Y.S. and Y.Z.; methodology, Y.S.; software, Y.S.; validation, Y.S., Y.Z. and Z.L.; formal analysis, Y.S.; investigation, Y.S.; resources, Y.S. and S.Y.; data curation, Y.S. and S.Z.; writing—original draft preparation, Y.S.; writing—review and editing, Y.S., Y.Z. and Z.L.; visualization, Y.S.; supervision, Y.Z.; project administration, Y.Z.; funding acquisition, Y.Z. All authors have read and agreed to the published version of the manuscript.

Funding: This research received no external funding.

Institutional Review Board Statement: Not applicable.

Informed Consent Statement: Not applicable.

Data Availability Statement: Not applicable.

Conflicts of Interest: The authors declare no conflict of interest.

References

1. Li, R.; Wang, H.; Liu, Z. Survey on Mapping Human Hand Motion to Robotic Hands for Teleoperation. *IEEE Trans. Circuits Syst. Video Technol.* **2022**, *32*, 2647–2665.
2. Liarokapis, M.V.; Artemiadis, P.K.; Kyriakopoulos, K.J. Mapping human to robot motion with functional anthropomorphism for teleoperation and telemanipulation with robot arm hand systems. In Proceedings of the IEEE/RSJ International Conference on Intelligent Robots & Systems, Tokyo, Japan, 3–7 November 2013; p. 2075.
3. Vongchumyen, C.; Bamrung, C.; Kamintra, W.; Watcharapupong, A. Teleoperation of Humanoid Robot by Motion Capturing Using KINECT. In Proceedings of the 2018 International Conference on Engineering, Applied Sciences, and Technology (ICEAST), Phuket, Thailand, 4–7 July 2018; IEEE: Piscataway, NJ, USA, 2018; pp. 1–4.
4. Wan, K.; Sawada, H. 3D measurement of human upper body for gesture recognition. In *Optomechatronic Computer-Vision Systems II*; SPIE: Bellingham, WA, USA, 2007.
5. Liu, Y.-T.; Zhang, Y.-A.; Zeng, M. Sensor to segment calibration for magnetic and inertial sensor based motion capture systems. *Measurement* **2019**, *142*, 1–9.
6. Yuan, Q.; Chen, I.M. Human velocity and dynamic behavior tracking method for inertial capture system. *Sens. Actuators A Phys.* **2012**, *183*, 123–131. [CrossRef]
7. Fan, B.; Li, Q.; Tan, T.; Kang, P.; Shull, P.B. Effects of IMU Sensor-to-Segment Misalignment and Orientation Error on 3D Knee Joint Angle Estimation. *IEEE Sens. J.* **2021**, *22*, 2543–2552.
8. Luinge, H.J.; Veltink, P.H. Measuring orientation of human body segments using miniature gyroscopes and accelerometers. *Med. Biol. Eng. Comput.* **2005**, *43*, 273–282. [CrossRef]
9. Seel, T.; Raisch, J.; Schauer, T. IMU-based joint angle measurement for gait analysis. *Sensors* **2014**, *14*, 6891–6909.
10. Favre, J.; Jolles, B.M.; Aissaoui, R.; Aminian, K. Ambulatory measurement of 3D knee joint angle. *J. Biomech.* **2008**, *41*, 1029–1035. [CrossRef]
11. Luinge, H.J.; Veltink, P.H.; Baten, C.T. Ambulatory measurement of arm orientation. *J. Biomech.* **2007**, *40*, 78–85.
12. El-Gohary, M.; Mcnames, J. Shoulder and Elbow Joint Angle Tracking With Inertial Sensors. *IEEE Trans. Bio-Med. Eng.* **2012**, *59*, 2635–2641.
13. Álvarez, D.; Alvarez, J.C.; González, R.C.; López, A.M. Upper limb joint angle measurement in occupational health. *Comput. Methods Biomech. Biomed. Eng.* **2016**, *19*, 159–170. [CrossRef]
14. Madgwick, S. *An Efficient Orientation Filter for Inertial and Inertial/Magnetic Sensor Arrays*; Report x-io and University of Bristol: Bristol, UK, 2010; Volume 25, pp. 113–118.
15. Mahony, R.; Hamel, T.; Pflimlin, J.M. Nonlinear Complementary Filters on the Special Orthogonal Group. *IEEE Trans. Autom. Control* **2008**, *53*, 1203–1218. [CrossRef]
16. Bachmann, E.R. Inertial and Magnetic Tracking of Limb Segment Orientation for Inserting Humans into Synthetic Environments. Doctoral Dissertation, Naval Postgraduate School, Monterey, CA, USA, 2000.
17. Shen, H.-M.; Lian, C.; Wu, X.-W.; Bian, F.; Yu, P.; Yang, G. Full-pose estimation using inertial and magnetic sensor fusion in structurized magnetic field for hand motion tracking. *Measurement* **2021**, *170*, 108697. [CrossRef]
18. Zihajehzadeh, S.; Loh, D.; Lee, M.; Hoskinson, R.; Park, E.J. A cascaded two-step Kalman filter for estimation of human body segment orientation using MEMS-IMU. In Proceedings of the 2014 36th Annual International Conference of the IEEE Engineering in Medicine and Biology Society, Chicago, IL, USA, 26–30 August 2014; IEEE: Piscataway, NJ, USA, 2014.
19. Young, A.D. Use of Body Model Constraints to Improve Accuracy of Inertial Motion Capture. In Proceedings of the International Conference on Body Sensor Networks, Singapore, 7–9 June 2010.
20. Zhang, Z.; Wong, L.W.C.; Wu, J.-K. 3D Upper Limb Motion Modeling and Estimation Using Wearable Micro-sensors. In Proceedings of the 2010 International Conference on Body Sensor Networks, Singapore, 7–9 June 2010; pp. 117–123.
21. Zhang, Z.-Q.; Wu, J.-K. A novel hierarchical information fusion method for three-dimensional upper limb motion estimation. *IEEE Trans. Instrum. Meas.* **2011**, *60*, 3709–3719. [CrossRef]
22. Tao, G.; Sun, S.; Huang, S.; Huang, Z.; Wu, J. Human modeling and real-time motion reconstruction for micro-sensor motion capture. In Proceedings of the 2011 IEEE International Conference on Virtual Environments, Human-Computer Interfaces and Measurement Systems Proceedings, Ottawa, ON, Canada, 19–21 September 2011; IEEE: Piscataway, NJ, USA, 2011.
23. Zhang, Z.Q.; Wong, W.C.; Wu, J.K. Ubiquitous Human Upper-Limb Motion Estimation using Wearable Sensors. *IEEE Trans. Inf. Technol. Biomed.* **2011**, *15*, 513–521. [CrossRef]
24. Roetenberg, D.; Luinge, H.J.; Baten, C.; Veltink, P.H. Compensation of magnetic disturbances improves inertial and magnetic sensing of human body segment orientation. *IEEE Trans. Neural Syst. Rehabil. Eng.* **2005**, *13*, 395. [CrossRef]
25. Fourati, H.; Manamanni, N.; Afilal, L.; Handrich, Y. Complementary Observer for Body Segments Motion Capturing by Inertial and Magnetic Sensors. *IEEE/ASME Trans. Mechatron.* **2014**, *19*, 149–157. [CrossRef]

26. Bharadwaj, R.; Parini, C.; Alomainy, A. Indoor tracking of human movements using UWB technology for motion capture. In Proceedings of the 2014 8th European Conference on Antennas and Propagation (EuCAP), The Hague, The Netherlands, 6–11 April 2014.
27. Queralta, J.P.; Almansa, C.M.; Schiano, F.; Floreano, D.; Westerlund, T. UWB-based system for UAV Localization in GNSS-Denied Environments: Characterization and Dataset. In Proceedings of the 2020 IEEE/RSJ International Conference on Intelligent Robots and Systems (IROS), Las Vegas, NV, USA, 25–29 October 2020.
28. Kim, D.-H.; Pyun, J.-Y. NLOS identification based UWB and PDR hybrid positioning system. *IEEE Access* **2021**, *9*, 102917–102929. [CrossRef]
29. Poulose, A.; Emeršič, Ž.; Eyobu, O.S.; Han, D.S. An accurate indoor user position estimator for multiple anchor UWB localization. In Proceedings of the 2020 International Conference on Information and Communication Technology Convergence (ICTC), Jeju Island, Republic of Korea, 21–23 October 2020; IEEE: Piscataway, NJ, USA, 2020; pp. 478–482.
30. Zampella, F.; Angelis, A.D.; Skog, I.; Zachariah, D.; Jiménez, A. A constraint approach for UWB and PDR fusion. In Proceedings of the International Conference on Indoor Positioning & Indoor Navigation, Montbeliard, France, 28–31 October 2013.
31. Corrales, J.A.; Candelas, F.A.; Torres, F. Hybrid tracking of human operators using IMU/UWB data fusion by a Kalman filter. In Proceedings of the Third ACM/IEEE International Conference on Human-Robot Interaction, Amsterdam, The Netherlands, 12–15 March 2008; IEEE: Piscataway, NJ, USA, 2008; p. 193.
32. Zihajehzadeh, S.; Yoon, P.K.; Park, E.J. A magnetometer-free indoor human localization based on loosely coupled IMU/UWB fusion. In Proceedings of the 2015 37th Annual International Conference of the IEEE Engineering in Medicine and Biology Society (EMBC), Milan, Italy, 25–29 August 2015; IEEE: Piscataway, NJ, USA, 2015.
33. Kok, M.; Hol, J.D.; Schon, T.B. Indoor Positioning Using Ultrawideband and Inertial Measurements. *IEEE Trans. Veh. Technol.* **2015**, *64*, 1293–1303. [CrossRef]
34. Zihajehzadeh, S.; Park, E.J. A novel biomechanical model-aided IMU/UWB fusion for magnetometer-free lower body motion capture. *IEEE Trans. Syst. Man Cybern. Syst.* **2016**, *47*, 927–938. [CrossRef]
35. Zihajehzadeh, S.; Yoon, P.K.; Kang, B.S.; Park, E.J. UWB-Aided Inertial Motion Capture for Lower Body 3-D Dynamic Activity and Trajectory Tracking. *IEEE Trans. Instrum. Meas.* **2015**, *64*, 3577–3587. [CrossRef]
36. Zhang, H.; Zhang, Z.; Gao, N.; Xiao, Y.; Meng, Z.; Li, Z. Cost-Effective Wearable Indoor Localization and Motion Analysis via the Integration of UWB and IMU. *Sensors* **2020**, *20*, 344. [CrossRef] [PubMed]
37. Peppoloni, L.; Filippeschi, A.; Ruffaldi, E.; Avizzano, C.A. A novel 7 degrees of freedom model for upper limb kinematic reconstruction based on wearable sensors. In Proceedings of the 2013 IEEE 11th International Symposium on Intelligent Systems and Informatics (SISY), Subotica, Serbia, 26–28 September 2013; IEEE: Piscataway, NJ, USA, 2013; pp. 105–110.
38. Tian, Y.; Meng, X.; Tao, D.; Liu, D.; Feng, C. Upper limb motion tracking with the integration of IMU and Kinect. *Neurocomputing* **2015**, *159*, 207–218. [CrossRef]
39. Roux, E.; Bouilland, S.; Godillon-Maquinghen, A.P.; Bouttens, D. Evaluation of the global optimisation method within the upper limb kinematics analysis. *J. Biomech.* **2009**, *35*, 1279–1283. [CrossRef] [PubMed]

sensors

MDPI

Article

FCCD-SAR: A Lightweight SAR ATR Algorithm Based on FasterNet

Xiang Dong, Dong Li * and Jiandong Fang

The College of Information Engineering, Information and Communication Engineering, Inner Mongol University of Technology, Hohhot 010080, China
* Correspondence: lidong@imut.edu.cn

Abstract: In recent times, the realm of remote sensing has witnessed a remarkable surge in the area of deep learning, specifically in the domain of target recognition within synthetic aperture radar (SAR) images. However, prevailing deep learning models have often placed undue emphasis on network depth and width while disregarding the imperative requirement for a harmonious equilibrium between accuracy and speed. To address this concern, this paper presents FCCD-SAR, a SAR target recognition algorithm based on the lightweight FasterNet network. Initially, a lightweight and SAR-specific feature extraction backbone is meticulously crafted to better align with SAR image data. Subsequently, an agile upsampling operator named CARAFE is introduced, augmenting the extraction of scattering information and fortifying target recognition precision. Moreover, the inclusion of a rapid, lightweight module, denoted as C3-Faster, serves to heighten both recognition accuracy and computational efficiency. Finally, in cognizance of the diverse scales and vast variations exhibited by SAR targets, a detection head employing DyHead's attention mechanism is implemented to adeptly capture feature information across multiple scales, elevating recognition performance on SAR targets. Exhaustive experimentation on the MSTAR dataset unequivocally demonstrates the exceptional prowess of our FCCD-SAR algorithm, boasting a mere 2.72 M parameters and 6.11 G FLOPs, culminating in an awe-inspiring 99.5% mean Average Precision (mAP) and epitomizing its unparalleled proficiency.

Keywords: synthetic aperture radar (SAR); automatic target recognition (ATR); deep learning; lightweight; fasterNet

1. Introduction

Synthetic aperture radar (SAR) [1] has achieved extensive utilization in diverse fields such as reconnaissance detection, geological exploration, disaster detection, and public area security screening. Its ability to operate round the clock and in all weather conditions makes it a vital tool. Automatic target recognition (ATR) [2] plays a crucial role in SAR image interpretation, encompassing the recognition of target regions of interest and inference of target class attributes. This paper focuses on SAR-ATR, which holds immense practical value and theoretical significance, providing valuable insights for image recognition [3], target recognition [4], image matching [5], and other remote sensing applications.

SAR achieves high-resolution imaging through linear FM signals and matched filtering techniques for distance direction, while the azimuth direction employs motion-based virtual aperture synthesis. It possesses remarkable attributes such as all-day [6], all-weather operability [7], high penetration capability [8], long-range observation [9], and high-resolution ground imaging [10]. The wealth of surface electromagnetic scattering information offered by SAR has contributed to its widespread adoption in various domains, including ocean exploration [11], forestry census [12], topographic mapping [13], land resource survey, and traffic control, as well as military applications like battlefield reconnaissance, radar guidance, and strike effect evaluation [14]. Enhancing image quality plays a crucial role in SAR applications, encompassing image super-resolution, denoising, deblurring, and contrast enhancement. These techniques not only improve visual effects but also enhance feature extraction quality, thereby facilitating subsequent image understanding and interpretation. However, traditional SAR imaging algorithms face challenges in effectively detecting and imaging moving targets due to their complex scattering characteristics and motion features.

With the rapid advancements in deep learning, convolutional neural networks (CNNs) have been increasingly employed for ship recognition in SAR images, exhibiting promising outcomes. However, deeper CNNs focus on accuracy at the expense of real-time performance, speed, and compatibility with resource-constrained embedded platforms. Hence, there is a pressing need to develop lightweight models that strike a balance between speed and accuracy, enabling real-time ship target recognition in SAR images and seamless deployment on embedded platforms [15]. In the realm of traditional machine learning, feature extraction and classification algorithms are commonly used for target recognition. Feature extraction techniques include edge, texture, shape, and polarization characteristics, enabling the extraction of target feature information from SAR images. The wavelet transform, Gabor filter, grayscale co-occurrence matrix, and principal component analysis are popular feature extraction algorithms. Following feature extraction, classifiers such as support vector machines (SVMs) [16], artificial neural networks, and decision trees are utilized for target classification. Traditional machine learning methods also encompass feature matching-based approaches, such as polarized scattering similarity and adaptive local orientation patterns. Du proposed the Fast C&W algorithm to counter attacks on SAR target recognition by deep convolutional neural networks [17]. Peng proposed a Speckle-variant attack algorithm for the adversarial attack on SAR target recognition of deep convolutional neural networks [18].

While traditional machine learning methods can achieve satisfactory results in SAR target recognition, they have limitations. Subjectivity and incompleteness arise from the need for manual feature selection during extraction. Furthermore, the performance of traditional machine learning methods is constrained by the capabilities of feature extraction and classifiers.

In recent years, the emergence of deep learning has prompted researchers to explore its application in SAR target recognition. Two-stage target recognition methods generally offer higher accuracy compared to single-stage methods. However, two-stage algorithms often exhibit slower training and recognition speeds compared to their single-stage counterparts. To address this limitation, researchers have increasingly turned to single-stage algorithms to ensure real-time recognition. However, single-stage methods are more susceptible to false recognitions and localization errors, particularly for small targets. Thus, there is a need to enhance the performance of single-stage algorithms in detecting small targets in real-time applications.

Existing target recognition algorithms primarily cater to optical images, focusing on accuracy improvement. Few detectors have been specifically tailored for SAR images. Directly applying target recognition algorithms designed for optical images to SAR images may yield suboptimal results due to differences in imaging mechanisms, target characteristics, and resolution disparities. Therefore, it is crucial to develop target recognition algorithms that consider the unique complexities and characteristics of SAR images.

Given the extensive SAR image data and high feature dimensionality, traditional deep learning models often suffer from a large number of parameters and high computational complexity. Hence, the research focus has shifted toward lightweight networks. Lightweight networks refer to neural network structures that achieve computational efficiency by reducing parameters and computations. In SAR automatic target recognition, lightweight networks offer improved computational speed and memory efficiency without compromising classification accuracy.

In conclusion, the demand for a lightweight and readily deployable single-stage target recognition algorithm for SAR image target recognition, especially on embedded platforms, is becoming increasingly urgent. This paper introduces the FCCD-SAR method, which addresses the SAR image target recognition challenge by striking a well-balanced approach. Our approach significantly improves recognition accuracy while minimizing the number of parameters and floating-point operations (FLOPs) required. The main contributions of this paper can be summarized as follows.

(1) To facilitate the development of a SAR image target recognition algorithm that is well-suited for embedded platforms, we adopt a more rational approach by employing the FasterNet algorithm for dataset lightening and feature extraction. This enables better alignment with the unique characteristics of SAR image data. Moreover, we effectively reduce the number of parameters while preserving satisfactory performance.

(2) To enhance the extraction of scattering information from targets and improve target recognition accuracy, we propose the utilization of a lightweight upsampling operator called CARAFE. This operator exhibits a wide perception field during reconfiguration, allowing for effective improvement in the recognition performance of SAR targets. Additionally, this design enables a reduction in the number of parameters and floating-point operations (FLOPs) required while maintaining high recognition performance.

(3) To further improve the model's recognition accuracy and computational efficiency, a faster and lighter module C3-Faster is used to reduce the number of parameters and computation while ensuring recognition accuracy.

(4) For the characteristics of multi-scale and large-scale variation of SAR targets, DyHead's attention-based mechanism detection head is used to better detect feature information at different scales adequately and improve the recognition effect of SAR targets.

(5) To obtain the ultimate network model, a pruning operation is introduced to prune the network structure to obtain the minimum optimal network model with guaranteed accuracy.

2. Related Work

Numerous publicly available datasets exist for SAR target recognition, with the MSTAR ten-class classification dataset being the most renowned. Various algorithms, including traditional and deep learning-based approaches, have been employed for SAR target recognition. Traditional algorithms often prove ineffective, relying heavily on manual parameter setting and design, lacking robustness, and exhibiting poor generalization to other SAR datasets. Additionally, their recognition speed and real-time performance fall short of engineering application requirements.

Consequently, the benefits of end-to-end deep learning algorithms have become increasingly evident. In recent years, researchers have shifted towards deep learning algorithms for SAR image target recognition, capitalizing on advancements in deep learning techniques. These algorithms eliminate the need for intricate manual feature extraction, instead focusing on designing robust network structures to effectively extract SAR target features. Convolutional neural networks (CNNs) have gained significant popularity in SAR image target recognition, particularly for ship targets. Pre-trained CNN [19] models have yielded promising results for feature extraction in SAR images, followed by classification using traditional classifiers. With the continuous evolution of deep learning techniques, researchers have explored complex deep neural network models such as RNNs and graph convolution neural networks (GCNNs) [20] for SAR target-recognition tasks.

The development of deep learning algorithms has introduced new methods for SAR target recognition. For instance, region extraction algorithms commonly used in target recognition, such as region-based convolutional neural networks (R-CNN) [21] and Single Shot MultiBox Detectors (SSD) [22], have been adapted for SAR target recognition. Additionally, novel network architectures and techniques have been proposed, such as the feature pyramid network for target recognition [23] and the fully convolutional attention block algorithm for SAR target recognition [24]. However, these methods often rely on deep network structures without considering practical engineering applications, leading to imbalanced parameters, FLOPs, and recognition accuracy.

Improving the accuracy and robustness of target recognition can be achieved by fusing SAR images with data from multiple sources, such as optical and infrared images. Multi-source data fusion plays a crucial role in SAR automatic target recognition [25], utilizing information from various sources for comprehensive analysis [26] and feature extraction to enhance accuracy and robustness. Data-level fusion and feature-level fusion are two main aspects of multi-source data fusion [27].

Given the growing interest in lightweight SAR target-recognition models, the focus has shifted toward networks with reduced parameters and computational requirements. Lightweight networks offer improved computational speed and memory efficiency without sacrificing classification accuracy [28]. Notable lightweight designs include a lossless lightweight CNN proposed by Zhang [29] and a modified convolutional random vector function link network [30] for SAR target recognition. However, existing models often fail to strike the appropriate balance between accuracy and lightweight design and neglect the need to tailor recognition models specifically for SAR image target recognition datasets.

Therefore, to address the requirements of real-world engineering applications, we have devised an innovative SAR target recognition algorithm that is both lightweight and highly precise. This algorithm has been specifically designed to cater to SAR image target recognition datasets.

3. Materials and Methods

In this paper, we propose a lightweight SAR ART algorithm based on FasterNet, the FCCD-SAR. As a consequence, we strike the best balance between accuracy and lightweight design. The FCCD-SAR model mainly consists of the following modules and strategies: the state-of-the-art target-recognition benchmark framework YOLOV5, the faster neural network backbone network FasterNet, the lightweight upsampling operator CARAFE that solves the problems of some general modules and operators, the faster lightweight module C3- Faster, and DyHead, which uses an attention mechanism to unify different target-detection heads. Compared to the current state-of-the-art methods, such as YOLOV8 and YOLOV7, YOLOV8 has just been released recently, and its model is still in the stage of frequent modification, thus it is not stable enough. YOLOV7, on the other hand, has a slightly lower inference speed than YOLOV5 and requires more memory resources. In contrast, YOLOV5, after many official modifications, has a more stable performance, a more mature network, and a faster inference speed, and at the same time, it is more economical in terms of memory consumption. Therefore, YOLOV5 was chosen as the benchmark framework for this study.

3.1. Architectural Overview of FCCD-SAR Network

The schematic representation in Figure 1 depicts the holistic network architecture of our FCCD-SAR model. Figure 2 shows the basic YOLOV5 network framework diagram, which facilitates the comparison with the improved structure. The model depicted in the figure comprises five distinct components: input, backbone, neck, head, and output. Notably, the input image initially undergoes processing through FasterNet [31], a custom-designed lightweight backbone network. This backbone network is adept at extracting the discrete scattering features of SAR images more rationally.

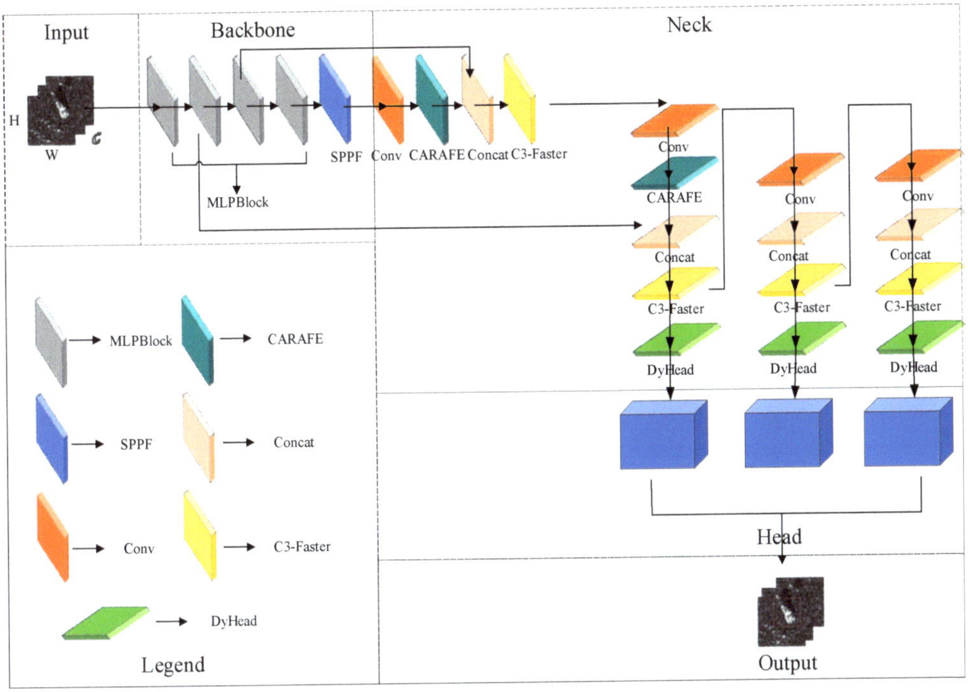

Figure 1. The overall network architecture of the FCCD-SAR model. There are five parts in the network structure, and the contents of each part are shown in the figure. SPPF: Spatial Pyramid Pooling-Fast.

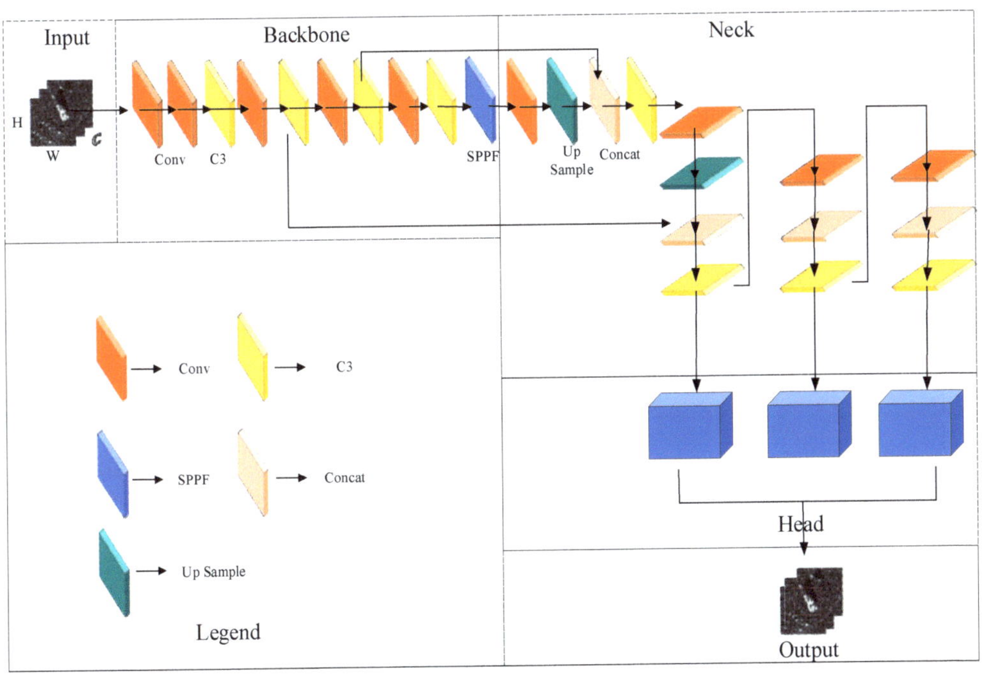

Figure 2. Basic YOLOV5 network framework.

Then, the backbone network into the neck meets the lightweight, universal upsampling operator CARAFE [32], which is used to achieve significant improvements in different tasks while introducing only a small number of parameters and computational costs. Then, before extracting the small target scale features output, a faster lightweight module C3-Faster is introduced to combine the respective advantages of CNN and self-attention to complement each other, which can improve the recognition performance of SAR targets while reducing the number of parameters and FLOPs. Finally, the processed image flows into DyHead [33], our chosen attention-based detection head. DyHead incorporates attention mechanisms across scale-aware feature layers, spatial locations for spatial perception, and output channels for task perception. This innovative approach greatly enhances the expressiveness of the model's target-detection head without imposing an additional computational burden. We will discuss the detailed improvements in these four areas later on.

3.2. *Faster and Better Neural Networks: FasterNet*

To design fast neural networks, much work has focused on reducing the number of FLOPs. However, we observe that this reduction in FLOPs does not necessarily lead to a similar degree of reduction in latency. This mainly stems from the inefficiency of low FLOPS per second.

To achieve faster networks, we revisited the popular operators and demonstrated that such low FLOPS are mainly due to frequent memory accesses of the operators, especially deep convolution. Therefore, we adopted a new partially convolutional (PConv) that can extract spatial features more efficiently by reducing both redundant computations and memory accesses.

To optimize the backbone network, we incorporated FasterNet-T0, the smallest version of FasterNet, and retained its MLPBlock architecture. Additionally, we made improvements by eliminating unnecessary MLPBlocks through stacking. The resulting lightweight backbone, FasterBackbone, was specifically designed to efficiently extract scattering features from SAR datasets.

Figure 3 provides an overview of the structural details of FasterBackbone, while Table 1 presents the specific parameters used. Through extensive experimental validation using the SAR dataset, we demonstrated the remarkable feature extractability of the backbone we designed.

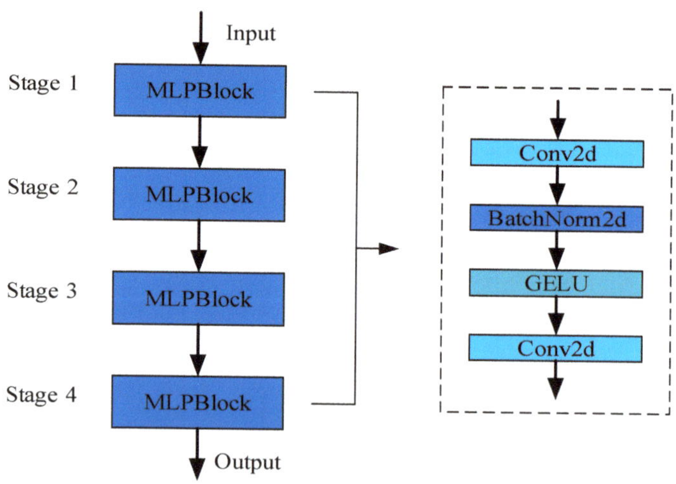

Figure 3. Details of the FasterBackbone network structure. The FasterBackbone consists of a total of four MLPBlocks, and the specific structure of the MLPBlock is shown in Figure 3. GELU: Gaussian Error Linear Unit.

Table 1. Parameters of FasterBackbone network structure.

	Out Channel	Kernel Size	Stride	Expand
MLPBlock	40	3	1	3
MLPBlock	80	3	1	3
MLPBlock	160	3	1	3
MLPBlock	320	3	1	3
SPPF	1024	3	1	-

3.3. Lightweight Upsampling Operator: CARAFE

The upsampling operation was achieved through feature recombination, which involves the dot product between the upsampling kernel and the corresponding neighborhood pixels in the input feature map. The fundamental network structure, with a small receptive field, ignores some useful information, and therefore, the receptive field needs to be enlarged. The upsampling operation CARAFE, on the other hand, can have a large receptive field during reorganization and guides the reorganization process based on the input features. Meanwhile, the whole CARAFE operator structure is small, which meets the lightweight requirement. Specifically, the input feature map is utilized to predict unique upsampling kernels for each position, followed by feature recombination based on these predicted kernels. CARAFE demonstrates significant performance improvements across various tasks while only introducing minimal additional parameters and computational overhead.

CARAFE consists of two primary modules: the upsampling kernel prediction module and the feature recombination module, as depicted in Figure 4. Assuming an upsampling multiplier of σ and an input features map with dimensions $H \times W \times C$, the process begins by predicting the upsampling kernel through the upsampling kernel prediction module. Subsequently, the feature recombination module is employed to complete the upsampling procedure, resulting in an output feature map with dimensions $\sigma H \times \sigma W \times C$.

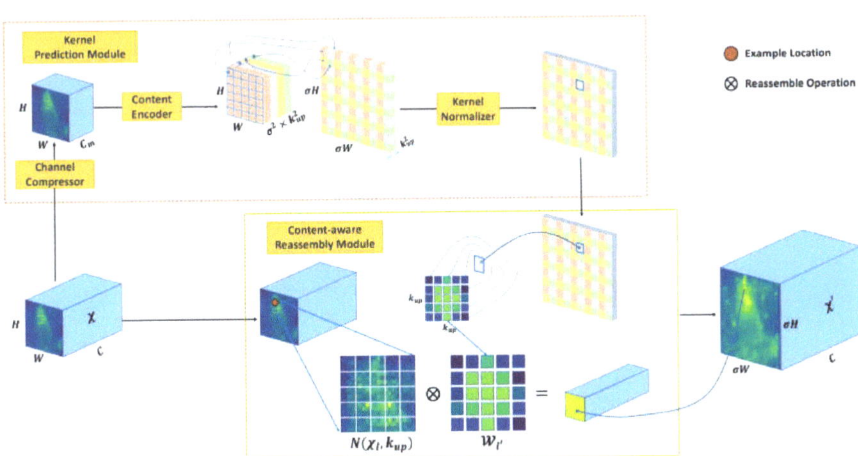

Figure 4. The overall framework of CARAFE. CARAFE is composed of two key components, i.e., kernel pre-diction module and content-aware reassembly module. A feature map with size C × H × W is upsampled by a factor of σ (=2) in this figure.

Given an input feature map of shape $H \times W \times C$, our initial step involves channel compression, reducing the channel number to C_m using a 1×1 operation. The primary objective of this compression is to alleviate the computational burden on subsequent steps. Following that, we proceeded with content encoding and upsampling kernel prediction, assuming a specific upsampling kernel size of $k_{up} \times k_{up}$. It is worth noting that a larger upsampling kernel offers a broader perceptual field, but it also entails a higher computational cost. To incorporate distinct upsampling kernels for each position in the output feature map, it is necessary to predict the shape of the upsampling kernel as $\sigma H \times \sigma W \times k_{up} \times k_{up}$. In the initial step, after compressing the input feature map, we employed a convolutional layer with $k_{encoder} \times k_{encoder}$ channels to predict the upsampling kernel. The number of input channels is C_m, and the number of output channels is $\sigma^2 k_{up}^2$. Following this, we expanded the channel dimension across the spatial dimension, resulting in an upsampling kernel with the shape $\sigma H \times \sigma W \times k_{up}^2$.

At each location within the output feature map, we performed a mapping back to the corresponding region in the input feature map. This region, centered on the location, encompassed a region of size $k_{up} \times k_{up}$. Subsequently, we computed the dot product between this region and the predicted upsampling kernel specific to that point, resulting in the output value. It is worth noting that different channels at the same location shared the same upsampling kernel.

3.4. Faster and Lighter Modules: C3-Faster

In recent years, the fields of computer vision (CV) have witnessed a surge of interest in convolutional neural networks (CNNs) and self-attention networks (SNNs). CNNs have achieved remarkable breakthroughs in CV domains, including image classification, target recognition, and target tracking, consistently attaining state-of-the-art performance across diverse datasets. Concurrently, the rapid development of vision transformers has led to the emergence of transformer-based models with various self-attention mechanisms that have begun to surpass CNNs in several vision tasks, thereby redefining the performance benchmarks in these areas.

ACmix offers a compelling fusion of convolution and self-attention, making it a suitable approach for enhancing hybrid representation learning in SAR image target recognition. With the challenge of detecting small targets in SAR images in mind, we opted to replace the original YOLOV5 C3 module with C3-Faster, provided by FasterNet. Faster-Block and BottleNeck structures are shown in Figure 5. Among them, Faster-Block has one more partial convolution than BottleNeck for spatial fusion and one Drop path to reduce the amount of calculation. Figure 5 showcases the design of this faster, lightweight module. By incorporating the lightweight C3-Faster, we further enhanced the speed of target recognition, addressing the need for efficient and swift target identification.

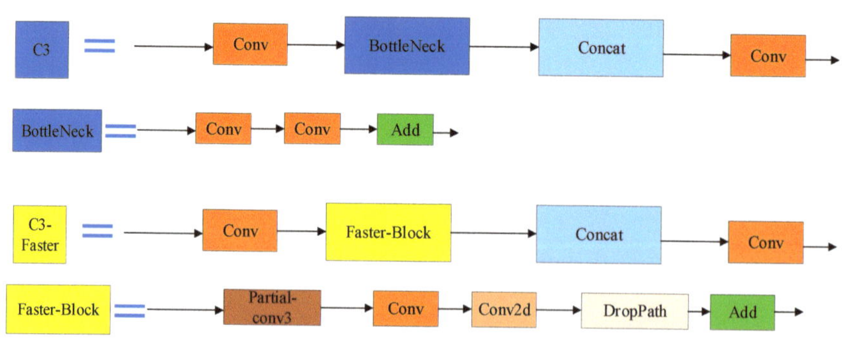

Figure 5. Comparison of C3-Faster of FCCD-SAR and the C3 structure of YOLOV5. Faster-Block and BottleNeck structure are shown in Figure 5.

3.5. Detection Head Based on Attention Mechanism: DyHead

In Figure 1, the third component showcased our attention mechanism-based detection head called DyHead, which was tailored for the SAR image dataset. DyHead introduced a novel dynamic head framework that unifies various target detection heads using an attention mechanism. By leveraging attention between feature levels for scale perception, spatial locations for spatial perception, and output channels for task perception, this approach substantially enhances the expressiveness of the model's target detection head without imposing additional computational burden.

DyHead is a fusion of three attention mechanisms: scale-aware attention π_L, spatial attention π_S, and channel attention π_C. These attention mechanisms are stacked together to form a single block. The final head consists of multiple blocks, each incorporating this stack of attention mechanisms.

With the feature tensor $F \in R^{L\times H\times W\times C}$ at hand, we can describe the generalized form of self-attentiveness as follows:

$$W(F) = \pi(F) \times F \tag{1}$$

The simplest approach would be to employ a fully connected layer, but directly learning the attention function across all dimensions would result in excessive computational requirements and prove impractical due to the high dimensionality. Instead, we tackled this challenge by breaking down the attention function into three sequential attentions, each targeting a single dimension:

$$W(F) = \pi_C(\pi_S(\pi_L(F) \times F) \times F) \times F \tag{2}$$

Scale-aware attention π_L: to address the fusion of features at different scales based on their semantic significance, we began by introducing scale-aware attention:

$$\pi_L(F) \times F = \sigma\left(f\left(\frac{1}{SC}\sum_{S,C} F\right)\right) \times F \tag{3}$$

In this context, $f(\cdot)$ corresponds to a linear function that utilizes 1×1 convolutional approximation, while $\sigma(x)$ represents a hard-sigmoid activation function.

Spatial-aware Attention π_S: continuing with our exploration, we then introduced another module called spatial location-aware attention to emphasize the discriminative capabilities of various spatial locations. Given the large extent of S, we decoupled it into two stages: first, we employed deformation convolution to achieve sparse attention learning, and then we integrated features across different scales to complete the process:

$$\pi_S(F) \times F = \frac{1}{L}\sum_{l=1}^{L}\sum_{k=1}^{K} w_{l,k} \times F(l; p_k + \Delta_{p_k}; c) \times \Delta_{m_k} \tag{4}$$

In this scenario, K represents the number of sparsely sampled positions. The remaining parameter information is analogous to that in deformation convolution, $\omega_{l,k}$ is an importance factor to add bias, and Δm_k is an importance factor for adaptive weighting and thus, it is omitted for brevity.

Task-aware attention π_C: to facilitate collaborative learning with the enhanced generalizability of goal representation capabilities, we devised a task-aware attention mechanism. This attention mechanism dynamically adjusts feature channels to assist various tasks as needed:

$$\pi_C(F) \times F = \max\left(\alpha^1(F) \times F_c + \beta^1(F), \alpha^2(F) \times F_c + \beta^2(F)\right) \tag{5}$$

The hyperparameter plays a crucial role in controlling the activation threshold, akin to DyReLU. α and β are used as rescale and reshift, respectively. By sequentially implementing the aforementioned attention mechanism, we can stack multiple instances of it. The configuration of the DynamicHead is illustrated in Figure 6, providing a visual representation of its structure.

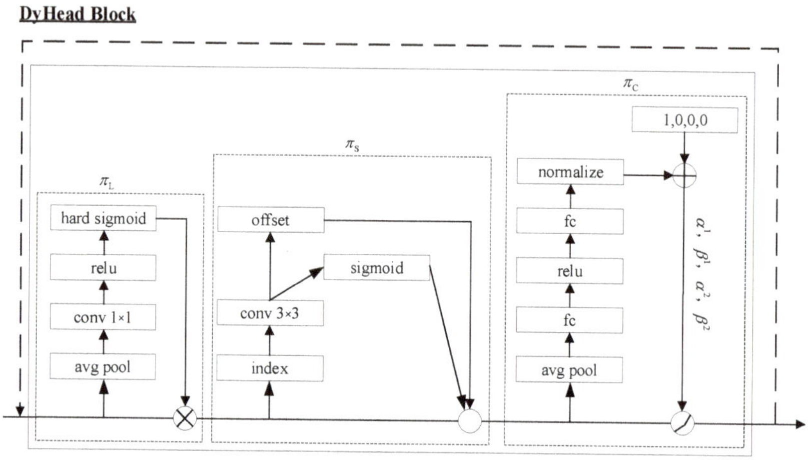

Figure 6. A detailed design of Dynamic Head.

3.6. Pruning

The deployment of CNNs in practical applications is often hindered by their high computational requirements. In this study, we propose a straightforward and efficient approach called network pruning, which involves the sparsity of network channels. This method is particularly well-suited for CNN architectures, as it minimizes the training overhead and yields models that can be deployed without the need for specialized hardware or software acceleration while still maintaining high performance. By training on thick networks and automatically filtering and removing redundant channels during the training process, we can generate streamlined networks that achieve comparable accuracy levels.

This process involves applying L1 regularization to the scaling factor within the Batch Normalization (BN) layer and iteratively adjusting the scaling factor. By converging the scaling factor towards zero, we can identify and remove unimportant channels. This can be visualized in Figure 7, where the regularization gradually reduces the scaling factor values, leading to the elimination of unnecessary channels.

By applying L1 regularization to the scaling factors, each corresponding to a specific convolutional channel or neuron in the fully connected layer, we can effectively discriminate and prune unimportant channels in subsequent operations. Although the additional regularization term has a minimal impact on model performance, it can potentially enhance training accuracy. While pruning unimportant channels may initially lead to a temporary performance drop, this can be rectified through subsequent fine-tuning.

The pruned network obtained after the pruning process exhibits a more compact size, reduced running time, and decreased computational operations compared to the original network.

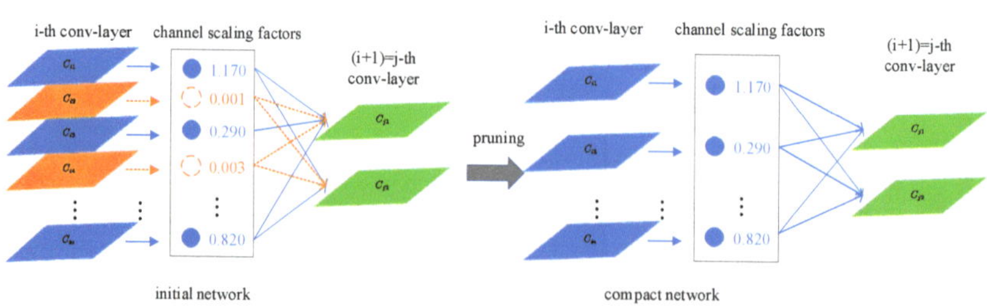

Figure 7. Pruning process diagram.

For each channel in the network, a scaling factor γ is introduced, which is multiplied by the output of that channel. Both the network and these scale factors undergo training, and sparse regularization is continuously applied to the scaling factors throughout the training process. Eventually, channels with significantly small scaling factor values are pruned, and the network is further fine-tuned. Here are the specific details of the process:

$$L = \sum_{(x,y)} l(f(x,W),) + \lambda \sum_{\gamma \in \Gamma} g(\gamma) \tag{6}$$

In the training process, the loss function is defined as a combination of terms. The first term on the left represents the loss for the normal training of the CNN, where (x, y) denotes the input and target. The second term $g(\cdot)$ introduces a sparsity penalty on the scaling factor. The balance factor λ normalizes the latter term. In our experiments, the L1 paradigm, $g(s) = |s|$ is chosen for the later sparsification training. To optimize the non-smoothed L1 penalty term, we use the subgradient descent method. Alternatively, the non-smoothed L1 penalty can be replaced with a smoothed L1 penalty to avoid the need for subgradients at non-smoothed points.

Channel pruning involves removing the input-output connectivity related to the channel, leading to a narrower network. The scaling factors serve as channel selectors, and when optimized with the network, they facilitate the removal of unimportant channels without significantly affecting the generalization performance.

BN finds extensive application in contemporary CNNs, facilitating rapid model convergence and enhancing generalization performance. We draw inspiration from BN's technique of normalizing activation values and employ it as a basis to devise a straightforward yet efficient approach for combining channel scaling factors. Specifically, the BN layer employs mini-batch statistics to normalize the activation values within a given segment. Considering z_{in} as the input and z_{out} as the output of the BN layer, with B representing the present mini-batch. The BN layer executes the subsequent transformation:

$$z = \frac{z_{in} - \mu_B}{\sqrt{\sigma_B^2 + \varepsilon}}; z_{out} = \gamma z + \beta \tag{7}$$

where μ_B represents the mean and σ_B represents the variance of the inputs in B. γ and β denote trainable affine transformation parameters responsible for normalizing the activation values and subsequently linearly transforming them to an arbitrary scale.

The conventional approach is to add a BN layer after the convolutional layer with a channel scaling/offset factor. This allows us to directly use the γ parameter in the BN layer as the scaling factor for network pruning. It offers the advantage of avoiding any additional overhead in the network and is an efficient way to implement channel pruning.

4. Results

In our experimental setup, we employed the MSTAR standard 10-class classification dataset to showcase the performance of FCCD-SAR. Additionally, we conducted a comparative analysis between FCCD-SAR and existing recognition methods, which reveals the superior performance of the former.

4.1. Dataset Introduction and Experimental Settings

4.1.1. MSTAR Dataset

To precisely evaluate the effectiveness and performance of the proposed model, we utilized the renowned MSTAR dataset, as illustrated in Figure 8. This dataset comprises measured SAR ground stationary target data made available by the MSTAR program and supported by the Defense Advanced Research Projects Agency. It encompasses SAR target images obtained from various vehicle targets at varying azimuths. The MSTAR dataset consists of ten ground targets under standard operating conditions (SOC), including artillery (2S1, ZSU234), armored vehicles (BRDM2, BTR60, BTR70, BMP2, D7, ZIL131), and tanks (T62, T72). Notably, the T72 in the BRDM2 armored vehicle category encompasses three variant types (9563, 9566, C21), while the T72 in the tank category includes three variant types as well (132, 812, S7). Please refer to Table 2 for specific parameters.

Figure 8. Introduction to the MSTAR dataset.

Table 2. Basic information and experimental settings of experimental datasets.

	Images Size for Training	Train:Valid	Number of Class	Batch Size	Epoch
MSTAR	640 × 640	8:2	10	16	300
SSDD	640 × 640	8:2	1	16	300

4.1.2. SSDD Dataset

The SAR ship dataset SSDD was then selected and used to validate the generalization in comparison experiments to avoid model overfitting. SSDD is the first publicly available dataset specialized in SAR image ship target recognition at home and abroad, which can be used for training and testing to examine the algorithm, and has been used by thirty levels of colleges and research institutes.

SSDD was obtained by downloading publicly available SAR images on the Internet and cropping the target area to a size of 640 × 640 pixels and by manually labeling the ship target positions. The dataset is shown in Figure 9. The models were trained using identical parameters, including YOLOV5, a batch size of 16, and a training image size of 640 × 640. The experiments were performed.

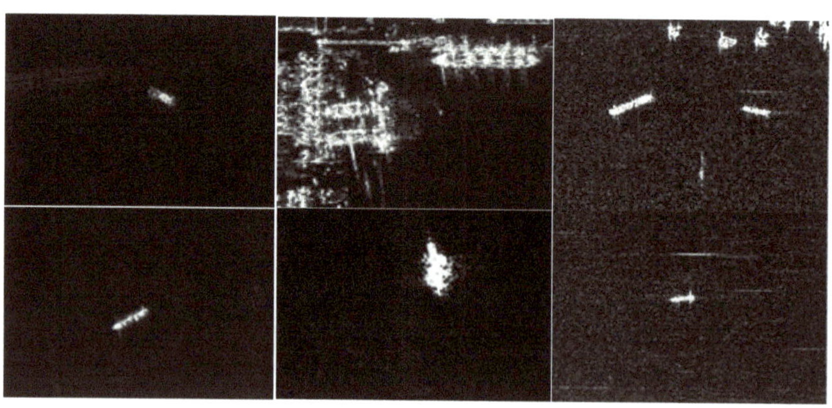

Figure 9. Examples of some of the samples in the SSDD sample.

4.2. Experimental Index

In this experiment, we employ the visual object classes (VOCs) evaluation criteria to assess the recognition performance of our proposed method. To quantify the accuracy of our method, we utilized the average precision at 50% intersection-over-union (mAP50) index. The mAP was calculated by considering both precision (P) and recall (R) values.

Precision measures the proportion of accurate predictions among the samples predicted as positive instances. It was determined based on the number of true positives (*TPs*) and false positives (*FPs*) using the following formula:

$$P = \frac{TP}{TP + FP} \tag{8}$$

The recall is a measure of the probability that the predicted positive instances cover all true positive (*TP*) samples among the actual positive samples. This probability is computed using the *TP* and false negatives (*FN*) according to the following formula:

$$R = \frac{TP}{TP + FN} \tag{9}$$

Accordingly, the mAP is determined by computing the integrated area under the precision-recall (*PR*) curve using the given formula:

$$mAP = \int_0^1 P(R)dR \tag{10}$$

The *F1* score is a commonly employed evaluation metric that provides a comprehensive assessment of the model's performance by incorporating both precision and recall indicators. It combines accuracy and recall measures to gauge the overall quality of the model:

$$F1 = 2 \times \frac{P \times R}{P + R} \tag{11}$$

In the realm of deep learning, it is customary to evaluate the model's parameter size, FLOPs, and model volume while designing a model, considering the specific requirements of the application.

(1) Parameters: Parameters serve as a measure of the model's size, akin to the space complexity of an algorithm. The number of parameters is determined by the video memory size. This encompasses all layers of the model, including convolutional, BN, and fully connected layers, along with the total number of weight parameters in the visual network components.

(2) FLOPs: An abbreviation for floating-point operations, FLOPs represent the quantity of floating-point operations, which can be regarded as a measure of the computational workload. It can be employed to gauge the complexity of an algorithm or model.

4.3. Experimental Results

We conducted experimental validation on the MSTAR dataset and performed compatibility experiments to verify the effectiveness and performance of the methods presented in this paper, showcasing the efficacy of each proposed module and improved method. Lastly, we compared our proposed FCCD-SAR with existing SAR target-recognition methods to establish the superior performance of our method.

4.3.1. Comparison of Parameters for CARAFE

We compared different parameters of CARAFE and selected the optimal CARAFE parameters for our method. The results of this comparison are presented in Table 3:

Table 3. Recognition results with various encoder kernel sizes $k_{encoder}$ and reassembly kernel sizes k_{up}.

$k_{encoder}$	k_{up}	Params (M)	FLOPs (G)	mAP (%)
1	3	7.07	16.1	98.8
1	5	7.08	16.1	98.8
3	3	7.11	16.1	98.9
3	5	7.19	16.3	99.1
5	5	7.21	16.4	99.0

Upon analyzing the data in the table, we observed variations in the number of parameters, FLOPs, and mAP values corresponding to different choices of $k_{encoder}$ and k_{up}. After comparing the data in the table, we ultimately selected $k_{encoder}$ as 3, k_{up} as 5 to achieve the optimal results.

4.3.2. Ablation Experiments

Ablation experiments were conducted on the MSTAR dataset to validate the effectiveness of our proposed module and improved method. The experiments focused on the following aspects:

1. Replacing the DarkNet53 backbone of the YOLOV5 benchmark with the more efficient and faster neural network, FasterNet;
2. Substituting the original upsampling operator with the lightweight CA-RAFE upsampling operator;
3. Exchanging the original C3 module with the faster and lightweight C3-Faster module;
4. Incorporating DyHead, a recognition head based on the attention mechanism. The experimental results are presented in Table 4.

Table 4. Results of the ablation experiments on the MSTAR dataset.

Model	FasterNet	CARAFE	C3-Faster	DyHead	Params (M)	FLOPs (G)	mAP (%)
YOLOV5s	×	×	×	×	7.05	16.0	98.1
FCCD-SAR	√	×	×	×	5.57	11.3	98.8
FCCD-SAR	√	√	×	×	5.86	12.0	99.1
FCCD-SAR	√	√	√	×	5.84	11.9	99.1
FCCD-SAR	√	√	√	√	6.00	12.2	99.5

From the results of the ablation experiments, the following findings are evident:

1. By adopting the improved and more efficient neural network FasterNet, the FLOP is reduced from 16.0 G to 11.3 G, resulting in a 0.7% improvement in mAP and a reduction in parameters from 7.05 M to 5.57 M. These results validate that our FasterBackbone exhibits superior feature extraction capabilities with fewer parameters and FLOPs. It also proves the correctness of choosing FasterNet as the main network, and FasterNet has a strong lightweight property.
2. Substituting the original upsampling operator with the lightweight CARAFE upsampling operator leads to a 0.6% improvement in mAP, accompanied by a slight increase in FLOPs and parameters. The experimental results demonstrate that CARAFE enhances accuracy while reducing the model's size. The upsampling operator CARAFE has excellent recognition performance and has the advantage of model lightweight.
3. By replacing the original C3 module with the faster and lightweight C3-Faster module, FLOPs are reduced while maintaining mAP stability. The number of parameters is reduced from 5.86 M to 5.84 M, and the computation is reduced from 12.0 G to 11.9 G. It is also proved that C3-Faster is better than C3 in structure and more in line with the requirements of lightweight.
4. The incorporation of DyHead, a detection head based on the attention mechanism, yields a 0.4% improvement in mAP. After using DyHead, the number of parameters and FLOPs have a small increase, but compared with this, it is an increase of mAP, which meets the expected goal. Experimental results show that the addition of the DyHead attention mechanism can identify more detailed image features.

In conclusion, the employed FCCD-SAR method showcases notable improvements in reducing the model size and minimizing FLOPs and parameters, all while maintaining a high level of accuracy. These findings substantiate the lightweight nature of our proposed model. Figure 10 presents the visualization comparison results of the ablation experiment.

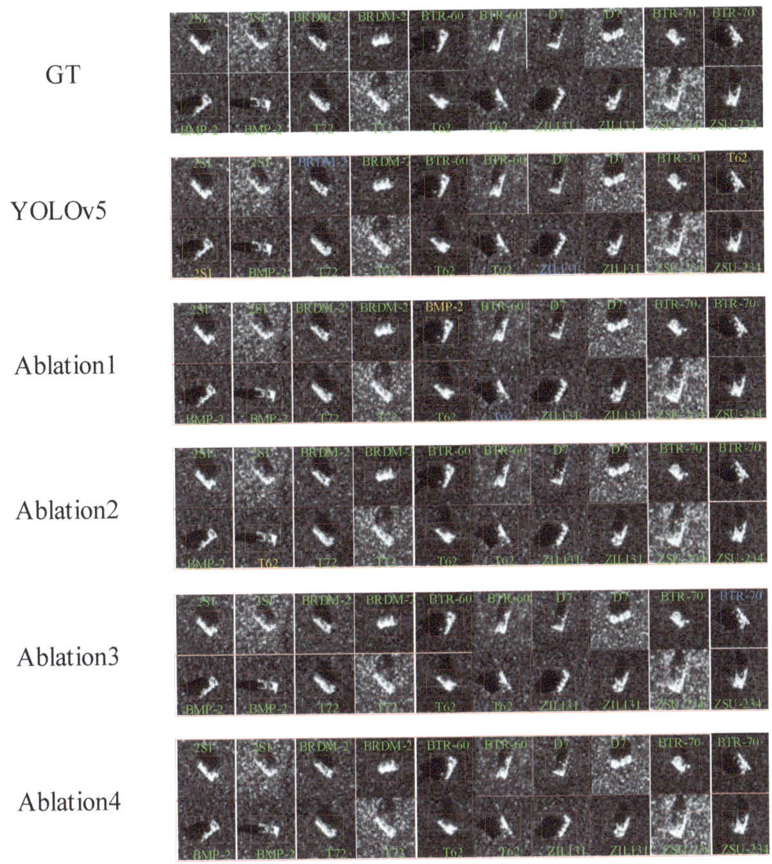

Figure 10. Comparison results of ablation experiment visualization on the MSTAR dataset. The blue rectangular box indicates the missed recognition target, and the yellow rectangular box indicates the false recognition target.

4.3.3. Comparison of Network Model Pruning Results

Pruning experiments were performed on the proposed network structure on the MSTAR dataset to compare the pruning results and select the best one.

The experimental results presented in Table 5 demonstrate that as the pruning degree increases, there is a corresponding decrease in the number of parameters and computational requirements. However, accuracy degradation becomes evident when the pruning magnitude exceeds 50%. Therefore, we selected a pruning rate of 50% to strike a balance between achieving a favorable reduction in parameters and computation while maintaining training accuracy.

Table 5. Pruning results for FCCD-SAR.

Method	Params (M)	FLOPs (G)	mAP (%)
FCCD-SAR (Baseline)	6.00	12.2	99.5
FCCD-SAR (20%Pruned)	5.12	10.98	99.5
FCCD-SAR (40%Pruned)	3.91	8.54	99.5
FCCD-SAR (50%Pruned)	3.34	7.32	99.5
FCCD-SAR (60%Pruned)	2.72	6.11	99.2
FCCD-SAR (80%Pruned)	1.50	3.66	98.6

4.3.4. Comparative Experiment

Comparison results with the latest target-recognition methods on the MSTAR dataset: To further validate the robust detectability of our proposed FCCD-SAR model, we conducted experiments to compare it with commonly used single-stage and two-stage target recognition methods on optical images, including YOLOV3, YOLOV5, Faster R-CNN, Cascade-RCNN, and other methods. Compared to the single-stage target recognition method with fewer parameters and lower computational effort, the present experimental method has fewer parameters and less computational effort and is more accurate. Compared to the more accurate two-stage target recognition method, this experimental method is more accurate than it. Meanwhile, this experimental method had fewer parameters and fewer computations, which could satisfy the lightweight requirement. The parameters for each method were set to be approximately the same, ensuring fairness in the comparison experiments. The results are presented in Table 6. Our proposed method achieves the lowest computational and parametric requirements while surpassing the accuracy and exhibiting the highest overall performance among the compared methods. Visualization of the predictions for each recognition method is shown in Figure 11. Our method demonstrated superior target-recognition performance and a lower false recognition miss rate compared to other target recognition methods.

Table 6. Comparison with the latest target-recognition methods on the MSTAR dataset.

Method	Params (M)	FLOPs (G)	P (%)	mAP (%)	Inference Time (ms)
YOLOV5s	7.05	16.0	97.8	98.1	4.1
YOLOV3	8.7	13.0	96.9	97.2	5.7
Faster-RCNN	41.12	91.41	96.5	96.4	82.6
Cascade-RCNN	69.17	119.05	96.9	96.8	98.5
MobileNetV3	4.29	7.2	96.3	96.6	4.4
FCCD-SAR	2.72	6.11	99.5	99.5	3.4

In order to verify the generalization and avoid overfitting, the article added the SAR ship dataset SSDD to corroborate the model performance. Under the SSDD dataset, the experimental results are shown in Figure 12. The experimental results are shown in Table 7. According to the experimental results, it can be found that the model proposed in this paper has certain generalization, and no overfitting occurs, which fully proves the excellent effect of the proposed model.

Table 7. Comparison with the latest target-recognition methods on the SSDD dataset.

Method	Params (M)	FLOPs (G)	P (%)	mAP (%)	Inference Time (ms)
YOLOV5s	7.06	16.2	96.8	97.8	4.2
YOLOV3	8.9	13.3	96.2	96.8	6.7
Faster-RCNN	44.34	92.11	95.2	94.3	85.1
Cascade-RCNN	71.32	120.48	96.1	95.9	100.8
MobileNetV3	4.58	8.2	93.5	93.6	5.3
FCCD-SAR	3.32	7.4	98.8	98.7	3.6

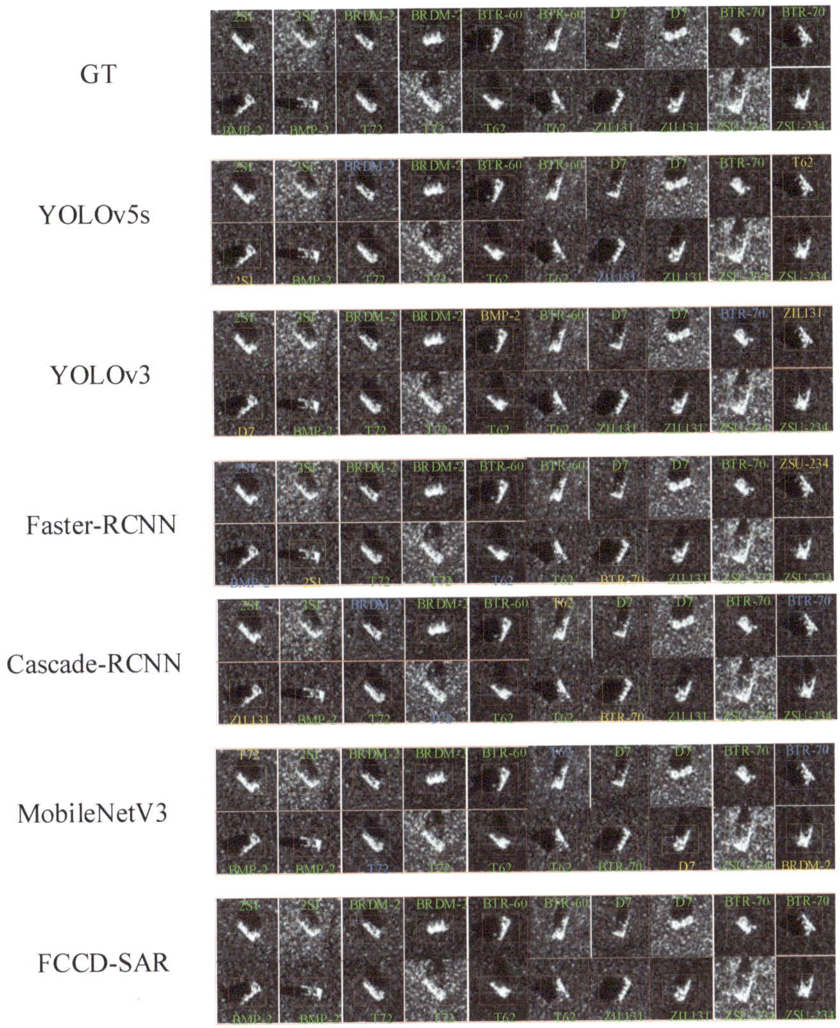

Figure 11. Comparison of visualization recognition results under MSTAR dataset with the latest method. The blue rectangular box indicates the missed recognition target and the yellow rectangular box indicates the false recognition target.

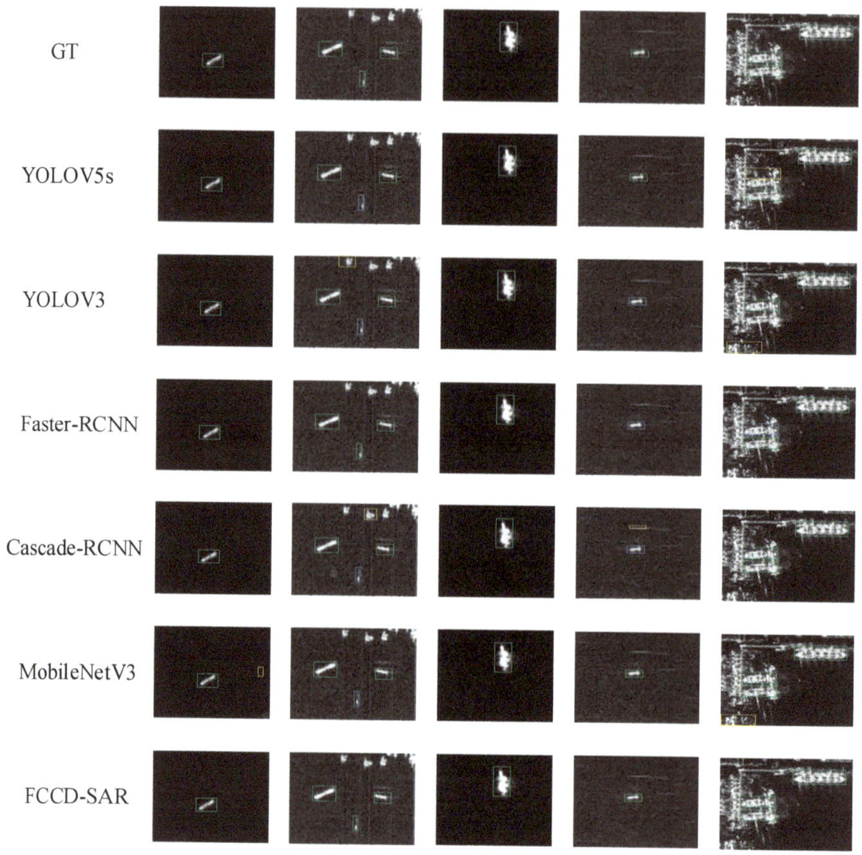

Figure 12. Comparison of visualization recognition results under the SSDD dataset with the latest method. The blue rectangular box indicates the missed recognition target and the yellow rectangular box indicates the false recognition target.

5. Discussion

This paper presented FCCD-SAR, a lightweight algorithm for SAR target recognition based on FasterNet. The method was specifically designed for deployment on embedded devices, taking into consideration the lightweight requirements. Moreover, it incorporates the unique feature information characteristics of SAR images.

In this study, the lightweight benchmark model YOLOv5 was initially introduced, and subsequently, FasterNet, a more efficient and faster neural network, was used to replace the main network. The choice of FasterNet was motivated by its compatibility with the unique characteristics of the SAR image dataset, striking a balance between speed and accuracy.

To minimize both the model size and computational effort, we employed the lightweight upsampling operator CARAFE. CARAFE performs a dot product between the upsampling kernel and the pixels in the surrounding neighborhood of each position within the input feature map. This operation allows for a broader perceptual field during recombination and guides the recombination process using input features. As a result, it enhances the recognition performance of SAR targets while simultaneously reducing the number of parameters.

To improve both the recognition accuracy and computational efficiency of the model, we incorporated the C3-Faster module, which is faster and lighter. This module effectively reduced the number of parameters and computational requirements by selectively discarding unimportant information while maintaining the required level of recognition accuracy.

The attention mechanism-based detection head, DyHead, was incorporated to handle the multi-scale features of SAR targets. DyHead is a dynamic head framework that utilizes an attention mechanism to unify various target detection heads. It leverages attention mechanisms across feature levels for scale perception, spatial locations for spatial perception, and output channels for task perception. By employing this approach, the model's target detection head achieved enhanced expressiveness and improved target recognition accuracy without increasing the computational effort.

To obtain the optimal model, we employed a pruning technique to reduce the network's complexity while preserving its accuracy. Subsequently, we evaluated the proposed method on the MSTAR dataset, and the results demonstrated its exceptional performance, achieving an MAP of 99.5%. Notably, the number of parameters was merely 2.72 M, and the FLOPs amounted to 6.11 G, showcasing the model's efficiency.

Author Contributions: Conceptualization, X.D. and D.L.; methodology, X.D.; software, X.D.; validation, X.D. and D.L.; formal analysis, X.D. and D.L.; investigation, X.D.; resources, X.D.; data curation, X.D.; writing, original draft preparation, X.D.; writing, review and editing, D.L. and J.F.; visualization, X.D. and D.L.; supervision, D.L. and J.F.; project administration, D.L.; funding acquisition, D.L. All authors have read and agreed to the published version of the manuscript.

Funding: This research was funded by the Inner Mongolia Autonomous Region Natural Science Foundation Project, grant number 2022QN06004.

Institutional Review Board Statement: Not applicable.

Informed Consent Statement: Not applicable.

Data Availability Statement: The original data will be shared at the author's discretion. Please contact the corresponding author directly.

Conflicts of Interest: The authors declare no conflict of interest.

References

1. Yang, G.; Li, H.C.; Yang, W.; Fu, K.; Sun, Y.J.; Emery, W.J. Unsupervised change detection of SAR images based on variational multivariate Gaussian mixture model and Shannon entropy. *IEEE Geosci. Remote Sens. Lett.* **2018**, *16*, 826–830. [CrossRef]
2. Wang, C.; Huang, Y.; Liu, X.; Pei, J.; Zhang, Y.; Yang, J. Global in local: A convolutional transformer for SAR ATR FSL. *IEEE Geosci. Remote Sens. Lett.* **2022**, *19*, 1–5. [CrossRef]
3. Wan, H.; Chen, J.; Huang, Z.; Xia, R.; Wu, B.; Sun, L.; Yao, B.; Liu, X.; Xing, M. AFSar: An anchor-free SAR target detection algorithm based on multiscale enhancement representation learning. *IEEE Trans. Geosci. Remote Sens.* **2021**, *60*, 1–14. [CrossRef]
4. Zhang, F.; Wang, Y.; Ni, J.; Zhou, Y.; Hu, W. SAR target small sample recognition based on CNN cascaded features and AdaBoost rotation forest. *IEEE Geosci. Remote Sens. Lett.* **2019**, *17*, 1008–1012. [CrossRef]
5. Zhou, Z.; Chen, J.; Huang, Z.; Wan, H.; Chang, P.; Li, Z.; Yao, B.; Wu, B.; Sun, L.; Xing, M. FSODS: A Lightweight Metalearning Method for Few-Shot Object Detection on SAR Images. *IEEE Trans. Geosci. Remote Sens.* **2022**, *60*, 1–17. [CrossRef]
6. Tsokas, A.; Rysz, M.; Pardalos, P.M.; Dipple, K. SAR data applications in earth observation: An overview. *Expert Syst. Appl.* **2022**, *205*, 117342. [CrossRef]
7. Jiang, J.; Li, Y.; Yuan, Y.; Zhu, Y. Generalized Persistent Polar Format Algorithm for Fast Imaging of Airborne Video SAR. *Remote Sens.* **2023**, *15*, 2807. [CrossRef]
8. Zhang, B.; Xu, G.; Zhou, R.; Zhang, H.; Hong, W. Multi-channel back-projection algorithm for mmwave automotive MIMO SAR imaging with Doppler-division multiplexing. *IEEE J. Sel. Top. Signal Process.* **2022**, *17*, 445–457. [CrossRef]
9. Wang, J.; Liang, X.D.; Chen, L.Y.; Li, Y.L. First demonstration of airborne MIMO SAR system for multimodal operation. *IEEE Trans. Geosci. Remote Sens.* **2021**, *60*, 1–13. [CrossRef]
10. Du, W.; Zhang, F.; Ma, F.; Yin, Q.; Zhou, Y. Improving SAR target recognition with multi-task learning. In Proceedings of the IGARSS 2020–2020 IEEE International Geoscience and Remote Sensing Symposium, Waikoloa, HI, USA, 26 September 2020; pp. 284–287. [CrossRef]
11. Zhang, L.; Leng, X.; Feng, S.; Ma, X.; Ji, K.; Kuang, G.; Liu, L. Domain knowledge powered two-stream deep network for few-shot SAR vehicle recognition. *IEEE Trans. Geosci. Remote Sens.* **2021**, *60*, 1–15. [CrossRef]
12. Feng, S.; Ji, K.; Zhang, L.; Ma, X.; Kuang, G. SAR target classification based on integration of ASC parts model and deep learning algorithm. *IEEE J. Sel. Top. Appl. Earth Obs. Remote Sens.* **2021**, *14*, 10213–10225. [CrossRef]

13. Wang, S.; Wang, Y.; Liu, H.; Sun, Y. Attribute-guided multi-scale prototypical network for few-shot SAR target classification. *IEEE J. Sel. Top. Appl. Earth Obs. Remote Sens.* **2021**, *14*, 12224–12245. [CrossRef]
14. Zhao, J.; Datcu, M.; Zhang, Z.; Xiong, H.; Yu, W. Contrastive-regulated CNN in the complex domain: A method to learn physical scattering signatures from flexible PolSAR images. *IEEE Trans. Geosci. Remote Sens.* **2019**, *57*, 10116–10135. [CrossRef]
15. Wang, K.; Qiao, Q.; Zhang, G.; Xu, Y. Few-Shot SAR Target Recognition Based on Deep Kernel Learning. *IEEE Access* **2022**, *10*, 89534–89544. [CrossRef]
16. Wei, Q.R.; He, H.; Zhao, Y.; Li, J.A. Learn to Recognize Unknown SAR Targets From Reflection Similarity. *IEEE Geosci. Remote Sens. Lett.* **2020**, *19*, 1–5. [CrossRef]
17. Du, C.; Huo, C.; Zhang, L.; Chen, B.; Yuan, Y. Fast C&W: A fast adversarial attack algorithm to fool SAR target recognition with deep convolutional neural networks. *IEEE Geosci. Remote Sens. Lett.* **2021**, *19*, 1–5. [CrossRef]
18. Peng, B.; Peng, B.; Zhou, J.; Xia, J.; Liu, L. Speckle-variant attack: Toward transferable adversarial attack to SAR target recognition *IEEE Geosci. Remote Sens. Lett.* **2022**, *19*, 1–5. [CrossRef]
19. Du, L.; Dai, H.; Wang, Y.; Xie, W.; Wang, Z. Target discrimination based on weakly supervised learning for high-resolution SAR images in complex scenes. *IEEE Trans. Geosci. Remote Sens.* **2019**, *58*, 461–472. [CrossRef]
20. Li, C.; Du, L.; Li, Y.; Song, J. A novel SAR target recognition method combining electromagnetic scattering information and GCN. *IEEE Geosci. Remote Sens. Lett.* **2022**, *19*, 1–5. [CrossRef]
21. Ren, S.; He, K.; Girshick, R.; Sun, J. Faster r-cnn: Towards real-time object detection with region proposal networks. *arXiv* **2015**, arXiv:1506.01497.
22. Song, Y.; Li, J.; Gao, P.; Li, L.; Tian, T.; Tian, J. Two-stage cross-modality transfer learning method for military-civilian SAR ship recognition. *IEEE Geosci. Remote Sens. Lett.* **2022**, *19*, 1–5. [CrossRef]
23. Lin, T.Y.; Dollár, P.; Girshick, R.; He, K.; Hariharan, B.; Belongie, S. Feature pyramid networks for object detection. In Proceedings of the IEEE Conference on Computer Vision and Pattern Recognition (CVPR), Honolulu, HI, USA, 21–26 July 2017; pp. 2117–2125. [CrossRef]
24. Li, R.; Wang, X.; Wang, J.; Song, Y.; Lei, L. SAR target recognition based on efficient fully convolutional attention block CNN. *IEEE Geosci. Remote Sens. Lett.* **2020**, *19*, 1–5. [CrossRef]
25. Li, S.; Pan, Z.; Hu, Y. Multi-Aspect Convolutional-Transformer Network for SAR Automatic Target Recognition. *Remote Sens.* **2022**, *14*, 3924. [CrossRef]
26. Zhang, C.; Wang, Y.; Liu, H.; Sun, Y.; Hu, L. SAR target recognition using only simulated data for training by hierarchically combining CNN and image similarity. *IEEE Geosci. Remote Sens. Lett.* **2022**, *19*, 1–5. [CrossRef]
27. Xu, R.; Li, D. Automatic Target Recognition Technology of SAR Images Based on 2DPCA+ PNN. In Proceedings of the 2020 4th International Conference on Electrical, Automation and Mechanical Engineering, Beijing, China, 21–22 June 2020; Volume 1626, No. 1. p. 012108. [CrossRef]
28. Yang, Y.; Li, D. Improving YOLOv5's lightweight helmet wear detection algorithm. *J. Comput. Eng. Appl.* **2022**, *58*, 201–207. [CrossRef]
29. Zhang, F.; Liu, Y.; Zhou, Y.; Yin, Q.; Li, H.C. A lossless lightweight CNN design for SAR target recognition. *Remote Sens. Lett.* **2020**, *11*, 485–494. [CrossRef]
30. Dai, Q.; Zhang, G.; Fang, Z.; Xue, B. SAR target recognition with modified convolutional random vector functional link network. *IEEE Geosci. Remote Sens. Lett.* **2021**, *19*, 1–5. [CrossRef]
31. Chen, J.; Kao, S.H.; He, H.; Zhuo, W.; Wen, S.; Lee, C.H.; Chan, S.H.G. Run, Don't Walk: Chasing Higher FLOPS for Faster Neural Networks. In Proceedings of the IEEE/CVF Conference on Computer Vision and Pattern Recognition (CVPR), Vancouver, BC, Canada, 18–22 July 2023; pp. 12021–12031. [CrossRef]
32. Wang, J.; Chen, K.; Xu, R.; Liu, Z.; Loy, C.C.; Lin, D. Carafe: Content-aware reassembly of features. In Proceedings of the IEEE/CVF International Conference on Computer Vision (ICCV), Seoul, Republic of Korea, 27 October–3 November 2019; pp. 3007–3016. [CrossRef]
33. Dai, X.; Chen, Y.; Xiao, B.; Chen, D.; Liu, M.; Yuan, L.; Zhang, L. Dynamic head: Unifying object detection heads with attentions. In Proceedings of the IEEE/CVF Conference on Computer Vision and Pattern Recognition (CVPR), Virtual, 20–25 June 2021; pp. 7373–7382. [CrossRef]

sensors

MDPI

Article

Designing UAV Swarm Experiments: A Simulator Selection and Experiment Design Process

Abhishek Phadke [1,2,*], F. Antonio Medrano [1,2], Chandra N. Sekharan [2] and Tianxing Chu [1,2]

1 Conrad Blucher Institute for Surveying and Science, Texas A&M University-Corpus Christi, Corpus Christi, TX 78412, USA
2 Department of Computer Science, Texas A&M University-Corpus Christi, Corpus Christi, TX 78412, USA
* Correspondence: aphadke@islander.tamucc.edu

Abstract: The rapid advancement and increasing number of applications of Unmanned Aerial Vehicle (UAV) swarm systems have garnered significant attention in recent years. These systems offer a multitude of uses and demonstrate great potential in diverse fields, ranging from surveillance and reconnaissance to search and rescue operations. However, the deployment of UAV swarms in dynamic environments necessitates the development of robust experimental designs to ensure their reliability and effectiveness. This study describes the crucial requirement for comprehensive experimental design of UAV swarm systems before their deployment in real-world scenarios. To achieve this, we begin with a concise review of existing simulation platforms, assessing their suitability for various specific needs. Through this evaluation, we identify the most appropriate tools to facilitate one's research objectives. Subsequently, we present an experimental design process tailored for validating the resilience and performance of UAV swarm systems for accomplishing the desired objectives. Furthermore, we explore strategies to simulate various scenarios and challenges that the swarm may encounter in dynamic environments, ensuring comprehensive testing and analysis. Complex multimodal experiments may require system designs that may not be completely satisfied by a single simulation platform; thus, interoperability between simulation platforms is also examined. Overall, this paper serves as a comprehensive guide for designing swarm experiments, enabling the advancement and optimization of UAV swarm systems through validation in simulated controlled environments.

Keywords: UAV; swarm; robotics; simulation

1. Introduction

UAV (Unmanned Aerial Vehicle) swarms refer to a collective group of UAVs that operate in a coordinated manner to achieve a common goal. UAV swarms are characterized by their ability to exhibit coordinated behaviors, where individual UAVs collaborate and interact with one another to accomplish tasks efficiently and effectively. This enables UAV swarms to perform complex operations for applications in surveillance [1,2], agriculture [3], military [4], search and rescue missions [5], and environmental monitoring [6,7].

The use of swarms offers several advantages over single UAVs, including increased robustness, redundancy, scalability, and enhanced mission capabilities. By leveraging swarm intelligence and advanced coordination algorithms, UAV swarms have the potential to revolutionize various industries and applications, opening up new possibilities for autonomous and collaborative aerial operations. Current swarm development can be broadly classified by their topology [8,9], by type of swarm agents such as homogeneous [10] or heterogeneous [11], and by application-specific usage such as remote sensing [12]. Of course, more advanced swarm descriptions and classifications exist that are created with a specific development goal in mind.

The environment that these swarms work in is prone to disruptions, which can impede their operation. With the degree of close-knit topology built into these swarms, damage or failure of swarm agents can result in a compromise in mission progress [13]. Additionally, swarms have a multilayered architecture of various components working cohesively to bring about acceptable operation. The creation of resilient UAV swarms requires the integration of multiple components [14]. Researchers across the globe are creating novel methods for engineering resiliency in UAV swarms [15]. For safety and reliability, it is necessary to perform rigorous testing of swarm systems in all phases of development. A wide range of general robotic or UAV specific simulation platforms exists that may be explored for this. Choosing a suitable simulator is important, as different simulation environments offer different performance, model detail, and built-in features, all of which may affect the success and the merit of a simulation-based study. Despite numerous options to choose from, there is a lack of descriptive research that categorizes and indexes them for the convenience of future experiment designers.

There are a multitude of tools available in the robotics development scenario, such that it is almost impossible to keep track of all of them without studies like this one to track and describe use-case scenarios. Each simulator tool and the platforms they are built on have their strengths and weaknesses, which the developer must be aware of before choosing a simulator. However recent trends in development have shifted focus in terms of the compatibility of these tools, such that a lot of these tools can be used in conjunction with each other. It is now entirely possible to connect integral as well as third-party plugins from different platforms to talk to each other for data sharing and system design. This leads to advantages such as distributing simulation tasks among various platforms. While this may not necessarily increase simulation performance, it opens up a wider array of capabilities and features during system design and control. The domain of mobile robot simulation, design, and development is a growing field along with its associated subcategories. The authors in [16] portray swarm robotics as a subclass of mobile robots.

The objectives of this study are to present a descriptive case study of UAV swarm simulation tools that the authors have used to design experiments and validate results. However, not all simulation platforms were capable enough to support the expectations of all experiments. Hence, a variety of tools were needed to perform them. Sometimes interfacing the various tools for real-time and post-processed data exchange was also necessary. Additionally, this study highlights the experimental design process required to conduct both real-world and simulation tests of several UAV swarm experiments. While making progress on specific objectives in UAV swarm development that require the design of a variety of experiments, the authors have examined and used a range of simulators capable of expressing UAV swarms. The experiment designs were aimed at incorporating resiliency into UAV swarms.

The three objectives for resilient UAV swarm development currently being worked on as part of building resilient UAV swarms by the authors are as follows:

(1) Creating inter-agent and global policies for path planning, swarm movement, and collision avoidance using techniques such as artificial potential fields and bioinspired pheromone maps.
(2) Creating swarm agent-specific SAR (Search and Rescue) frameworks that focus on improving operational swarm resilience rather than external operations.
(3) Examining the impacts to swarm dynamics and performance on the introduction of heterogeneous agents in the UAV swarm.

This study describes a concise experiment design strategy executed while creating resilient swarm protocols for various UAV swarm scenarios. These experiments were planned and categorized in an experimental series each having unique objectives and outcomes. The major contributions of this research paper include a thorough examination of existing simulation platforms, aiding researchers in selecting the most suitable tool for their swarm experimentation needs. Additionally, the outlined experimental design process will provide a valuable framework to assist users in selecting the appropriate simulation tool for developing specific objectives for UAV swarms. Presently, surveys concerning simulation platforms offer general overviews of these platforms without delving into specific case scenarios. The designed experiment series herein are implemented for the choice of simulation platforms, and sample experimental results are shown.

The novelty of this research lies in its systematic approach to designing experiments for UAV swarm systems. While the development of UAV swarm technology has gained traction due to its diverse applications, there has been a lack of standardized experimental procedures to validate and optimize the performance of these systems. This research bridges the gap by offering a well-defined and rigorous experimental design process, encompassing the selection of appropriate simulation platforms and the formulation of specific objectives. By addressing the need for robust experimental validation before deploying UAV swarms in dynamic environments, this study contributes to the advancement and resilience of swarm technology, paving the way for its successful integration in real-world applications. Researchers and practitioners can utilize the insights from this paper to conduct evidence-based evaluations, ensuring the reliability and efficiency of UAV swarm systems across various scenarios and challenges.

Our findings are condensed in this brief article below and organized in the following manner. Section 2 outlines previous work on simulation platforms and lists reasons to re-examine them. Section 3 briefly defines swarm characteristics such as network topology, deployment strategies, and composition. Section 4 describes the experimental series and their targeted development objectives. Section 5 shows implementation scenarios of the selected simulation platforms for every experiment series. Section 6 outlines future work considerations and Section 7 has a concluding statement.

2. Background and Motivation

A structured framework design for important UAV swarm processes, such as experiment design and validation, is of paramount importance for several reasons. Firstly, it provides a systematic and organized approach to tackle complex problems and research questions. By following a well-defined framework, researchers can establish clear objectives, identify variables, and establish a coherent plan, ensuring that their experiments yield reliable and meaningful results. Secondly, a structured framework enhances the reproducibility and comparability of experiments. With a detailed and standardized methodology, other researchers can replicate the study to validate its findings or build upon the existing knowledge, fostering scientific progress and collaboration. Furthermore, a well-designed framework helps in minimizing bias and errors in the experimental process [17]. By carefully considering potential confounding factors and controlling variables, researchers can enhance the accuracy and validity of their conclusions, strengthening the overall credibility of their work. Lastly, a structured approach aids in identifying limitations and potential pitfalls early in the research process. By incorporating rigorous validation procedures and statistical analyses, researchers can gain a deeper understanding of the robustness of their results, acknowledge any shortcomings, and suggest areas for further investigation. Thus, a structured framework design experiment design and validation is a cornerstone of reliable and impactful research. It fosters clarity, reproducibility, objectivity, and efficiency, ultimately advancing knowledge and facilitating evidence-based decision making in various domains.

A search for recent articles focused on UAV simulator review yielded only a few results and are outlined in Table 1. A comparison of three simulation tools for UAV use is described in [18]. While this is a more concise approach, the authors in [19] present a broader review of the various platforms available. Article [20] further expands the review process by considering 17 different simulation platforms to review. Article [16] presents a broad view of swarm robotics, simulators, hardware, and behavior.

Table 1. Recent works focusing on robotic simulators arranged by their published year.

Reference	Published Year	Description
[18]	2018	Comparing the performance of three popular robot simulators.
[21]	2019	Survey on UAV simulators, their features, and architecture.
[16]	2021	Survey on robotic simulators, platforms, and frameworks, with a distinction between the three terms.
[20]	2021	An overview of robotic simulators suitable for use in education.
[19]	2022	Review on swarm-focused simulators, real-life hardware, and applications.
This study	2023	An updated and comprehensive examination of simulation platforms capable of handling UAVs with an accompanying experimental design process for swarm-based research objectives.

During the course of the literature review, it was observed that multiple tools mentioned in past literature are now deprecated and, thus, were not included in Table 1. As technology rapidly evolves, previous studies may not reflect the current state-of-the-art in terms of available simulators and their capabilities. Tools such as USARSim [22] have pre-2020 development dates and are no longer updated. Additionally, tools such as VREP [23] have changed names, whereas Microsoft AirSim [24] while still having workable versions has been archived in favor of future projects. Additionally, as observed in [25], many researchers create custom scenario-based simulation packages for testing certain components. These lack code reusability and further support. Even currently acceptable broad-range simulation platforms may not satisfy all requirements for the objective study [26].

Therefore, there is a pressing need to create an updated and comprehensive list of UAV simulators that encompasses the latest advancements in the field. This list will serve as a valuable resource for researchers, developers, and practitioners, providing them with accurate and relevant information to choose the most suitable simulator for their specific needs. Researchers can stay informed about the latest options and make informed decisions when it comes to simulation-based studies and the development of UAV systems.

Article [27] in its literature review survey concludes that MATLAB was the most widely used simulation platform for UAV swarm experiments. However, their methodology compares items such as Java against Visual Studio, which are a programming language and an IDE, respectively. Our search parameters were more concentrated and considered simulation platforms only. Based on our analysis, we have summarized our findings in the following figures and tables. Figure 1 shows the top six platforms that were used by researchers working on aerial robotic swarms in the past five years. Table 2 presents additional information about simulation platforms examined such as OS support and cross-platform availability. Table 3 lists notable publications on swarm robotic development in the past three years (2021–2023) along with the simulation platform used for experiments and result validation.

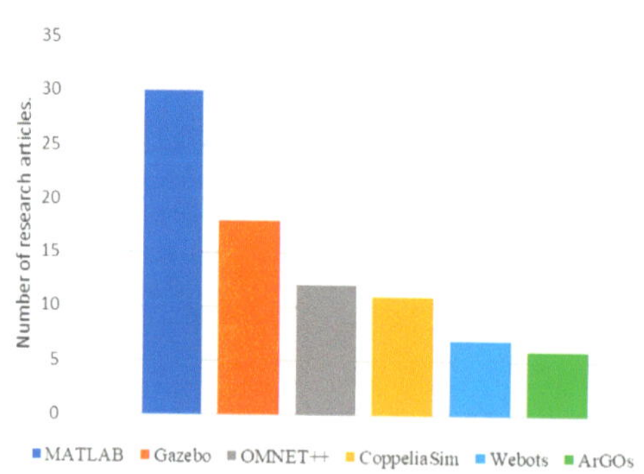

Figure 1. The number of studies examined using different simulation platforms over the past five years (2019–2023).

Table 2. Simulation tools examined.

Name	OS Support [1]	UAV Specific?	Possibility of Cross-Platform Connectivity [2]	Notable Publication by Platform Creators	Remarks
Gazebo	W *, M *, L	No	Yes	—	—
Webots	W, M, L	No	Yes	—	Independent simulation platform with support for various robot platforms.
CoppeliaSim	W, M, L	No	Yes	[23]	Formerly known as VREP.
UAV toolbox (MATLAB)	W, M, L *	No	Yes	—	Although MATLAB is not UAV specific, the UAV toolbox is designed for UAV development.
RflySim	W	No	Yes	[28]	Independent simulation platform inspired by PX4 and MATLAB simulation platform.
ARGoS	M *, L	No	Yes+	[29]	Multiphysics robot simulator.
OMNET++	W, M, L	No	Yes	—	Discrete event simulation platform focused on networking and communication.
AVENS	W	Yes	Yes	[30]	Works with the OMNET++ network simulator and X-Plane flight simulator.
MORSE	W *, M *, L *	No	Yes	[31]	—

[1] A * near the OS label indicates that direct binaries of the tool may not be available; however, alternatives such as docker deployments or community supported guides do exist. [2] A + near the yes indicates that the software has indirect passing of model data for cross-platform connectivity or the possibility of self-developed plug-ins.

Table 3. Recent work on UAV swarm development and the simulation platform they used is categorized by published year (Range 2021 to 2023).

Reference	Year Published	Study Description	Platform Used [1]
[32]	2021	Formation control of heterogeneous UAV and USV swarms.	CoppeliaSim
[33]	2021	Architecture of UAV swarm to find a load and transport it to its destination cooperatively.	CoppeliaSim
[34]	2021	Adaptive formation control for UAV swarms with multiple leaders and switching topologies.	Gazebo
[35]	2021	Control of UAV agents in a swarm using vision-based approaches.	Gazebo
[36]	2021	Development of control layers to enable the autonomous and cooperative navigation of a swarm of UAVs.	Gazebo
[37]	2021	Optimized area coverage by autonomous multi-UAV.	Gazebo + MATLAB
[38]	2021	Approach to address coverage and flocking problems in multi-UAV.	Gazebo
[39]	2021	Bioinspired neural network for cooperative planning of multi-UAV.	Gazebo
[40]	2021	Safe allocation of UAV swarm mission resources based on random labels.	OMNET++
[41]	2022	Coverage and path planning algorithm for swarms to detect points of interest and collect information from them.	OMNET++

Table 3. *Cont.*

Reference	Year Published	Study Description	Platform Used [1]
[42]	2022	A distributed UAV swarm formation system optimized by a hybrid evolutionary algorithm.	ARGoS
[43]	2022	Approach to optimized UAV swarm for improved intruder detection.	ArGoS
[44]	2022	A framework for simulating cooperative UAV swarms performing joint missions.	OMNET++
[45]	2022	Task decomposition and task correlation of UAV swarm	OMNET++
[46]	2022	Multiple UAVs collaborating for improved field phenotyping.	Gazebo
[47]	2022	Resource balancing for MAV (Mobile Aerial Vehicle).	Gazebo
[48]	2022	Development of a PSO-based threat avoidance and reconnaissance FANET.	Gazebo
[49]	2022	Entrap multiple targets using a robot swarm.	MATLAB + CoppeliaSim
[50]	2023	A decentralized method for multiple UAVs to explore separate areas.	Gazebo
[51]	2023	A novel IDS (Intrusion Detection System) that identifies deviation in normal UAV behaviors as means of indicating external threats.	Gazebo
[52]	2023	A multiple tracking methodology for aircraft at low altitudes.	MATLAB
[53]	2023	Collision avoidance strategy for D2D (Device to Device) communications in UAV networks.	MATLAB
[54]	2023	Evaluation of routing protocols in UAV ad hoc networks in SAR scenarios	MATLAB
[1]	2023	Autonomous cooperative mission planning for multiple UAVs conducting surveillance missions.	MATLAB
[55]	2023	Trajectory planning and formation maintenance in swarm using MPC and standoff algorithm.	MATLAB
[56]	2023	An improved PSO for optimized base station placement for UAVs.	CoppeliaSim

[1] With regards to the use of MATLAB for experimental design, all cited references may not necessarily use the UAV toolbox, Simulink, or other add-ons, choosing to rely only on the main program. However, they are included in the table since the standalone MATLAB platform may be sufficient for their validation requirements.

3. Generalized Swarm Process Consideration during Development

It is vital to understand swarm characteristics before experiment design. Based on the three research objectives highlighted in the introduction, the experiment series will be created to observe and validate the performance of developed resilient mechanisms for the major swarm processes as discussed below.

3.1. Network Topologies and Communication

Network topologies and communication play a crucial role in the coordination and effectiveness of UAV swarm systems. The selection and design of appropriate network topologies significantly impact the swarm's ability to exchange information, make collective decisions, and execute coordinated actions. Various topologies, such as centralized, decentralized, and hybrid architectures [57], can be employed based on the specific requirements and constraints of the swarm application.

Centralized topologies, with a central controller, facilitate efficient communication and decision making but may suffer from single points of failure. Decentralized topologies distribute the decision-making across multiple UAVs, promoting resilience [58] and scalability but introducing challenges in synchronization and coordination. Figure 2 shows a centralized communication topology on the left and a decentralized communication topology on the right.

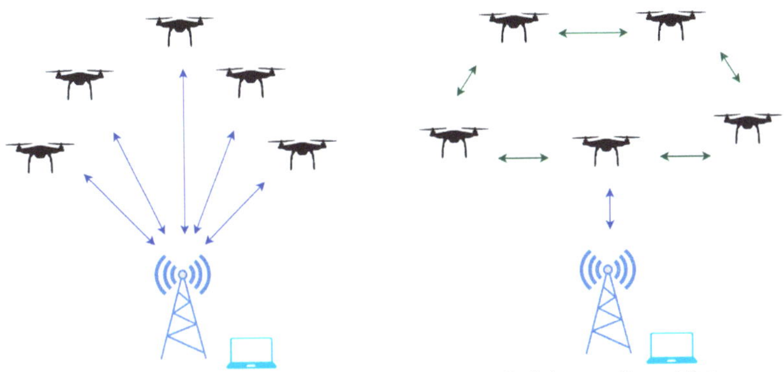

Figure 2. Centralized topology (**left**) and decentralized topology (**right**) for a swarm of UAV agents.

Hybrid topologies combine centralized and decentralized elements to strike a balance between efficiency and robustness. Communication protocols and mechanisms, such as direct or indirect communication, broadcast, or multi-hop routing, determine how UAVs exchange information and collaborate within the swarm. Efficient communication protocols must address issues such as packet loss, latency, and bandwidth constraints to ensure reliable and timely data transmission. Additionally, incorporating advanced techniques like adaptive routing [59], dynamic network reconfiguration, and cognitive radio systems [60] can further enhance the swarm's communication capabilities. Future research should focus on developing efficient network topologies and communication protocols tailored to the specific needs of UAV swarm systems, considering factors such as scalability, robustness, energy efficiency [61], and adaptability to dynamic environments. By improving network topologies and communication, UAV swarms can achieve enhanced coordination, cooperation, and performance, enabling them to tackle a wide range of complex tasks and operate effectively in diverse scenarios.

3.2. Deployment Strategies

Deployment strategies [62] play a critical role in the effective utilization of UAV swarms over an ROI (Region of Interest). The choice of strategy depends on various factors, including the mission objectives, environmental conditions, available resources, and desired outcomes [63]. One commonly employed strategy is the grid-based deployment approach, where the ROI is divided into a grid pattern and UAVs are strategically positioned at predefined grid points. While the decomposed grid may be any polygon [64], four-sided cells are the most common. A collection of such cells is called a subgrid [65]. Depending on a use case scenario these subgrids may be of different shapes and dimensions. Grid-based decomposition ensures comprehensive coverage of the entire area and facilitates systematic exploration of the ROI. Deploying agents to assigned grids or subgrids is particularly useful in scenarios that require even distribution of surveillance or data collection efforts, such as monitoring large agricultural fields or conducting wide-scale search and rescue operations.

Another deployment strategy is targeted deployment, which involves placing UAVs at specific locations within the ROI based on the mission requirements. This strategy allows for a more focused approach, concentrating the swarm's resources on critical areas of interest. For instance, in disaster management scenarios, UAVs can be deployed near disaster-stricken regions or potential danger zones to gather detailed information, assess damage, or provide real-time situational awareness to aid response teams in their decision-making process. Figure 3 shows examples of deployment strategies that were visualized during the experiment design process. Predefined agent deployment assumes the initial placement of agents at selected points in the ROI; randomized deployment uses a single take-off point and is then relined on a randomized search by agents backed by consensus for target detection. On the left, the different colors indicate that each agent was preassigned its subgrid to search. On the right, randomized deployment shows all agents flying from a single take-off point, with no pre-assignment of subgrids.

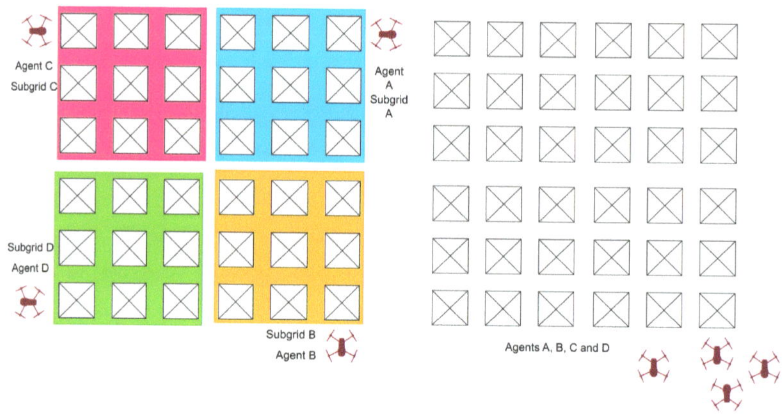

Figure 3. Predefined agent deployment (**left**); single point deployment (**right**).

Figure 4 shows the two possible subgrid selections highlighted in yellow and green. These are just examples. In practice, the size and shape can be defined by the operator depending on factors such as agent capability, sensor range, coverage, or ROI geographical features. These subgrids can be selected initially and stay constant or may change dynamically across the mission timeline.

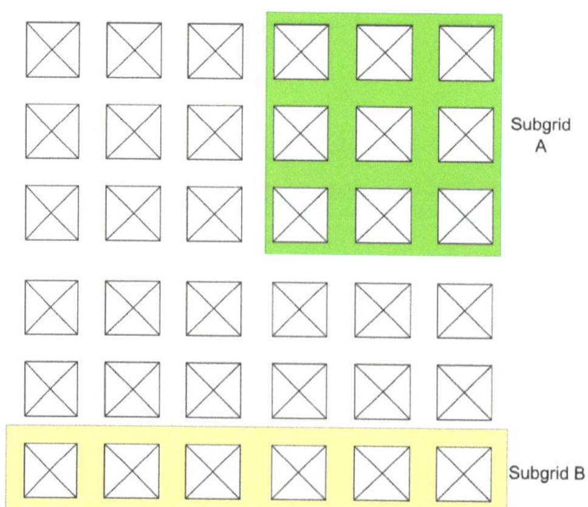

Figure 4. Possible subgrid selections are shown as A or B.

Randomized deployment strategies introduce variability and unpredictability into the swarm's positions. By using algorithms that generate random or pseudo-random deployment patterns, the swarm can explore the ROI in a non-deterministic manner. This strategy can be beneficial in scenarios where it is essential to reduce predictability, enhance resilience, or counteract potential adversarial efforts to anticipate the swarm's behavior. Randomized deployment strategies can be particularly useful in surveillance and security applications, where diverse and unpredictable movement patterns can make it more challenging for adversaries to identify and track the swarm's activities.

In certain dynamic scenarios, adaptive deployment strategies come into play. These strategies involve dynamically adjusting the swarm's configuration based on real-time information and changing mission requirements. By leveraging feedback from sensors, communication networks, or external data sources, the swarm can adapt its deployment pattern, alter its flight paths, or redistribute resources [66] within the ROI. Adaptive deployment strategies enhance the swarm's flexibility, responsiveness, and ability to prioritize and allocate resources based on evolving situational demands. This adaptability is especially valuable in scenarios with dynamic events, such as monitoring traffic congestion, tracking fast-moving targets, or responding to emerging threats in real-time.

The choice of deployment strategy depends upon careful consideration of mission objectives, environmental factors, available resources, and the desired level of coverage and redundancy. By selecting an appropriate deployment strategy, UAV swarms can maximize their effectiveness, optimize data collection or surveillance efforts, and accomplish the intended mission goals within the ROI.

3.3. *Swarm Formation Control Strategies*

Swarm formation control strategies are fundamental to achieving coordinated behavior and desired spatial arrangements within UAV swarms. These strategies involve designing algorithms and mechanisms that enable the swarm agents to self-organize and maintain specific formations while dynamically adapting to changes in the environment or mission requirements. Various approaches, such as potential fields, artificial potential functions, behavior-based methods, distributed consensus algorithms, and network-to-distance strategies can be employed to govern swarm formation control.

Potential field-based methods utilize attraction and repulsion forces to guide agents toward desired positions while avoiding collisions. Artificial potential functions [67] define a mathematical representation of the desired formation and drive the swarm toward it. Behavior-based methods focus on defining individual agent behaviors and interactions that collectively result in the desired formation. Distributed consensus algorithms [32,68] enable agents to reach a consensus on their positions and orientations through local interactions and information exchange. These strategies typically rely on local sensing, communication, and decision-making capabilities to achieve the desired formations without relying on centralized control.

A network-to-distance formation controller in a UAV swarm operates on the principle that agents in close proximity should have a stronger network link between them. This relationship allows the network strength to serve as a representation of the distance constraints between two agents. By setting a network strength threshold, the controller can establish and maintain the desired formation among the UAVs. The controller continuously monitors the network connectivity and measures the strength of communication links between neighboring agents. When the network strength between two agents exceeds the predefined threshold, it indicates that the agents are within the desired distance range. In this case, the controller maintains their current positions to preserve the formation. However, if the network strength falls below the threshold it signifies that the agents are too far apart, potentially violating the desired formation. The controller then initiates appropriate control actions, such as adjusting the agents' velocities or orientations, to bring them back within the desired distance range. By leveraging the network strength as a surrogate for distance, the network-to-distance formation controller enables coordinated movement and formation control in a UAV swarm. It allows the swarm to dynamically adapt to environmental changes or mission requirements. This controller facilitates robust and scalable coordination among the UAVs, enabling them to collectively perform complex tasks and accomplish objectives efficiently and effectively.

The choice of formation control strategy depends on factors such as the desired formation shape, scalability, robustness, and computational complexity. By enhancing swarm formation control strategies, UAV swarms can achieve precise spatial arrangements, coordinated movements, and cooperative behaviors, enabling them to effectively tackle a wide range of tasks, including surveillance, mapping, and collaborative sensing in diverse real-world scenarios.

3.4. *Swarm Composition and Vehicle Characteristics*

Heterogeneous UAV swarms, composed of UAVs with different capabilities, have gained significant attention in the field of swarm robotics due to their potential to enhance swarm performance in various applications [69]. By combining UAVs with diverse functionalities, such as different sensor payloads, communication capabilities, or task-specific capabilities, heterogeneous swarms offer increased versatility and adaptability compared to homogeneous swarms [70]. The inclusion of heterogeneous agents is predicted to surpass the performance set by homogeneous agent swarms as well as pave the way for coevolutionary abilities [71].

One significant advantage of heterogeneous UAV swarms is their ability to efficiently accomplish complex tasks through task specialization. By assigning specific roles or functions to different types of UAVs within the swarm, the workload can be distributed effectively, leading to improved task completion times. For instance, UAVs equipped with high-resolution cameras can focus on detailed surveillance and target identification, while UAVs with heavy-lift capabilities can handle payload transport or deployment tasks. This division of labor enhances the overall efficiency and enables the swarm to tackle tasks that would be challenging for a homogeneous swarm. A UGV (Unmanned Ground Vehicle) and UAV combination can work together to produce survey maps from different perspectives [72] or cooperative surveillance [73].

Moreover, heterogeneous UAV swarms demonstrate improved coverage capabilities [74]. The diverse range of sensors and payloads in heterogeneous swarms enables them to gather more comprehensive and diverse data about the environment. This enhanced coverage facilitates better situational awareness, enabling the swarm to make informed decisions and respond effectively to dynamic environmental changes. Additionally, the heterogeneous nature of the swarm allows for optimized resource allocation, as UAVs with specific capabilities can be strategically deployed in areas where their expertise is most valuable. This targeted deployment ensures efficient resource utilization and maximizes the coverage area, ultimately improving the overall effectiveness of the swarm.

Another critical aspect influenced by heterogeneous UAV swarms is fault tolerance. The inclusion of different types of UAVs with redundant or complementary capabilities enhances the swarm's resilience in the face of individual UAV failures. In the event of a malfunction or system failure, other UAVs within the heterogeneous swarm can compensate for the loss by taking over the failed UAV's responsibilities. This fault-tolerant behavior increases the reliability and robustness of the swarm, ensuring the continuity of mission execution even in challenging or unpredictable scenarios. Figure 5 shows two possible scenarios where cooperative behavior between heterogeneous agents is enabled. Scenario A shows applications where UAV and UGV work together to produce multi-perspective maps of environments. Scenario B shows a UWSV (Unmanned Water Surface Vehicle) equipped with a landing platform [75]. Such deployments have been popularly used in marine search and rescue scenarios where the landing platform allows UAVs to recharge or synchronize data between flights, thus extending their capability and flight time.

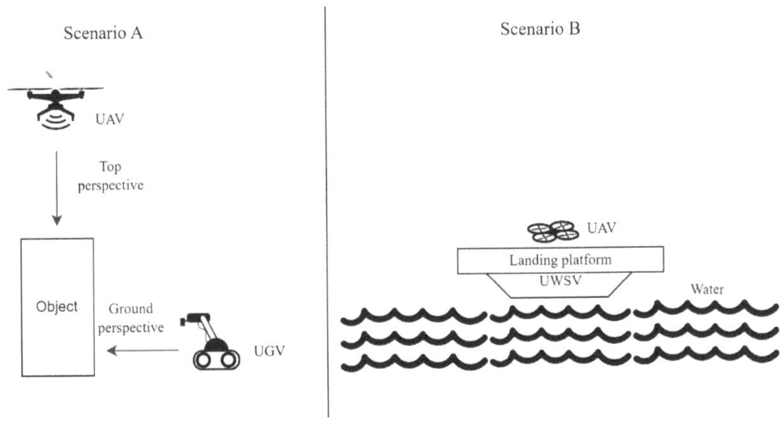

Figure 5. Two scenarios showing cooperation operation between heterogeneous robots.

However, the integration of heterogeneous UAVs also presents challenges that must be carefully addressed. Communication and coordination between different UAV types become crucial factors in maintaining the coherence and efficiency of the swarm. Effective communication protocols and coordination mechanisms must be developed to enable seamless collaboration and exchange of information between heterogeneous UAVs. Furthermore, the heterogeneity of the swarm necessitates sophisticated task allocation and decision-making algorithms to optimize resource allocation and task assignments based on individual UAV capabilities and mission objectives. The incorporation of UAVs with diverse capabilities enhances the swarm's efficiency, adaptability, and resilience. However, addressing communication, coordination, and task allocation challenges become imperative to fully harness the potential of heterogeneous UAV swarms.

A UAV-specific simulation platform may not be the best choice for the design and validation of heterogeneous swarms, particularly if the participation of swarm agents over a variety of operational spaces is needed. This highlights the need for platforms with cross-platform interfacing as well as generalized tools that are capable of simulating a wide range of robots. A systematic experimental design process should be tailored specifically for validating the resilience and performance of UAV swarm systems exhibiting the above swarm process and application-specific usage.

4. Simulator Selection and Experimental Design

A set of experiments were formulated to consider the assessment of the aforementioned simulation platforms and the specific research objectives. While complete experiment descriptions are not included in this article, a series categorization was created for each one. The selection of the most appropriate simulation platform was made for each experiment series to ensure an optimal environment for conducting the experiments. The experiment series and their characteristics along with the selected platform are outlined in Table 4.

Each experiment series addresses specific aspects of UAV behavior, communication, and performance. ES1 focuses on observing fast swarm phenomena, specifically flocking maneuvers and 2D path planning. The goal is to study the collective behavior of swarms and their ability to navigate efficiently in two-dimensional space. Experiments observing swarm phenomena such as flocking maneuvers and 2D path planning were initiated here. MATLAB was the choice of simulation tool used due to its robust environment for algorithm development and simulation, with specialized toolboxes for robotics and control systems. Developers also have the option to transfer created deployments into in-house higher-level modules such as Simulink and UAV toolbox. Cross-platform abilities allow interfacing with community-created packages such as the RYZE TELLO application to control Tello

EDU swarms or other simulations such as CoppeliaSim. ES2 was used for observing FANET (Flying Ad Hoc Networks) and MANET (Mobile Ad Hoc Networks) topology and performance for developing inter-agent communication and routing protocols [76,77]. The choice of simulation platform was OMNET++ and associated plugins.

Table 4. Experiment series with descriptions, target observations, and simulation platforms used.

Experiment Series (ES)	Description	Target Mechanisms Developed and Observed	Simulation Tools Used
1	Basic swarm phenomenon observation	Flocking maneuvers, 2D path planning.	MATLAB UAV Scenario Designer
2	Studying FANET topology and performance.	Examine inter-agent communication process, equipment range, and routing protocols.	OMNET++ and associated plugins.
3	Photorealistic Single UAV design and observation.	Examine the addition of various cameras, sensors on UAV, and data collection using simulated environments.	Webots, MATLAB UAV toolbox, & Simulink
4	Examining basic movement operations of UAV swarm.	Establishing and defining inter-swarm policies, agent deployment, and defining swarm topology	CoppeliaSim
5	Implementing ground terrain features as well as realistic obstacles with varied agents.	Defining buildings, trees, and realistic heterogeneous agents	CoppeliaSim, Webots

ES1 and ES2 focus on critical swarm developments and on producing fast accurate results using bare simulation mechanisms. Use of high-level graphical components was limited because scenario design for swarms in a pure MATLAB based implementation is overly complex. Additional support from toolboxes, external simulator platforms, and graphic engines would be required. These methods were instead used starting from ES3. Additionally, the introduction of realistic environments such as 3D UAVs and environment models may increase the computational load, shifting the focus away from crucial observations by creating unnecessary bottlenecks. ES3 involved designing and observing photorealistic single UAVs in associated environments. Different cameras and sensors were added to the UAV, and their impact on the overall performance and capabilities of the UAV was examined through simulations. The choice of platform was the UAV toolbox in MATLAB accompanied by Simulink and the MATLAB inspired RflySim platform.

ES4 focused on examining basic movement and specific operations of the UAV swarm. Inter-swarm communication policies and swarm topology developed in ES2 were used to understand coordination and cooperation among UAVs in a swarm. The choice of platforms used was Webots and CoppeliaSim. Webots, much like CoppeliaSim has excellent support for the creation of UAV swarms and associated topologies. While the newer versions of CoppeliaSim have a generic UAV frame with modifiable physical characteristics and the ability to mount sensors such as vision and ultrasonic sensors, GPS, and LiDAR, Webots has two specific UAV models. They are Crazieflie [78] and the DJI Mavic 2 Pro [79]. Both allow the use of MATLAB scripts to function as agent-specific or global swarm controllers. ES5 implemented ground terrain features and realistic static and dynamic obstacles.

In this series, ground terrain features and realistic obstacles were implemented in the simulations. The focus was on defining buildings, trees, and heterogeneous agents to create a realistic environment for UAVs. This allowed for the study of UAV navigation and obstacle avoidance strategies. CoppeliaSim has so far proven to be the best choice for this series with the ability to define random and specific behavior for obstacles, including physics, as well as porting control and network parameters from previous ES. Figure 6 shows the swarm development process flow using the above-mentioned experiment series.

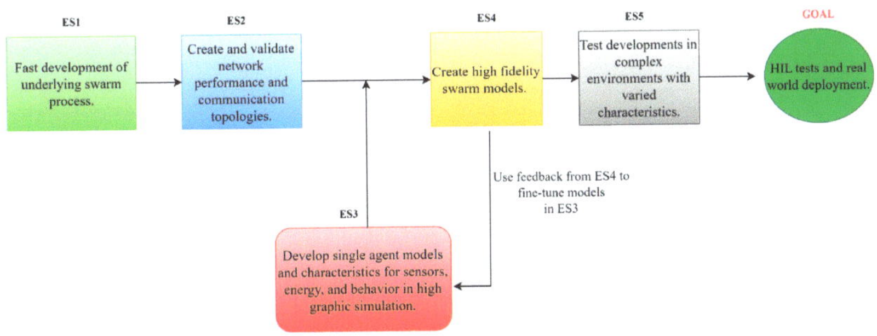

Figure 6. A suggested workflow using the above-proposed experiment series to design a comprehensive robust swarm.

5. Experiment Descriptions

This section shows the preliminary experiment progress in the various simulation platforms mentioned above. Quick 2D swarm simulations in MATLAB such as agent consensus and formation control in application scenarios such as target convergence using distance-to-network formation controller and pheromone techniques [80]. Figure 7 shows a swarm of five agents moving to a target and carrying out an encircling process. While basic results are possible using pure MATLAB, supported applications such as the UAV Scenario Designer extend functionality.

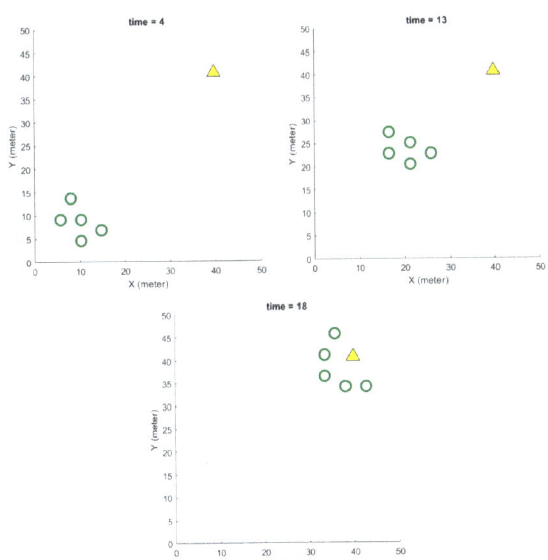

Figure 7. Using MATLAB to model five simple agents navigating with two programmed characteristics: formation control using network strength and target convergence using basic pheromone strategy moving on a fixed target in a 2D map. Triangles are targets, circles are agents.

Figure 8 shows a swarm of three homogeneous fixed-wing agents that were programmed to survey a city block. The experiment examined the relationship between trajectory point selection and flying altitude on agent fuel efficiency. Using minimum overhead graphics to examine essential swarm constraints is made possible using the ES1 process.

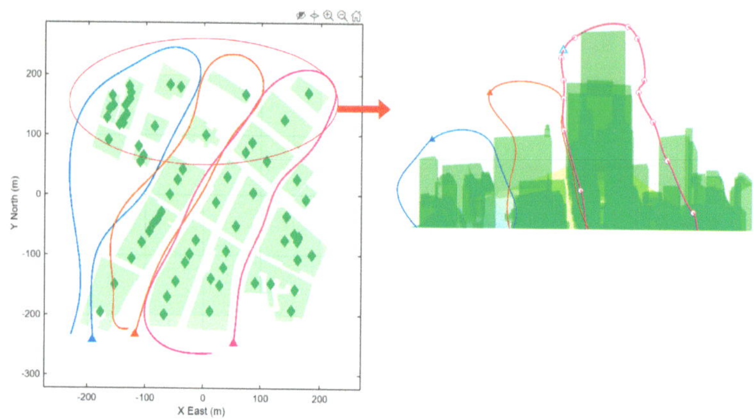

Figure 8. A MATLAB simulation of a top-down view of three agents and selected trajectory for a city block survey, and a sectional 3D view of the agents at their respective highest trajectory section.

The UAV swarm development domain is large enough that multiple components are needed and requires separate development and testing. It is easier to test standalone components in isolated scenarios before implementation with other components. ES2 involves the creation and testing of the underlying swarm network. FANET topologies may be based on previous MANET deployments. However, the former has been specifically designed to undergo more severe dynamic changes in node movement than a MANET. Additionally, MANET may be constricted to movement in two dimensions, with the rate of altitude change being very slow. OMNET++ allows the modeling of mobile nodes implemented with UAV characteristics and network underlay. Simulation of network node phenomenon for clustering, multi-hop data forwarding, and edge computing are necessary operations for a robust swarm network [81]. Figure 9 shows a swarm of 3, 7, and 10 agents that were created to optimize area coverage under a shared network load for data transfer.

Routing protocols form the basis of the network creation for UAV swarms for reliable node-to-node packet transmission, security, and energy efficiency. AODV (Ad Hoc On-Demand Distance Vector) is a self-starting routing protocol that has a degree of tolerance toward node failures [82–84]. A request–response cycle is initiated for hops. Although lacking in implicit security, it is used as the basis for future development of routing protocols [84]. Figure 10 shows interactions between a UAV agent acting as a leader and a set of mobile UAV swarm agents.

For ES3, a single high-fidelity 9 DOF (Degree Of Freedom) drone equipped with a fisheye camera is simulated flying in a city block scenario. The UAV Toolbox works with MATLAB and Simulink platforms the implementation of mathematical constraints governing the aircraft model as well as simulation environment parameters. In this particular experiment, the weather was adjusted to fog and light rain. The Simulink model makes it possible to implement additional sensors such as LiDAR and ultrasonic sensors or tweak fuel level, GPS parameters, and aircraft framework physics. A randomly generated traffic light overlay was added as the targeted data for collection in the form of image data from the onboard sensor. Figure 11 shows the UAV agent simulation from different angles and a feed from the image sensor.

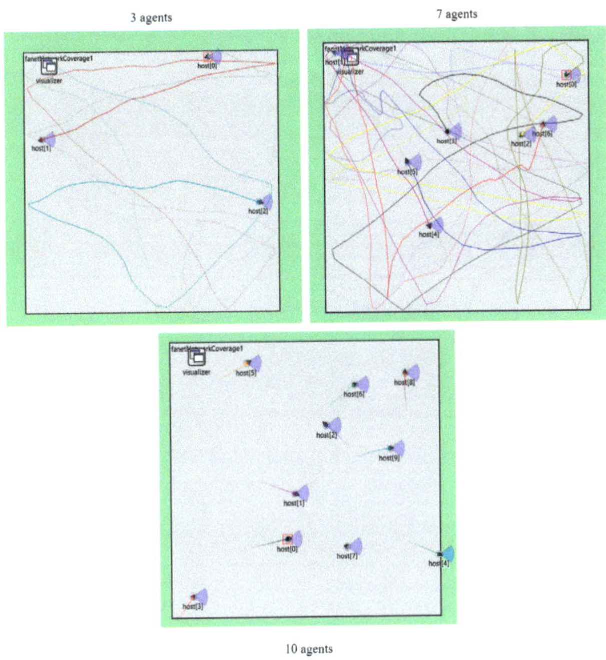

Figure 9. A swarm of three, seven, and ten UAV agents for area coverage using OMNET++.

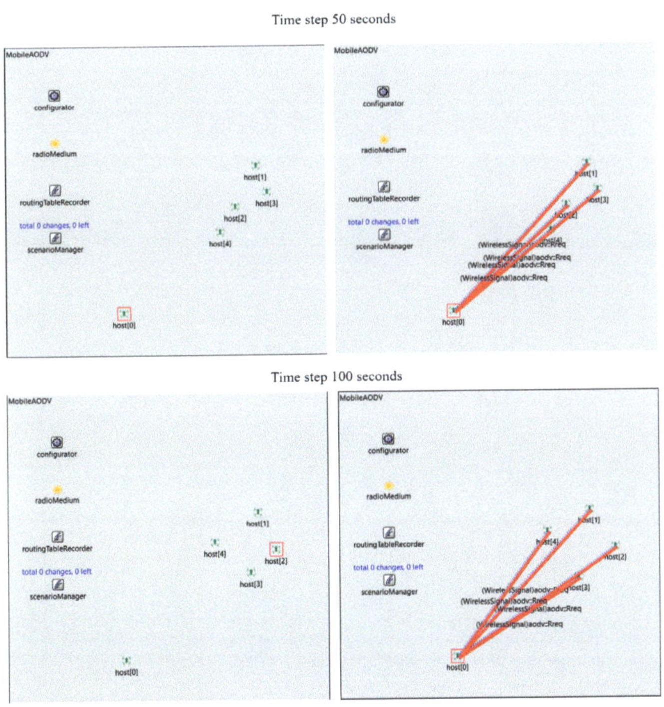

Figure 10. A mobile AODV routing protocol for a set of mobile and non-mobile agents using OMNET++.

Figure 11. MATLAB simulation using the UAV Toolbox sub-figures on top show different angles of the simulated UAV and the bottom shows the feed from the fisheye image sensor.

ES4 experiments focused on swarm topology and inter-agent policies discussed above. Formation controllers can be further fine-tuned to adjust the allowable intrusion into an agent's sphere of minimum distance. Low tolerance controllers record even the least amount of intrusion by one agent into another agent. These incidents are then noted by swarm control and corrective action for agent movement is taken. There is a fine tradeoff between the set tolerance and the computation power required. Low tolerance controllers promise lower collision rates; however, even minimum intrusions are examined. This results in more computational power used to examine intrusions and take corrective action. Higher tolerance controllers may not realize all intrusions except the most severe, freeing up computation power but opening up the risk for unpredictable collisions. Inaccuracies in measurement can be supplemented by additional methods. This can work for any sensor reading including GNSS (Global Navigation Satellite System) readings. A GNSS positioning system can be supplemented with localized passive beacons that transmit exact information, whereas drifts in agent movement can be compensated using formation control and additional sensors. Figure 12 shows this simulated approach.

Figure 12. A 3-agent swarm simulated in MATLAB with a downward sensor for passive beacons and a limited-range lateral sensor to detect other agents.

Figures 13 and 14 show three and fourteen agents in a swarm, respectively. A minimum and maximum distance between agents are set as means to complement other sensors and swarm formation methods. Based on the tolerance level set and the level of intrusion by one agent into the space of another, the controller alerts the possibility of an inter-agent collision such that suitable adjustments can be made to avoid it.

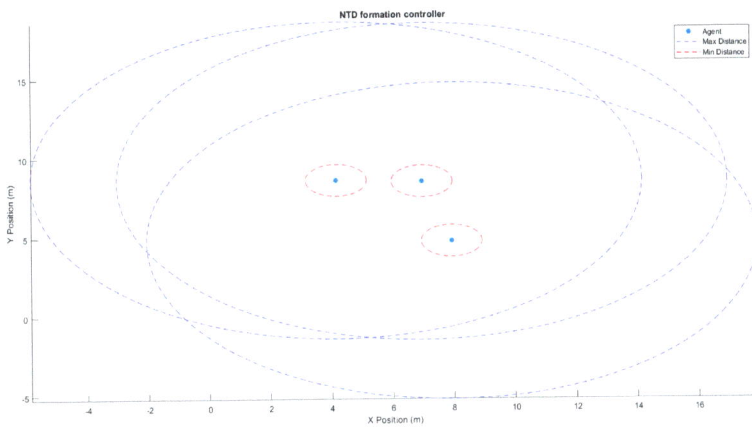

Figure 13. Three agents in an airspace with maximum and minimum inter-agent distances are set for formation control.

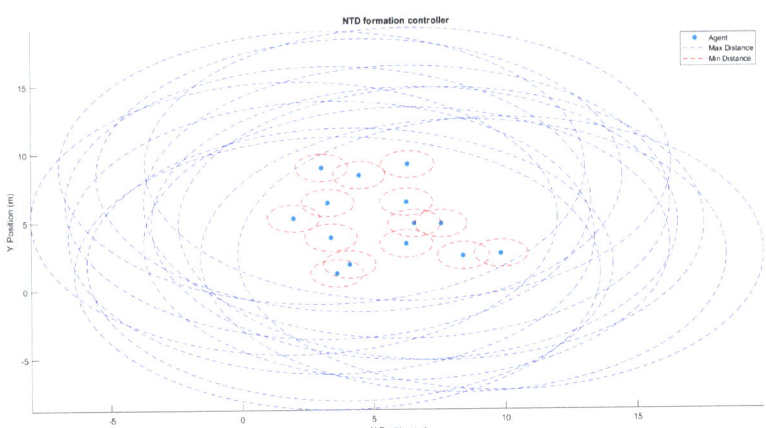

Figure 14. Fourteen agents flying the airspace with maximum and minimum inter-agent distances set for formation control.

An ES5 experiment has been designed in Webots for examining the application-specific functionality of a homogeneous swarm in a realistic environment. Here a swarm of low-level UAVs indicated by the green oval in Figure 15 work to locate a person in distress. An external agent controlled by an operator is also shown to provide an additional POV (Point Of View). Floating windows show an agent POV that has located the person in distress, a second POV from a different agent in the swarm, and the view of the manually controlled Mavic 2 camera following the swarm.

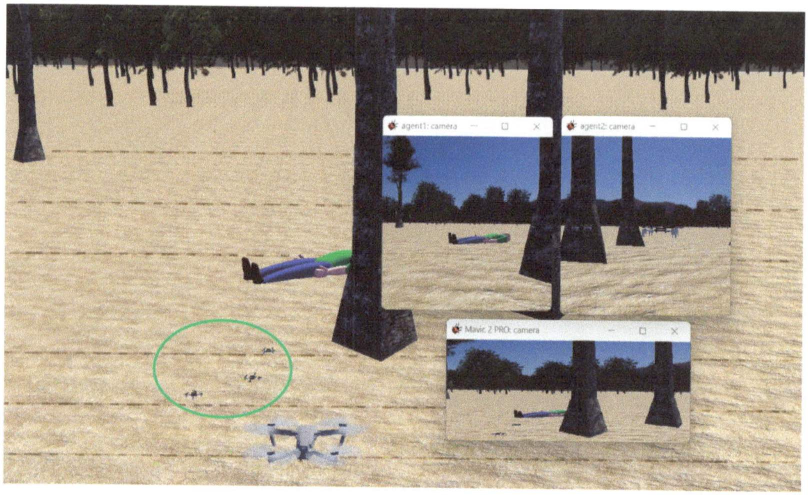

Figure 15. A distributed low-cost UAV swarm working towards targeted search and rescue simulated in a hyper-realistic environment in Webots.

An additional pass-on module to MATLAB tracked agent positions to collect and visualize position data as a means of creating an efficient distance to network strength formation controller for maintaining inter-agent distances. The developed approach uses the concept of proportional changes in inter-agent network strengths as a result of changes in distance between them. APF (Artificial Potential Field) is used to create attraction and repulsion forces. The figure shows the Crazieflie [78] agents in a swarm converging on the person. Target search and detection applications such as this open a range of possibilities in disaster management and emergency response [5,85,86].

The inclusion of agents with different capabilities in swarms is necessary to push the limitations of performance set by homogeneous swarms. A heterogeneous swarm can be recognized by various factors such as different operational spaces the swarm agents work in, the different nature of agents, or their hardware. An example of a swarm with all three properties was simulated in a CoppeliaSim simulation in Figure 16. An agent with a close-range sensor for obstacle detection works with an agent that has a long-range vision sensor; a ground-based wheeled robot is also a part of the swarm, and a tree is included as a sample for a static obstacle and a walking person is a dynamic obstacle. The aerial agents work in conjunction to detect obstacles and share information across the swarm. The ground-based robot is assigned different functions depending on the scenario such as acting as a platform for the safe landing of aerial agents or finding safe spots for the agents to land on the ground. The green spheres indicate possible zones for movement as mapped by the ground robot. Efficient information exchange pathways and the demonstration of heterogeneous agents with extended parameters are one of the targets for ES5. Simulation quality is enhanced with features such as obstacle variety, realistic scenery, and a mixed-sensor array. To design and test varied systems such as this, it is necessary for the right choice of simulation platforms that can support it.

Experiments beginning from ES3 use simulation platforms where it might be possible to introduce variability in experiments scenarios using weather effects. Rain, wind, and snow exert significant influence on UAV operations, underscoring the need for comprehensive testing. Rain can compromise UAV sensors and electronics, potentially disrupting communication and navigation systems. Wind poses challenges in maintaining stable flight paths and can increase energy consumption. Snow accumulation can alter UAV weight and aerodynamics, affecting maneuverability. The sun's azimuth angles collectively shape the complex interplay of factors affecting UAV operations. The sun's azimuth angle determines the direction of sunlight and casts shadows, impacting visual perception and sensor data accuracy. UAVs' ability to interpret their environment and execute tasks can be compromised by glare, changing illumination, and altered shadow patterns. Properly accounting for the sun's angle is crucial for accurate navigation, object recognition, and obstacle avoidance. All of these factors collectively impede control precision and reduce flight endurance, necessitating robust algorithms and hardware designs to ensure reliable UAV performance in adverse weather conditions. Integrating these environmental variables into UAV operation testing provides a comprehensive understanding of their combined impact, enabling the development of more robust and adaptable UAV systems capable of functioning effectively under a wide range of real-world conditions.

Figure 16. A heterogeneous swarm scenario with two UAV and one ground robot using CoppeliaSim.

Figure 17 shows an example of how the presence of fog and changes in the sun's angles may influence object detection by vision sensors. Adverse weather conditions may require the implementation of add-on methods to ensure that the output of sensors remain coherent [87]. The performance of other sensors such as LiDAR may also be affected. Table 5 summarizes some simulation platforms and the various weather effects that they are capable of supporting. Weather conditions such as wind, rain, and fog, can significantly impact the performance of UAVs, challenging their navigation, communication, and coordination capabilities. By incorporating weather effects, researchers can more accurately assess the resilience of swarm algorithms and control strategies under diverse environmental conditions. This approach not only provides a more comprehensive evaluation of UAV swarm behavior but also aids in identifying potential vulnerabilities and optimizing strategies for real-world deployment. Ultimately, the integration of weather variability lends a vital layer of complexity to UAV swarm experiments, promoting the development of more adaptable and robust swarm technologies.

The operating weather in UAV swarm experiments is an important factor as it enhances the realism and robustness of simulated scenarios. However, modifying graphical environment parameters may require additional development during the development phase of a platform as well as more computational resources while executing scenarios. The capability of a simulator to support different weather conditions can also be a crucial parameter that can assist in deciding which simulation platform to choose to perform experiment validation.

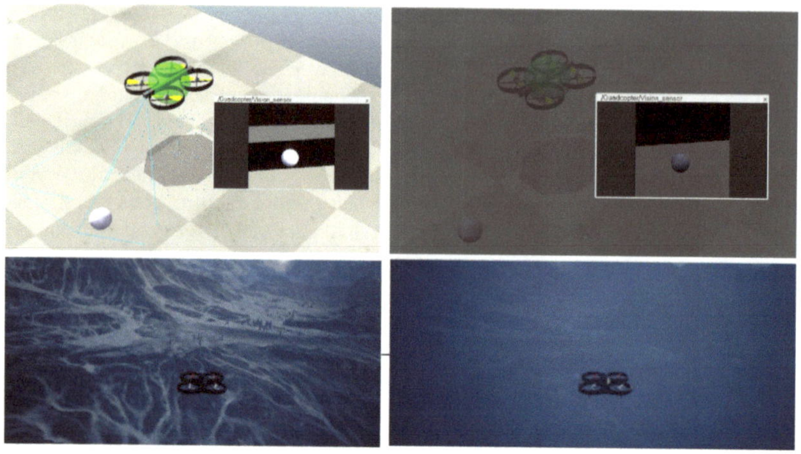

Figure 17. Top two figures show the effect of sun azimuth angle and fog affecting vision sensor capabilities. The bottom two figures show a UAV flying in clear weather (**left**) and in fog and snow weather (**right**) using Microsoft AirSim.

Table 5. A comparison of simulation platforms and the various weather effects they support.

Simulation Platform	Weather Effect	Additional Notes
CoppeliaSim	Limited to scripted animations	No built-in support for weather effects
Gazebo	Supports weather effects through plugins	Rain, fog, snow, and more can be simulated using community plugins
Webots	Simulates environmental conditions	Offers tools to adjust parameters like lightening, wind, and physics
UAV toolbox	Sun angle, time of day fog, and rain are possible by default	A slight variability in the mentioned weather factors is possible
Microsoft AirSim	A larger number of weather variations are possible.	In-built functions include controlling wind direction, rain, snow, dust, and fog.

6. Future Work

A future research direction is to bridge the gap between simulation platforms and real-world UAV systems by establishing interfaces that enable bidirectional communication and synchronization. This would allow for the testing and validation of simulation results in controlled environments using fly nets and motion capture cameras. By integrating motion capture systems that accurately track and trace the movement of UAV agents, the collected data can be fed back into the simulation platform to validate and refine the simulated behaviors and algorithms. Modern simulation platforms or their associated packages allow for the transfer and deployment of code on supported hardware. There has been an increase in the number of platforms that hardware manufacturers support out of the box. The Ryze Tello package screen for MATLAB shown in Figure 18 supports Tello EDU drones. Control data can be exported for use and analysis. Multiple Tello drones in a swarm can also be programmed [88]. The easy interface of Tello drones with MATLAB and code functionality with Python, along with their low cost and swarming capability, were some of the reasons that they were chosen as the platform for conducting all hardware in-loop experiments by the authors. The framework in this article is not hardware specific and could be implemented using other platforms. The technology landscape is constantly evolving, and newer hardware such as the Robolink Codrone Edu platform introduced in a 2022 drone version offers similar capabilities to the Tello [89].

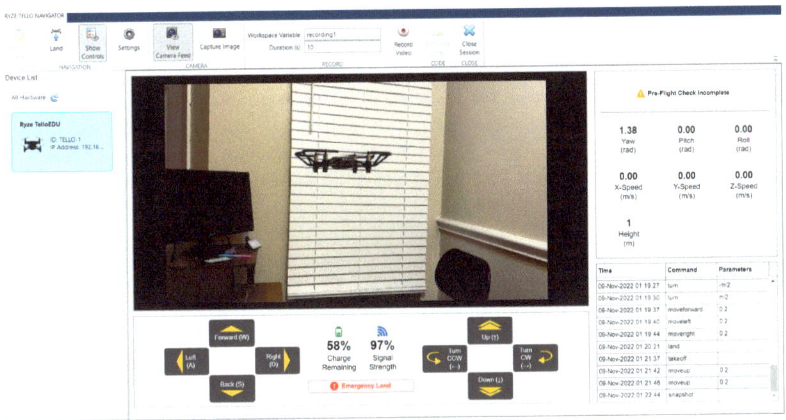

Figure 18. Ryze Tello package for Tello Edu drones.

This closed-loop approach enables a continuous feedback loop between simulation and real-world experiments, enhancing the fidelity and reliability of the simulation results. Furthermore, the controlled environment provided by fly nets ensures the safety of the experiments while enabling the examination of intricate swarm behaviors and complex interaction dynamics. Such an approach would not only enhance the accuracy and realism of simulation results but also facilitate the transferability of findings from simulation to real-world applications. The ability to update simulation platforms also invites the possibility of allowing them to work with older hardware that supported the use of simulation tools but lacked certain features. A popular example is the now discontinued Robolink Codrone Pro platform which does not have swarm functionality, as seen in Figure 19–left. This platform was replaced by the Robolink Codrone Edu UAV. Swarming on the old version is still possible by creating custom scripts that pass position data to intermediaries. Every agent sends information to a node that updates other agents. Using this method, a decentralized proof-of-concept was created for the pro drone platform to enable agent swarming. Manufacturers now are more aware of the value of creating easily implementable bridges between simulation platforms and hardware. The DJI Tello EDU platforms now support swarm mode using MATLAB packages, as seen in Figure 19–right.

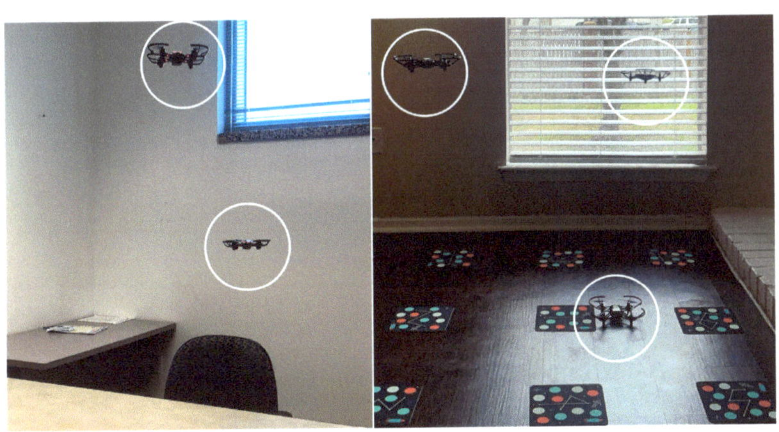

Figure 19. Two Robolink CoDrones flying in formation, (**left side**), and a swarm of three DJI Tello EDU drones including two active and one reserve agent (**right side**).

It is vital that the UAV designs themselves and the swarm experiments designed for them evolve to stay in line with new regulations enforced on the airspace that they fly in. The recently enforced rule by the FAA [90] mandates the presence of a RID module either as a modification to current hardware or by extension. The purpose of this module is to broadcast vital self-identifying information of the UAV and its controlling entity as a means to deter unauthorized activity, as well as to reduce load on conventional aircraft tracking measures. Simulation tools such as CoppeliaSim allow the creation of additional objects that can be fused to a prebuilt UAV and custom behavior can be defined for them. In this case, it would be the RID module periodically broadcasting information about the UAV such as position, task ID, and operator information. While this new rule has multiple concerns such as privacy concerns and extra costs, the data that it broadcasts can also be imbibed into existing swarm policies for resource tracking and management [91].

Future work should focus on developing standardized protocols and methodologies for integrating simulation platforms with motion capture systems and fly nets, enabling comprehensive and systematic validation of UAV swarm behaviors in controlled environments. Other than the experiment design itself, advances in the operation of UAV agents themselves should be explored. ANNs have been used to mitigate challenges with MBC design of flight controllers [92]. Similar approaches can be used to tweak experiment parameters that would enable observing a wider range of agent responses to introduced disruptions. This approach would significantly contribute to the advancement and practical applicability of UAV swarm research, fostering the development of robust and reliable autonomous systems for various domains, including surveillance, disaster response, and environmental monitoring.

7. Conclusions

The goal of the descriptive analysis presented here is to serve as a reference to other researchers who are currently working on similar swarm development experiments and are looking for various tools that might assist them in doing so. The contributions of this research lie in its thorough examination of simulation platforms and the proposed experimental design process. By providing a comprehensive guide for designing swarm experiments, the paper enables researchers to validate the robustness and efficiency of UAV swarm systems before field implementation. Also provided is a concise review of existing simulation platforms, assessing their suitability for swarm experimentation needs. This evaluation was used during the process of platform selection for designing swarm experiments. In future work, researchers can explore further advancements in simulation platforms, refine experimental design processes, and investigate novel swarm behaviors and coordination strategies. With continued research and evidence-based evaluation, UAV swarm systems can be further optimized, enabling their widespread implementation, and contributing to the advancement of autonomous aerial operations.

Author Contributions: Conceptualization, A.P. and C.N.S.; methodology, A.P.; software, A.P.; validation, F.A.M., A.P. and T.C.; investigation, A.P. and F.A.M.; resources, A.P.; writing—original draft preparation, A.P.; writing—review and editing, F.A.M., C.N.S. and T.C.; visualization, A.P. and C.N.S.; supervision, F.A.M. All authors have read and agreed to the published version of the manuscript.

Funding: This research received no external funding.

Data Availability Statement: Data sharing is not applicable.

Acknowledgments: The authors would like to thank Scott King, director of the iCORE (Innovation in Computing Research) lab at Texas A&M University–Corpus Christi for providing the Robolink CoDrones and members Evan Krell and Mahmoud Eldefrawy for experiment feedback that made this study possible.

Conflicts of Interest: The authors declare no conflict of interest.

Abbreviations

ANNs	Artificial Neural Networks
APF	Artificial Potential Field
AODV	Ad Hoc On demand Distance Vector
D2D	Device to Device
DOF	Degree of Freedom
ES	Experiment Series
FANET	Flying Ad Hoc Networks
GNSS	Global Navigation Satellite System
IDS	Intrusion Detection System
MAV	Micro Air Vehicle
MPC	Model Predictive Control
MANET	Mobile Ad Hoc Network
PSO	Particle Swarm Optimization
POV	Point Of View
ROI	Region of Interest
RID	Remote Identification
SAR	Search and Rescue
UAV	Unmanned Aerial Vehicle
UGV	Unmanned Ground Vehicle
USV	Unmanned Surface Vehicle
UAS	Unmanned Aircraft System
UWSV	Unmanned Water Surface Vehicle

References

1. Zhang, X.; Zhao, W.; Liu, C.; Li, J. Distributed Multi-Target Search and Surveillance Mission Planning for Unmanned Aerial Vehicles in Uncertain Environments. *Drones* **2023**, *7*, 355. [CrossRef]
2. Machovec, D.; Siegel, H.J.; Crowder, J.A.; Pasricha, S.; Maciejewski, A.A.; Friese, R.D. Surveillance mission scheduling with unmanned aerial vehicles in dynamic heterogeneous environments. *J. Supercomput.* **2023**, *79*, 13864–13888. [CrossRef]
3. Radoglou-Grammatikis, P.; Sarigiannidis, P.; Lagkas, T.; Moscholios, I. A Compilation of UAV Applications for Precision Agriculture. *Comput. Netw.* **2020**, *172*, 107148. [CrossRef]
4. Gans, N.R.; Rogers, J.G. Cooperative Multirobot Systems for Military Applications. *Curr. Robot. Rep.* **2021**, *2*, 105–111. [CrossRef]
5. Cao, Y.; Qi, F.; Jing, Y.; Zhu, M.; Lei, T.; Li, Z.; Xia, J.; Wang, J.; Lu, G. Mission Chain Driven Unmanned Aerial Vehicle Swarms Cooperation for the Search and Rescue of Outdoor Injured Human Targets. *Drones* **2022**, *6*, 138. [CrossRef]
6. Bakacsy, L.; Tobak, Z.; van Leeuwen, B.; Szilassi, P.; Biró, C.; Szatmári, J. Drone-Based Identification and Monitoring of Two Invasive Alien Plant Species in Open Sand Grasslands by Six RGB Vegetation Indices. *Drones* **2023**, *7*, 207. [CrossRef]
7. Guan, S.; Sirianni, H.; Wang, G.; Zhu, Z. sUAS Monitoring of Coastal Environments: A Review of Best Practices from Field to Lab. *Drones* **2022**, *6*, 142. [CrossRef]
8. Wheeb, A.H.; Nordin, R.; Samah, A.A.; Alsharif, M.H.; Khan, M.A. Topology-Based Routing Protocols and Mobility Models for Flying Ad Hoc Networks: A Contemporary Review and Future Research Directions. *Drones* **2021**, *6*, 9. [CrossRef]
9. Park, M.; Lee, S.; Lee, S. Dynamic Topology Reconstruction Protocol for UAV Swarm Networking. *Symmetry* **2020**, *12*, 1111. [CrossRef]
10. Kent, T.; Richards, A.; Johnson, A. Homogeneous Agent Behaviours for the Multi-Agent Simultaneous Searching and Routing Problem. *Drones* **2022**, *6*, 51. [CrossRef]
11. Deng, H.; Huang, J.; Liu, Q.; Zhao, T.; Zhou, C.; Gao, J. A Distributed Collaborative Allocation Method of Reconnaissance and Strike Tasks for Heterogeneous UAVs. *Drones* **2023**, *7*, 138. [CrossRef]
12. Flores Pena, P.; Luna, M.A.; Ale Isaac, M.S.; Ragab, A.R.; Elmenshawy, K.; Martin Gomez, D.; Campoy, P.; Molina, M. A Proposed System for Multi-UAVs in Remote Sensing Operations. *Sensors* **2022**, *22*, 9180. [CrossRef] [PubMed]
13. Phadke, A.; Medrano, A. A Resilient Multi-UAV System of Systems (SoS). *Acad. Lett.* **2021**, *2*, 2771–9359. [CrossRef]
14. Phadke, A.; Medrano, F.A. Towards Resilient UAV Swarms—A Breakdown of Resiliency Requirements in UAV Swarms. *Drones* **2022**, *6*, 340. [CrossRef]
15. Phadke, A.; Antonio Medrano, F.; Chu, T. Engineering resiliency in UAV swarms–A bibliographic analysis. *J. Physics Conf. Ser.* **2022**, *2330*, 012007. [CrossRef]
16. Dias, P.G.F.; Silva, M.C.; Rocha Filho, G.P.; Vargas, P.A.; Cota, L.P.; Pessin, G. Swarm Robotics: A Perspective on the Latest Reviewed Concepts and Applications. *Sensors* **2021**, *21*, 2062. [CrossRef]

17. Scholtes, M.; Westhofen, L.; Turner, L.R.; Lotto, K.; Schuldes, M.; Weber, H.; Wagener, N.; Neurohr, C.; Bollmann, M.H.; Kortke, F. et al. 6-Layer Model for a Structured Description and Categorization of Urban Traffic and Environment. *IEEE Access* **2021**, *9*, 59131–59147. [CrossRef]
18. Pitonakova, L.; Giuliani, M.; Pipe, A.; Winfield, A. Feature and Performance Comparison of the V-REP, Gazebo and ARGoS Robot Simulators. In *Towards Autonomous Robotic Systems*; Springer: Cham, Switzerland, 2018; pp. 357–368.
19. Calderón-Arce, C.; Brenes-Torres, J.C.; Solis-Ortega, R. Swarm Robotics: Simulators, Platforms and Applications Review. *Computation* **2022**, *10*, 80. [CrossRef]
20. Tselegkaridis, S.; Sapounidis, T. Simulators in Educational Robotics: A Review. *Educ. Sci.* **2021**, *11*, 11. [CrossRef]
21. Mairaj, A.; Baba, A.I.; Javaid, A.Y. Application specific drone simulators: Recent advances and challenges. *Simul. Model. Pract Theory* **2019**, *94*, 100–117. [CrossRef]
22. Carpin, S.; Lewis, M.; Wang, J.; Balakirsky, S.; Scrapper, C. USARSim: A robot simulator for research and education. In Proceedings of the 2007 IEEE International Conference on Robotics and Automation, Roma, Italy, 10–14 April 2007; pp. 1400–1405.
23. Rohmer, E.; Singh, S.P.N.; Freese, M. CoppeliaSim (formerly V-REP): A Versatile and Scalable Robot Simulation Framework. In Proceedings of the IEEE/RSJ International Conference on Intelligent Robots and Systems, Tokyo, Japan, 3–7 November 2013; pp. 1321–1326. [CrossRef]
24. Shah, S.; Dey, D.; Lovett, C.; Kapoor, A. AirSim: High-Fidelity Visual and Physical Simulation for Autonomous Vehicles. In Proceedings of the 11th Conference on Field and Service Robotics, Zürich, Switzerland, 13–15 September 2017.
25. Arafat, M.Y.; Moh, S. Routing Protocols for Unmanned Aerial Vehicle Networks: A Survey. *IEEE Access* **2019**, *7*, 99694–99720. [CrossRef]
26. Stepanyants, V.; Romanov, A. Analysis of Requirements for Next-Generation Complex Urban and Transportation System Simulation. In Proceedings of the 2022 Dynamics of Systems, Mechanisms and Machines (Dynamics), Omsk, Russia, 15–17 November 2022; pp. 1–5. [CrossRef]
27. Poudel, S.; Arafat, M.Y.; Moh, S. Bio-Inspired Optimization-Based Path Planning Algorithms in Unmanned Aerial Vehicles: A Survey. *Sensors* **2023**, *23*, 3051. [CrossRef] [PubMed]
28. Quan, Q.; Dai, X.; Wang, S. *Multicopter Design and Control Practice: A Series Experiments Based on MATLAB and Pixhawk*; Springer: Berlin/Heidelberg, Germany, 2020.
29. Pinciroli, C.; Trianni, V.; O'Grady, R.; Pini, G.; Brutschy, A.; Brambilla, M.; Mathews, N.; Ferrante, E.; Di Caro, G.; Ducatelle, F.; et al. ARGoS: A modular, parallel, multi-engine simulator for multi-robot systems. *Swarm Intell.* **2012**, *6*, 271–295. [CrossRef]
30. Marconato, E.A.; Rodrigues, M.; Pires, R.M.; Pigatto, D.F.; Filho, L.C.Q.; Pinto, A.S.R.; Branco, K.R.L.J.C. AVENS–A Novel Flying Ad Hoc Network Simulator with Automatic Code Generation for Unmanned Aircraft System. In Proceedings of the Hawaii International Conference on System Sciences (HICSS), Waikoloa Village, HI, USA, 4–7 January 2017.
31. Echeverria, G.; Lassabe, N.; Degroote, A.; Lemaignan, S.e. Modular Open Robots Simulation Engine: MORSE. In Proceedings of the International Conference on Robotics and Automation, Shanghai, China, 9–13 May 2011. [CrossRef]
32. Xue, K.; Wu, T. Distributed Consensus of USVs under Heterogeneous UAV-USV Multi-Agent Systems Cooperative Control Scheme. *J. Mar. Sci. Eng.* **2021**, *9*, 1314. [CrossRef]
33. Huang, K.; Chen, J.; Oyekan, J. Decentralised aerial swarm for adaptive and energy efficient transport of unknown loads. *Swarm Evol. Comput.* **2021**, *67*, 100957. [CrossRef]
34. Xie, Y.; Han, L.; Dong, X.; Li, Q.; Ren, Z. Bio-inspired adaptive formation tracking control for swarm systems with application to UAV swarm systems. *Neurocomputing* **2021**, *453*, 272–285. [CrossRef]
35. Schilling, F.; Schiano, F.; Floreano, D. Vision-Based Drone Flocking in Outdoor Environments. *IEEE Robot. Autom. Lett.* **2021**, *6*, 2954–2961. [CrossRef]
36. Madridano, Á.; Al-Kaff, A.; Flores, P.; Martín, D.; de la Escalera, A. Software Architecture for Autonomous and Coordinated Navigation of UAV Swarms in Forest and Urban Firefighting. *Appl. Sci.* **2021**, *11*, 1258. [CrossRef]
37. Hong, Y.; Jung, S.; Kim, S.; Cha, J. Autonomous Mission of Multi-UAV for Optimal Area Coverage. *Sensors* **2021**, *21*, 2482. [CrossRef]
38. Elmokadem, T.; Savkin, A. Computationally-Efficient Distributed Algorithms of Navigation of Teams of Autonomous UAVs for 3D Coverage and Flocking. *Drones* **2021**, *5*, 124. [CrossRef]
39. Godio, S.; Primatesta, S.; Guglieri, G.; Dovis, F. A Bioinspired Neural Network-Based Approach for Cooperative Coverage Planning of UAVs. *Information* **2021**, *12*, 51. [CrossRef]
40. Liu, L.; Qian, H.; Hu, F. Random Label Based Security Authentication Mechanism for Large-Scale UAV Swarm. In Proceedings of the 2019 IEEE International Conference on Parallel & Distributed Processing with Applications, Big Data & Cloud Computing, Sustainable Computing & Communications, Social Computing & Networking (ISPA/BDCloud/SocialCom/SustainCom), Xiamen, China, 16–18 December 2019; pp. 229–235. [CrossRef]
41. Bezas, K.; Tsoumanis, G.; Angelis, C.T.; Oikonomou, K. Coverage Path Planning and Point-of-Interest Detection Using Autonomous Drone Swarms. *Sensors* **2022**, *22*, 7551. [CrossRef] [PubMed]
42. Stolfi, D.H.; Danoy, G. An Evolutionary Algorithm to Optimise a Distributed UAV Swarm Formation System. *Appl. Sci.* **2022**, *12*, 10218. [CrossRef]
43. Stolfi, D.H.; Brust, M.R.; Danoy, G.; Bouvry, P. SuSy-EnGaD: Surveillance System Enhanced by Games of Drones. *Drones* **2022**, *6*, 13. [CrossRef]

44. Lamenza, T.; Paulon, M.; Perricone, B.; Olivieri, B.; Endler, M. Gradys-Sim–A Omnet++/Inet Simulation Framework for Internet of Flying Things. *arXiv* **2022**, arXiv:2202.08134.
45. Gu, X.; He, F.; Wang, R.; Chen, L.; Xu, J. Group Mobility Model for Complex Multimission Cooperation of UAV Swarm. *Int. J. Aerosp. Eng.* **2022**, *2022*, 5261663. [CrossRef]
46. Lee, H.S.; Shin, B.S.; Thomasson, J.A.; Wang, T.; Zhang, Z.; Han, X. Development of Multiple UAV Collaborative Driving Systems for Improving Field Phenotyping. *Sensors* **2022**, *22*, 1423. [CrossRef]
47. Campo, L.V.; Ledezma, A.; Corrales, J.C. MCO Plan: Efficient Coverage Mission for Multiple Micro Aerial Vehicles Modeled as Agents. *Drones* **2022**, *6*, 181. [CrossRef]
48. Guo, Y.; Tang, H.; Qin, R. A Low Complexity Persistent Reconnaissance Algorithm for FANET. *Sensors* **2022**, *22*, 9526. [CrossRef]
49. Wang, C.; Shi, Z.; Gu, M.; Luo, W.; Zhu, X.; Fan, Z. Revolutionary entrapment model of uniformly distributed swarm robots in morphogenetic formation. *Def. Technol.* **2022**. [CrossRef]
50. Gui, J.; Yu, T.; Deng, B.; Zhu, X.; Yao, W. Decentralized Multi-UAV Cooperative Exploration Using Dynamic Centroid-Based Area Partition. *Drones* **2023**, *7*, 337. [CrossRef]
51. Subbarayalu, V.; Vensuslaus, M.A. An Intrusion Detection System for Drone Swarming Utilizing Timed Probabilistic Automata. *Drones* **2023**, *7*, 248. [CrossRef]
52. Memon, S.A.; Son, H.; Kim, W.-G.; Khan, A.M.; Shahzad, M.; Khan, U. Tracking Multiple Unmanned Aerial Vehicles through Occlusion in Low-Altitude Airspace. *Drones* **2023**, *7*, 241. [CrossRef]
53. Shan, L.; Li, H.-B.; Miura, R.; Matsuda, T.; Matsumura, T. A Novel Collision Avoidance Strategy with D2D Communications for UAV Systems. *Drones* **2023**, *7*, 283. [CrossRef]
54. Wheeb, A.H.; Nordin, R.; Samah, A.A.; Kanellopoulos, D. Performance Evaluation of Standard and Modified OLSR Protocols for Uncoordinated UAV Ad-Hoc Networks in Search and Rescue Environments. *Electronics* **2023**, *12*, 1334. [CrossRef]
55. Li, B.; Song, C.; Bai, S.; Huang, J.; Ma, R.; Wan, K.; Neretin, E. Multi-UAV Trajectory Planning during Cooperative Tracking Based on a Fusion Algorithm Integrating MPC and Standoff. *Drones* **2023**, *7*, 196. [CrossRef]
56. Pasandideh, F.; Cesen, F.E.R.; Pereira, P.H.M.; Rothenberg, C.E.; de Freitas, E.P. An Improved Particle Swarm Optimization Algorithm for UAV Base Station Placement. *Wirel. Pers. Commun.* **2023**, *130*, 1343–1370. [CrossRef]
57. Davoli, L.; Pagliari, E.; Ferrari, G. Hybrid LoRa-IEEE 802.11s Opportunistic Mesh Networking for Flexible UAV Swarming. *Drones* **2021**, *5*, 26. [CrossRef]
58. Wolf, S.; Cooley, R.; Fantl, J.; Borowczak, M. Secure and Resilient Swarms: Autonomous Decentralized Lightweight UAVs to the Rescue. *IEEE Consum. Electron. Mag.* **2020**, *9*, 34–40. [CrossRef]
59. Airlangga, G.; Liu, A. A Study of the Data Security Attack and Defense Pattern in a Centralized UAV–Cloud Architecture. *Drones* **2023**, *7*, 289. [CrossRef]
60. Abubakar, A.I.; Ahmad, I.; Omeke, K.G.; Ozturk, M.; Ozturk, C.; Abdel-Salam, A.M.; Mollel, M.S.; Abbasi, Q.H.; Hussain, S.; Imran, M.A. A Survey on Energy Optimization Techniques in UAV-Based Cellular Networks: From Conventional to Machine Learning Approaches. *Drones* **2023**, *7*, 214. [CrossRef]
61. Nguyen, M.-N.; Nguyen, L.D.; Duong, T.Q.; Tuan, H.D. Real-Time Optimal Resource Allocation for Embedded UAV Communication Systems. *IEEE Wirel. Commun. Lett.* **2019**, *8*, 225–228. [CrossRef]
62. Kashino, Z.; Nejat, G.; Benhabib, B. A Hybrid Strategy for Target Search Using Static and Mobile Sensors. *IEEE Trans. Cybern.* **2020**, *50*, 856–868. [CrossRef]
63. Phadke, A.; Medrano, F.A.; Ustymenko, S. A Review of Vehicular Micro-Clouds. In Proceedings of the 2021 International Conference on Computational Science and Computational Intelligence (CSCI), Las Vegas, NV, USA, 15–17 December 2021; pp. 411–417. [CrossRef]
64. Cho, S.-W.; Park, J.-H.; Park, H.-J.; Kim, S. Multi-UAV Coverage Path Planning Based on Hexagonal Grid Decomposition in Maritime Search and Rescue. *Mathematics* **2021**, *10*, 83. [CrossRef]
65. Recchiuto, C.T.; Sgorbissa, A. Post-disaster assessment with unmanned aerial vehicles: A survey on practical implementations and research approaches. *J. Field Robot.* **2018**, *35*, 459–490. [CrossRef]
66. Rogers, A.; Eshaghi, K.; Nejat, G.; Benhabib, B. Occupancy Grid Mapping via Resource-Constrained Robotic Swarms: A Collaborative Exploration Strategy. *Robotics* **2023**, *12*, 70. [CrossRef]
67. Guo, Y.; Liu, X.; Jiang, W.; Zhang, W. Collision-Free 4D Dynamic Path Planning for Multiple UAVs Based on Dynamic Priority RRT* and Artificial Potential Field. *Drones* **2023**, *7*, 180. [CrossRef]
68. Zhang, Y.; Feng, W.; Shi, G.; Jiang, F.; Chowdhury, M.; Ling, S.H. UAV Swarm Mission Planning in Dynamic Environment Using Consensus-Based Bundle Algorithm. *Sensors* **2020**, *20*, 2307. [CrossRef]
69. Liu, H.; Chen, Q.; Pan, N.; Sun, Y.; Yang, Y. Three-Dimensional Mountain Complex Terrain and Heterogeneous Multi-UAV Cooperative Combat Mission Planning. *IEEE Access* **2020**, *8*, 197407–197419. [CrossRef]
70. Shi, W.; Wang, S.; Yue, H.; Wang, D.; Ye, H.; Sun, L.; Sun, J.; Liu, J.; Deng, Z.; Rao, Y.; et al. Identifying Tree Species in a Warm-Temperate Deciduous Forest by Combining Multi-Rotor and Fixed-Wing Unmanned Aerial Vehicles. *Drones* **2023**, *7*, 353. [CrossRef]
71. Gomes, J.; Mariano, P.; Christensen, A.L. Cooperative Coevolution of Partially Heterogeneous Multiagent Systems. In Proceedings of the International Conference on Autonomous Agents and Multiagent Systems, Istanbul, Turkey, 4–8 May 2015.

72. Mahendran, A.; Dewan, A.; Soni, N.; Krishna, K.M. UGV-MAV Collaboration for Augmented 2D Maps. In Proceedings of the Conference on Advances in Robotics, Pune, India, 4–6 July 2013; pp. 1–6. [CrossRef]
73. Stolfi, D.H.; Brust, M.R.; Danoy, G.; Bouvry, P. UAV-UGV-UMV Multi-Swarms for Cooperative Surveillance. *Front. Robot. AI* **2021**, *8*, 616950. [CrossRef] [PubMed]
74. Chen, J.; Du, C.; Zhang, Y.; Han, P.; Wei, W. A Clustering-Based Coverage Path Planning Method for Autonomous Heterogeneous UAVs. *IEEE Trans. Intell. Transp. Syst.* **2022**, *23*, 25546–25556. [CrossRef]
75. Chen, M.; Zeng, F.; Xiong, X.; Zhang, X.; Chen, Z. A maritime emergency search and rescue system based on unmanned aerial vehicle and its landing platform. In Proceedings of the 2021 IEEE International Conference on Electrical Engineering and Mechatronics Technology (ICEEMT), Qingdao, China, 2–4 July 2021; pp. 758–761. [CrossRef]
76. Chen, S.; Jiang, B.; Pang, T.; Xu, H.; Gao, M.; Ding, Y.; Wang, X. Firefly swarm intelligence based cooperative localization and automatic clustering for indoor FANETs. *PLoS ONE* **2023**, *18*, e0282333. [CrossRef]
77. Abdulhae, O.T.; Mandeep, J.S.; Islam, M. Cluster-Based Routing Protocols for Flying Ad Hoc Networks (FANETs). *IEEE Access* **2022**, *10*, 32981–33004. [CrossRef]
78. Bitcraze. Crazyflie 2.0 Product Specification and Manual. Available online: https://www.bitcraze.io/products/old-products/crazyflie-2-0/ (accessed on 15 June 2023).
79. DJI. DJI Mavic 2 Pro Product Page and Specifications. Available online: https://www.dji.com/mavic-2 (accessed on 15 June 2023).
80. Zhou, Y.; Song, D.; Ding, B.; Rao, B.; Su, M.; Wang, W. Ant Colony Pheromone Mechanism-Based Passive Localization Using UAV Swarm. *Remote Sens.* **2022**, *14*, 2944. [CrossRef]
81. You, W.; Dong, C.; Cheng, X.; Zhu, X.; Wu, Q.; Chen, G. Joint Optimization of Area Coverage and Mobile-Edge Computing With Clustering for FANETs. *IEEE Internet Things J.* **2021**, *8*, 695–707. [CrossRef]
82. Mansour, H.S.; Mutar, M.H.; Aziz, I.A.; Mostafa, S.A.; Mahdin, H.; Abbas, A.H.; Hassan, M.H.; Abdulsattar, N.F.; Jubair, M.A. Cross-Layer and Energy-Aware AODV Routing Protocol for Flying Ad-Hoc Networks. *Sustainability* **2022**, *14*, 8980. [CrossRef]
83. Maakar, S.K.; Khurana, M.; Chakraborty, C.; Sinwar, D.; Srivastava, D. Performance Evaluation of AODV and DSR Routing Protocols for Flying Ad hoc Network Using Highway Mobility Model. *J. Circuits Syst. Comput.* **2022**, *31*, 2250008. [CrossRef]
84. Tan, X.; Zuo, Z.; Su, S.; Guo, X.; Sun, X. Research of Security Routing Protocol for UAV Communication Network Based on AODV. *Electronics* **2020**, *9*, 1185. [CrossRef]
85. Zhang, N.; Nex, F.; Vosselman, G.; Kerle, N. Training a Disaster Victim Detection Network for UAV Search and Rescue Using Harmonious Composite Images. *Remote Sens.* **2022**, *14*, 2977. [CrossRef]
86. Khalil, H.; Rahman, S.U.; Ullah, I.; Khan, I.; Alghadhban, A.J.; Al-Adhaileh, M.H.; Ali, G.; ElAffendi, M. A UAV-Swarm-Communication Model Using a Machine-Learning Approach for Search-and-Rescue Applications. *Drones* **2022**, *6*, 372. [CrossRef]
87. Pikun, W.; Ling, W.; Jiangxin, Q.; Jiashuai, D. Unmanned aerial vehicles object detection based on image haze removal under sea fog conditions. *IET Image Process.* **2022**, *16*, 2709–2721. [CrossRef]
88. MathWorks MATLAB Hardware Team. *MATLAB Support Package for Ryze Tello Drones*; Publisher MathWorks: Natick, MA, USA, 2023; Available online: https://www.mathworks.com/help/supportpkg/ryzeio/ (accessed on 5 June 2023).
89. Robolink. Available online: https://www.robolink.com/products/codrone-edu (accessed on 26 July 2023).
90. FAA. Remote Identification of Unmanned Aircraft-Final Rule. 2023. Available online: https://www.faa.gov/sites/faa.gov/files/2021-08/RemoteID_Final_Rule.pdf (accessed on 15 June 2023).
91. Phadke, A.; Boyd, J.; Medrano, F.A.; Starek, M. Navigating the skies: Examining the FAA's remote identification rule for unmanned aircraft systems. *Drone Syst. Appl.* **2023**, *11*, 1–4. [CrossRef]
92. Gu, W.; Valavanis, K.P.; Rutherford, M.J.; Rizzo, A. UAV Model-based Flight Control with Artificial Neural Networks: A Survey. *J. Intell. Robot. Syst.* **2020**, *100*, 1469–1491. [CrossRef]

sensors

MDPI

Article

Research on a Method of Locating Civil Aviation Radio Interference Sources Based on Time Difference of Arrival and Frequency Difference of Arrival for Four Unmanned Aerial Vehicles

Chao Zhou [1,*], Xingyu Zhu [2], Renhe Xiong [2], Kun Hu [2], Feng Ouyang [3], Chi Huang [2] and Tao Huang [1]

1 Institute of Electronic and Electrical Engineering, Civil Aviation Flight University of China, Guanghan 618307, China; 15520664026@163.com
2 College of Air Traffic Management, Civil Aviation Flight University of China, Guanghan 618307, China; zhcnzxy@gmail.com (X.Z.); renhexiong@gmail.com (R.X.); 17854335326@163.com (K.H.); 13778880943@163.com (C.H.)
3 CAAC Academy of Flight Technology and Safety, Civil Aviation Flight University of China, Guanghan 618307, China; oyf202313668304477@163.com
* Correspondence: zc_cafuc@163.com

Abstract: Monitoring and analyzing radio interference sources play a crucial role in ensuring the safe operation of civil aviation navigation, communication, airport management, and air traffic control. Traditional ground monitoring methods are slow and inadequate for tracking aerial and mobile interference sources effectively. Although flight methods such as helicopters and airships can effectively monitor aerial interference, the flight approval process is time-consuming and expensive. This paper investigates a novel approach to locating civil aviation radio interference sources using four unmanned aerial vehicles (UAVs) to address this issue. It establishes a model for aerial positioning of radio interference sources with the four UAVs and proposes a method for time synchronization and data communication among them. The paper conducts simulations of the four-UAV time–frequency difference positioning method, analyzing the geometric accuracy dilution with different deployment configurations of the UAVs, positioning biases, and root mean square errors (RMSEs) under varying interference source movement speeds. The simulation results provide crucial data to support subsequent experiments.

Keywords: four UAVs; civil aviation; radio interference source; time–frequency difference positioning

1. Introduction

In the realm of aviation, radio technology plays a critical role in communication, navigation, surveillance, meteorology, and various other aspects. However, the unauthorized establishment of "black radio", "pseudo-base stations", and similar devices has become a growing concern. These devices operate at frequencies perilously close to civil aviation radio frequencies, leading to escalating and severe interference with civil aviation communications. Such radio interference can significantly diminish air traffic control communications and crucial avionics equipment [1–5], posing a serious threat to aviation safety. Over the years, reported cases of radio interference at airports worldwide have shown a marked increase. For instance, according to the CAAC East China Regional Administration, there were 334 radio interference incidents in East China in 2015, and the number has remained consistently high, with over 300 incidents reported annually since then [6]. Additionally, the airspace near the airports of the Civil Aviation Flight University of China's branches in Luoyang, Xinjin, and Suining has experienced multiple instances of radio interference, severely affecting normal flight training operations. In comparison to ground monitoring methods, air platform-based radio monitoring offers distinct advantages. Utilizing UAVs for monitoring circumvents airspace limitations and eliminates the complications and lengthy approval processes required for route clearance. Employing multiple UAVs provides several benefits, including heightened flexibility, increased positioning accuracy, extensive coverage, and rapid positioning. This approach proves particularly valuable in monitoring aerial and mobile interference sources, ultimately advancing the ranking of civil aviation radio interference sources.

In this paper, we present a novel approach for identifying the sources of civil aviation radio interference using a time–frequency difference positioning technique with four UAVs. By employing this method, we effectively mitigate the impact of multipath radio wave propagation caused by obstacles, while addressing the limitations of ground-based troubleshooting methods. The proposed method offers the capability not only to monitor aerial interference but also to effectively detect ground-to-air interference and ground interference. Furthermore, in comparison to existing unmanned aerial vehicle (UAV)-based methods for locating civil aviation interference sources, our approach demonstrates superior positioning accuracy, broader coverage area, faster positioning speed, and increased flexibility in tracking moving point targets.

2. Related Work

The concept of radio monitoring was initially introduced by Western countries, and subsequently, the Federal Aviation Administration (FAA) [7] implemented multiple fixed and relocatable stations to establish the Airport Radio Interference Monitoring System (AIMS) and Radio Interference Monitoring System (IMDS). The requirements for an interference source monitoring/direction finding system include a frequency coverage of 25–3000 MHz, a frequency scan rate of 1000 MHz/s, and a direction finding accuracy better than 2° [8]. In recent studies, various researchers proposed innovative techniques for different applications. Rakshit Ramesh et al. [9] proposed a new protocol and technique based on the time difference of arrival (TDOA) method for UAV positioning. Kilari et al. [10] introduced a linear programming initialization method to complement the TDOA algorithm. Mario Nicola et al. [2] presented a novel interference management concept capable of detecting intentional interference in navigation satellite system signals and determining its source. Sanat K Biswas et al. [11] explored the use of Kalman filters to efficiently geolocate and track dynamic and static RF interference sources based on real measurements from a geolocation system. Additionally, Adrien Perkins et al. [12] detailed the design, development, and flight testing of the JAGER visual navigation system. In China, radio monitoring networks are extensively employed to identify and exclude interference sources that may affect aviation, railroad, and telecommunication units. Notably, recent research by Hao Caiyong et al. [13] proposed a high-precision positioning method utilizing UAV assistance. Xu Bojian et al. [14] from Beijing Global Information Application Development Center performed an analysis based on radio frequency parameters and established a specimen database. Jin Ping et al. [15] from the School of Information Science and Engineering at Yanshan University introduced an improved MUSIC algorithm for localizing coherent interference cognitive users. Wang Guangyu [16] from the Technical Support Center of CAAC Northeast Regional Administration achieved precise positioning of radio interference sources. Additionally, Li Jinshan [17] from the School of Information Engineering and Automation at Kunming University of Science And Technology proposed an existing radio interference source positioning technique.

The positioning of radio wave sources plays a crucial role not only in military applications like electronic warfare but also in civilian domains such as navigation systems [18,19], internal security [20], and search and rescue missions [21,22]. The majority of methods analyzed in the literature apply to stationary interference sources. The potential of using widely available UAVs to enhance communication service quality and to extend coverage has been explored in the context of fifth-generation (5G) mobile networks and fixed interference source positioning systems, as presented in references [23–25], respectively. Among the more commonly used techniques for estimating the positions of mobile transmitters are TDOA and frequency difference of arrival (FDOA) measurements obtained from multiple sensors, as discussed in references [25,26]. In [27], target tracking techniques based on wireless sensor network (WSN) TDOA measurements are described. Sathyan et al. [28] also recommended the use of the extended Kalman filter (EKF). Similarly, Kim et al. [29] proposed a method involving correlated TDOA and Gaussian mixture. Kelner J M et al. [30] evaluated the effectiveness of signal Doppler frequency methods in locating mobile radiation sources using swarms of UAVs. The team led by Zhou Chao at the Civil Aviation Flight University of China (CAFUC) has undertaken extensive research in this field, constructing various UAV monitoring platforms dedicated to monitoring civil aviation radio interference sources [31–39].

The commonly employed techniques for localizing stationary radiation sources using UAV swarms primarily include the FDOA method, the positioning algorithm using a phase interferometer, the Dual Station Direction Finding (DF) cross-positioning algorithm, and the TDOA method for passive positioning techniques. Regarding the positioning of dynamic radiation sources by UAV swarms, the principal methods consist of the least squares method [40], spatial electromagnetic environment platform positioning [41], and radar positioning of dynamic radiation sources based on active positioning [42].

Although the time and angle positioning methods mentioned above can achieve a high level of positioning accuracy, they are constrained by the continuous operation of the target jammer and the complexity of the positioning algorithm. This paper introduces a fast and intuitive positioning method with a simple algorithm to determine the position of the interference source. In the designated measurement area, a matrix of several radio monitors is deployed, forming a radio monitoring network. This network continuously monitors the signal strength of the interference source within the area, measures the magnitude of its received power, and analyzes the received power magnitude data from the radio monitors. The proposed algorithm deduces the location of the interference source based on the data detected by the radio monitors, and its efficacy is validated through simulation.

3. Design of Four-UAV Time–Frequency Difference Positioning Method for Interference Sources

The technique of cross-location between two UAVs primarily involves measuring the arrival angles between the interfering source and the monitoring station. The ray, originating from the monitoring station and passing through the interfering source, intersects with another ray to determine the position of the interference source. However, it should be noted that the cross-location method is only suitable for non-moving radio interference sources since the positioning accuracy is not sensitive to the positional errors of UAVs. In the case of a moving interference source, this method cannot provide precise positioning. Moreover, according to mathematical principles, the trajectory of a moving point with a constant difference in distances from two fixed points forms a hyperbola. To determine a point in three-dimensional space, at least three difference in distances and four monitoring stations are required. Therefore, TDOA positioning requires a minimum of four unmanned aerial vehicles (see Figure 1).

3.1. *The Four-UAV Time–Frequency Difference Positioning Algorithm for Interference Sources*

The four UAVs can be operated independently using their own paired remote controls or ground control computers. Once the UAVs have established a synchronized time reference, the signal-receiving equipment on each UAV monitoring platform measures the arrival time of the same interference source signal separately. Subsequently, the arrival time information, along with position, speed, and other relevant data from all UAVs, is collected and sent to the ground station through the downlink. The interference source's position is then calculated using the TDOA–FDOA joint positioning method, and the location of the radio interference source is displayed on a map. The UAV ground station serves as the core component of the system, responsible for system control and data processing, enabling efficient cooperative control of the UAVs.

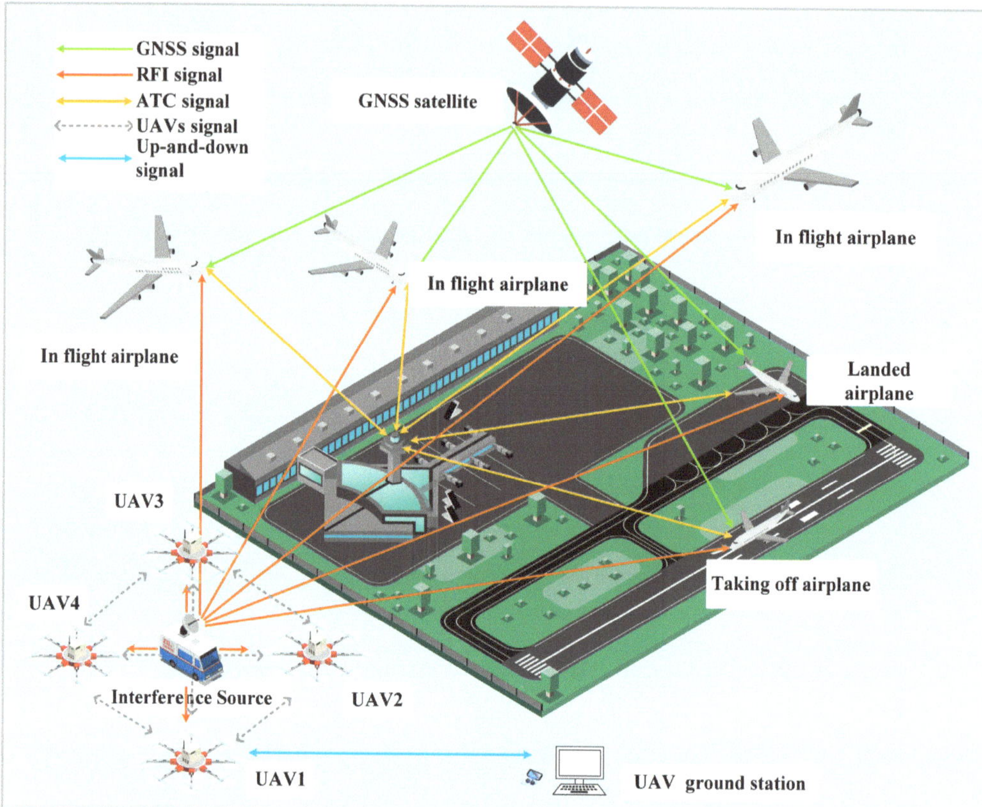

Figure 1. Four-UAV-based radio interference source positioning scenario for civil aviation.

In the data processing of a multi-UAV cooperative monitoring network, it is essential to ensure that the measurement values provided by each UAV can be transformed into the same reference station coordinate system for analysis and expression. The geodetic coordinate system represents the position of the UAV in terms of longitude L_i, latitude B_i, and geodetic height H_i, $i = 1, 2, \ldots, P$. In the Cartesian coordinate system, the position of the UAV is represented by variables X_i, Y_i, and Z_i. Given the geodetic coordinates of the UAV, the formula for calculating the Cartesian coordinates of UAV is as Equation (1):

$$\begin{cases} x_i = (N + H_i)\cos B_i \cos L_i \\ y_i = (N + H_i)\cos B_i \sin L_i \\ z_i = \left[N\left(1 - e_1^2\right) + H_i\right]\sin B_i \end{cases} \tag{1}$$

where e_1 represents the first eccentricity of the meridian ellipse and N denotes the curvature radius of the ellipsoidal surface along the prime vertical.

When the spatial Cartesian coordinate $U_i(x_i, y_i, z_i)$ of a UAV is known, the formula for calculating the geodetic coordinates of UAV is as Equation (2).

$$\begin{cases} L_i = \arctan \frac{y_i}{x_i} \\ B_i = \arctan \frac{(N+H_i)z_i}{(N+H_i-e_1^2N^2)\sqrt{x_i^2+y_i^2}} \\ H_i = \frac{\sqrt{x_i^2+y_i^2}}{\cos B_i} - N \end{cases} \tag{2}$$

Let $R = [x, y, z]^T$ represent the position of the moving civil aviation radio interference source and $\dot{R} = [\dot{x}, \dot{y}, \dot{z}]^T$ denote its moving speed. There are a total of four UAVs with their flight positions described in spatial Cartesian coordinates as $U_i = [x_i, y_i, z_i]$ and their flight speeds as $\dot{U}_i = [\dot{x}_i, \dot{y}_i, \dot{z}_i]$, $i = 1, 2, \ldots, 4$. Considering UAV U_1 as the master station, the distance difference, known as the range difference of arrival (RDOA) between the remaining UAVs and the master station concerning the mobile interference source can be expressed using Equation (3) as follows:

$$r_{i,1} = d_i - d_1 + n_{i,1} \tag{3}$$

where $d_i = \|R - U_i\|_2 = \sqrt{(x - x_i)^2 + (y - y_i)^2 + (z - z_i)^2}$ represents the actual distance between the i-th UAV and the mobile interference source and $n_{i,1}$ denotes the measurement error of the i-th time.

Expand Equation (3) as follows:

$$2(U_i - U_1)^T R + 2r_{i,1}d_1 = \left(U_i^T U_i - U_1^T U_1 - r_{i,1}^2\right) + 2d_i n_{i,1} \tag{4}$$

The time derivative of Equation (4) allows us to utilize the relevant information of FDOA:

$$\begin{aligned} &2(\dot{U}_i - \dot{U}_1)^T R + 2(U_i - U_1)^T \dot{R} + 2r_{i,1}\dot{d}_1 + 2\dot{r}_{i,1}d_1 \\ &= 2\left(\dot{U}_i^T U_i - \dot{U}_1^T U_1 - r_{i,1}\dot{r}_{i,1}\right) + 2d_i\dot{n}_{i,1} + 2\dot{d}_i n_{i,1} \end{aligned} \tag{5}$$

The time derivative of the actual distance $d_i = \|R - U_i\|_2$ between the i-th UAV and the mobile interference source is represented as Equation (6):

$$\dot{d}_i = \frac{(\dot{U}_i - \dot{U}_1)(U_i - U_1)}{d_i} \tag{6}$$

Based on the monitoring data, the TDOA information of the three UAVs is represented by $T_d = [t_{2,1}, t_{3,1}, t_{4,1}]$, the FDOA information is denoted by $F_d = [f_{2,1}, f_{3,1}, f_{4,1}]$, their RDOA is indicated as $r = [r_{2,1}, r_{3,1}, r_{4,1}]$ and the rate of change in RDOA is expressed as $\dot{r} = [\dot{r}_{2,1}, \dot{r}_{3,1}, \dot{r}_{4,1}]$. The relationship between these parameters can be expressed as Equation (7):

$$\begin{cases} r = c \times T_d = d + n \\ \dot{r} = c \times \frac{F_d}{f_0} = \dot{d} + \dot{n} \end{cases} \tag{7}$$

where f_0 represents the carrier frequency; c denotes the transmission speed of the radio interference source frequency; and $n = [n_{2,1}, n_{3,1}, n_{4,1}]$ and $\dot{n} = [\dot{n}_{2,1}, \dot{n}_{3,1}, \dot{n}_{4,1}]$ refer to the time noise vector and the frequency noise vector, respectively. In this algorithm, the noises are temporarily assumed to be zero-mean Gaussian white noise.

In the first step, we estimate by combining Equations (4) and (5), resulting in the following:

$$Q_1\theta_1 - h_1 = \varepsilon_1 \tag{8}$$

where $\theta_1 = \left[R^T, R_1, \dot{R}^T, \dot{R}_1\right]^T_{8\times 1}$, $B = 2 \times \text{diag}([d_2, d_3, d_4])$, $\dot{B} = 2 \times \text{diag}([\dot{d}_2, \dot{d}_3, \dot{d}_4])$, $h_1 = \begin{bmatrix} U_2^T U_2 - U_1^T U_1 - r_{2,1}^2 \\ \cdots \\ U_4^T U_4 - U_1^T U_1 - r_{4,1}^2 \\ 2(\dot{U}_2^T U_2 - \dot{U}_1^T U_1 - \dot{r}_{2,1} r_{2,1}) \\ \cdots \\ 2(\dot{U}_4^T U_4 - \dot{U}_1^T U_1 - \dot{r}_{M,1} r_{M,1}) \end{bmatrix}_{6\times 1}$, $Q_1 = \begin{bmatrix} (U_2 - U_1)^T & r_{2,1} & 0_{1\times 3} & 0 \\ \cdots & \cdots & \cdots & \cdots \\ (U_4 - U_1)^T & r_{4,1} & 0_{1\times 3} & 0 \\ (\dot{U}_2 - \dot{U}_1)^T & \dot{r}_{2,1} & (U_2 - U_1)^T & r_{2,1} \\ \cdots & \cdots & \cdots & \cdots \\ (\dot{U}_4 - \dot{U}_1)^T & \dot{r}_{4,1} & (U_4 - U_1)^T & r_{4,1} \end{bmatrix}_{6\times 8}$,
$\varepsilon_1 = B_1 \Delta\eta = \begin{bmatrix} B & 0_{3\times 3} \\ \dot{B} & B \end{bmatrix} \begin{bmatrix} n \\ \dot{n} \end{bmatrix}$.

Based on the preceding matrix, Equation (8) is transformed to the following:

$$\theta_1 = \left(Q_1^T T_1 Q_1\right)^{-1} Q_1^T T_1 h_1 \tag{9}$$

The obtained Equation (9) represents the weighted least squares estimate of the first step, where $T_1 = \left(B_1 Z^{-1} B_1^T\right)^{-1}$, Z denote the covariance matrix of the measurement noise $\Delta\eta$.

$$\text{cov}(\theta_1) = \left(Q_1^T T_1 Q_1\right)^{-1} \tag{10}$$

In the second-step estimate, the time–frequency difference joint positioning algorithm has two constraints on distance:

$$(R - U_1)^T (R - U_1) = d_1^2 \tag{11}$$

$$(\dot{R} - \dot{U}_1)^T (R - U_1) = \dot{d}_1 d_1 \tag{12}$$

Based on Equations (11) and (12), we can establish the following constraint model:

$$Q_2 \theta_2 - h_2 = \varepsilon_2 \tag{13}$$

Let $\theta_{1,R} = [\theta_1(1), \theta_1(2), \theta_1(3)]^T$, $\theta_{1,\dot{R}} = [\theta_1(5), \theta_1(6), \theta_1(7)]^T$, $\theta_2 = \begin{bmatrix} (R - U_1) \odot (R - U_1) \\ (\dot{R} - U_1) \odot (R - U_1) \end{bmatrix}$ $Q_2 = \begin{bmatrix} I_{3\times 3} & 0_{3\times 3} \\ 1_{1\times 3} & 0_{1\times 3} \\ 0_{3\times 3} & I_{3\times 3} \\ 0_{1\times 3} & 1_{1\times 3} \end{bmatrix}$,
$h_2 = \begin{bmatrix} (\theta_{1,R} - U_1) \odot (\theta_{1,R} - U_1) \\ \theta_1(4)^2 \\ \left(\theta_{1,\dot{R}} - U_1\right) \odot (\theta_{1,R} - U_1) \\ \theta_1(8)\theta_1(4) \end{bmatrix}$, where $I_{3\times 3}$ is a 3×3 identity matrix, $0_{3\times 3}$ is a 3×3 zero matrix, $1_{1\times 3}$ is a 1×3 matrix with all ones, $0_{1\times 3}$ is a 1×3 zero matrix, and $\odot$ represents the product of vectors.

Based on the preceding matrix, Equation (13) is transformed to the following:

$$\theta_2 = \left(Q_2^T T_2 Q_2\right)^{-1} Q_2^T T_2 h_2 \tag{14}$$

The obtained Equation (14) represents the weighted least squares estimation of the second step, where $T_2 = \left(B_2 \,\text{cov}(\theta_1) B_2^T\right)^{-1}$, $B_2 = \begin{bmatrix} 2\,\text{diag}(R - U_1) & 0 & 0_{3\times 3} & 0 \\ 0_{1\times 3} & 2d_1 & 0_{1\times 3} & 0 \\ \text{diag}(\dot{R} - \dot{U}_1) & 0 & \text{diag}(R - U_1) & 0_{3\times 1} \\ 0_{1\times 3} & d_1 & 0_{1\times 3} & \dot{d}_1 \end{bmatrix}$.

Estimate θ_2 and utilize Equations (15) and (16) to calculate the position information and velocity information of the mobile interference source:

$$R = V\left[\sqrt{\theta_2(1)}, \sqrt{\theta_2(2)}, \sqrt{\theta_2(3)}\right]^T + U_1 \quad (15)$$

$$\dot{R} = V\left[\frac{\theta_2(4)}{\sqrt{\theta_2(1)}}, \frac{\theta_2(5)}{\sqrt{\theta_2(2)}}, \frac{\theta_2(6)}{\sqrt{\theta_2(3)}}\right]^T + \dot{U}_1 \quad (16)$$

where $V = \text{diag}(\text{sgn}(\theta_{1,R} - U_1))$, where sgn is the positive or negative sign.

The first-step estimation and the second-step estimation mentioned above are repeated in a cycle until the difference between the two estimation results is smaller than the predefined threshold or the number of cycles reaches the set limit. At that point, the cycle is terminated, and the final result at the end of the cycle represents the most accurate estimate obtained using the algorithm.

Accordingly, the overall flow of the four-UAV time–frequency difference positioning interference source algorithm can be obtained: (See Algorithm 1).

Algorithm 1: The overall flow of the four-UAV time–frequency difference positioning interference source

Step 1: Input the UAV position and velocity information, followed by a conversion of the geodetic coordinates of the UAV into spatial Cartesian coordinates using a coordinate system transformation model specifically tailored for the algorithm;
Step 2: Construct a joint four-UAV TDOA–FDOA positioning model (Q_1 and h_1), and let T_1 be the identity matrix;
Step 3: Calculate the weighted least squares estimate for the first step based on the localization model: $\theta_1 = \left(Q_1^T T_1 Q_1\right)^{-1} Q_1^T T_1 h_1$;
Step 4: Calculate the coefficient matrix B_1 from θ_1, and reconstruct the weight matrix: $T_1 = \left(B_1 Z^{-1} B_1^T\right)^{-1}$;
Step 5: Repeat steps 3 and 4 until the absolute difference between the two results is smaller than the predefined threshold or the set number of cycles is reached. At this point, terminate the cycle, and obtain $\text{cov}(\theta_1) = \left(Q_1^T T_1 Q_1\right)^{-1}$;
Step 6: The constraint models for the joint four-UAV TDOA–FDOA positioning are constructed based on θ_1 (Q_2 and h_2);
Step 7: Construct the coefficient matrix B_2 of the constraint model weight matrix, and calculate the weight matrix: $T_2 = \left(B_2 \,\text{cov}(\theta_1) B_2^T\right)^{-1}$;
Step 8: Calculate the weighted least squares estimate for the second step based on the constrained model: $\theta_2 = \left(Q_2^T T_2 Q_2\right)^{-1} Q_2^T T_2 h_2$;
Step 9: Calculate R and $\dot{R}$ for mobile radio interference sources based on θ_2;
Step 10: Repeat steps 6 to 8 until the absolute difference between the two results is smaller than the predefined threshold or the set number of cycles is reached. At this point, terminate the cycle and obtain the final calculated interference sources R and $\dot{R}$;
Step 11: Transform the spatial Cartesian coordinates of the radio interference source into geodetic coordinates using the coordinate system conversion model and then output them.

3.2. Design of Four-UAV Time Synchronization

Due to the different clock behaviors on each UAV, ensuring meaningful time measurements necessitates adopting one UAV's time as the standard and synchronizing the time of the other three UAVs with it. In this paper, the utilized clock synchronization method involves transmitting the time difference between all UAVs and the GPS time to the ground station. Following processing using the time–frequency synchronization algorithm, the time difference information is inputted into the time–frequency difference positioning algorithm. Figure 2 depicts the schematic diagram of the four UAVs' time synchronization settings.

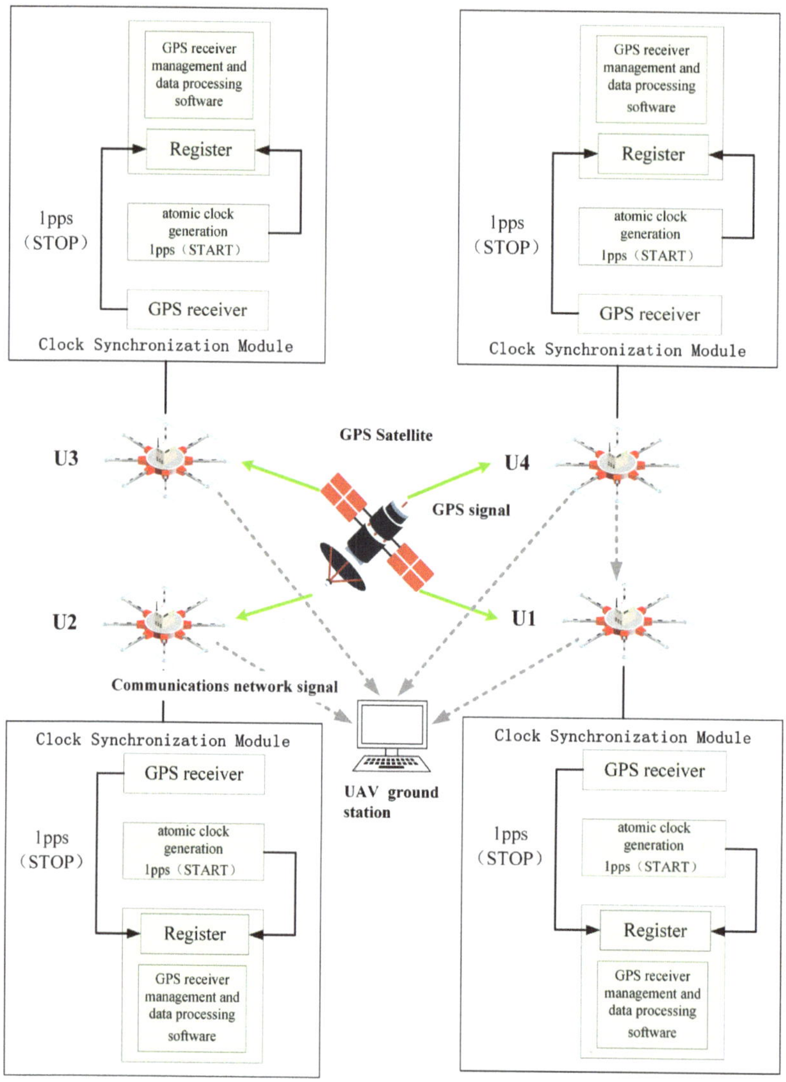

Figure 2. Schematic diagram of the four UAVs' time synchronization settings.

Let the clock time of UAV U_1 be t_{U_1}, the clock time of UAV U_2 be t_{U_2}, the clock time of UAV U_3 be t_{U_3} and the clock time of UAV U_4 be t_{U_4}, while the GPS time is denoted as t_{GPS}. The measurement method for the clock difference of the four UAVs is as follows: Under the same co-viewing schedule constraint, the GPS receivers of the four UAVs simultaneously receive the same GPS satellite signal. The output of the GPS receivers of all four UAVs produces a GPS time second pulse, which is then transmitted to the built-in counter of the GPS receivers, resulting in the GPS on-star time. By subtracting the received GPS on-star time from the local atomic clock seconds signal generated by the clock synchronization module, along with the time delays of the UAV monitoring platform equipment and the

GPS signal reaching the UAV monitoring platform, we obtain the difference between the clock time of each UAV and the GPS on-star time. This process is executed by the GPS receiver management and data processing software within the clock synchronization module. The resulting difference data are then transmitted from the UAVs to the ground for further processing. Consequently, subtracting the air–ground signal transmission delay from this difference yields the clock difference between the ground clock and the GPS clock.

$$\Delta t_{iGPS} = t_i - t_{GPS} - t_{Ri} - \tau_i - \kappa_i \quad (i = U_1, U_2, U_3, U_4) \tag{17}$$

where t_i represents the local time generated by the clock synchronization module of each UAV platform when the GPS signal is received by the UAV; t_{GPS} is the on-star time of the GPS satellite transmit signal; t_{Ri} denotes the equipment time delay of the corresponding UAV platform; τ_i is the transmission time delay of the GPS satellite signal reaching the UAV; and κ_i is the ground-to-air signal transmission time delay and is given by the following:

$$\tau_i = \frac{\sqrt{(x - x_i)^2 + (y - y_i)^2 + (z - z_i)^2}}{c} \quad (i = U_1, U_2, U_3, U_4) \tag{18}$$

where (x, y, z) represents the Cartesian coordinates of the GPS satellite and (x_i, y_i, z_i) denotes the Cartesian coordinates of each UAV. By utilizing Equations (17) and (18), it is possible to calculate the difference between the local clock of the four UAVs and the GPS clock at the ground station, resulting in the values of $\Delta t_{U_1,GPS}$, $\Delta t_{U_2,GPS}$, $\Delta t_{U_3,GPS}$ and $\Delta t_{U_4,GPS}$. By subtracting the values of $\Delta t_{U_1,GPS}$ from U_2, U_3, U_4 and U_1, we obtain the differences $\Delta t_{U_1,U_2}$, $\Delta t_{U_1,U_3}$, and $\Delta t_{U_1,U_4}$ between the clocks of the three UAVs U_2, U_3, and U_4 and that of UAV U_1.

$$\begin{aligned}\Delta t_{U_1,U_2} &= t_{U_1} - t_{U_2} \\ &= (\Delta t_{U_1,GPS} + \Delta t_{GPS} + \tau_{U_1} + t_{RU_1} + \kappa_{U_1}) - (\Delta t_{U_2,GPS} + \Delta t_{GPS} + \tau_{U_2} + t_{RU_2} + \kappa_{U_2}) \\ &= (\Delta t_{U_1,GPS} - \Delta t_{U_2,GPS}) + (\tau_{U_1} - \tau_{U_2}) + (t_{RU_1} - t_{RU_2}) + (\kappa_{U_1} - \kappa_{U_2})\end{aligned} \tag{19}$$

$$\begin{aligned}\Delta t_{U_1,U_3} &= t_{U_1} - t_{U_3} \\ &= (\Delta t_{U_1,GPS} + \Delta t_{GPS} + \tau_{U_1} + t_{RU_1} + \kappa_{U_1}) - (\Delta t_{U_3,GPS} + \Delta t_{GPS} + \tau_{U_3} + t_{RU_3} + \kappa_{U_3}) \\ &= (\Delta t_{U_1,GPS} - \Delta t_{U_3,GPS}) + (\tau_{U_1} - \tau_{U_3}) + (t_{RU_1} - t_{RU_3}) + (\kappa_{U_1} - \kappa_{U_3})\end{aligned} \tag{20}$$

$$\begin{aligned}\Delta t_{U_1,U_4} &= t_{U_1} - t_{U_4} \\ &= (\Delta t_{U_1,GPS} + \Delta t_{GPS} + \tau_{U_1} + t_{RU_1} + \kappa_{U_1}) - (\Delta t_{U_4,GPS} + \Delta t_{GPS} + \tau_{U_4} + t_{RU_4} + \kappa_{U_4}) \\ &= (\Delta t_{U_1,GPS} - \Delta t_{U_4,GPS}) + (\tau_{U_1} - \tau_{U_4}) + (t_{RU_1} - t_{RU_4}) + (\kappa_{U_1} - \kappa_{U_4})\end{aligned} \tag{21}$$

In this way, the three UAVs U_2, U_3, and U_4 can be synchronized on time based on the reference of the U_1 UAV's clock.

3.3. Design of Four UAVs' Data Communication

The communication system for monitoring civil aviation radio interference sources using four UAVs can be divided into three main parts: inter-aircraft link communication, UAV platform to ground station link communication (downlink), and ground station to UAV platform link communication (uplink). The inter-aircraft link is established based on a UAV self-assembling network architecture. All four UAVs are equipped with self-assembling network communication radios, enabling interaction and information exchange among them. Through the inter-aircraft link, each UAV shares its position, speed, and other relevant information with the other UAVs. For the downlink communication, data transmission radios are utilized. Each UAV is equipped with an antenna and a hardware front-end for the Software Defined Radio (SDR) platform, known as the Universal Software Radio Peripheral N321 (USRP N321) . This setup enables the UAVs to receive signals from radio interference sources. The received signals are then processed and transmitted to the ground station via data radios. In addition, the four UAVs transmit their collected information to the ground station using data radios. The uplink communication relies on wireless self-assembling radios. Each of the four UAVs is equipped with a self-assembling radio, similar to the inter-aircraft communication equipment. These radios receive control commands from the ground self-assembling transmitter radio, which is connected to a laptop computer functioning as the ground station. The laptop runs the necessary software for UAV flight, facilitating communication between UAVs, between UAVs and ground stations, and between ground stations and UAVs.

Figure 3 depicts the schematic design of the four UAVs' data communication. For the downlink design, each UAV is equipped with a digital transmission radio responsible for the real-time transmission of both the UAV's flight information and the USRP-converted digital signal information to the digital reception radio. This arrangement enables the ground station to display the radio spectrum monitored by the UAV platform and facilitates the monitoring of the UAV's flight status.

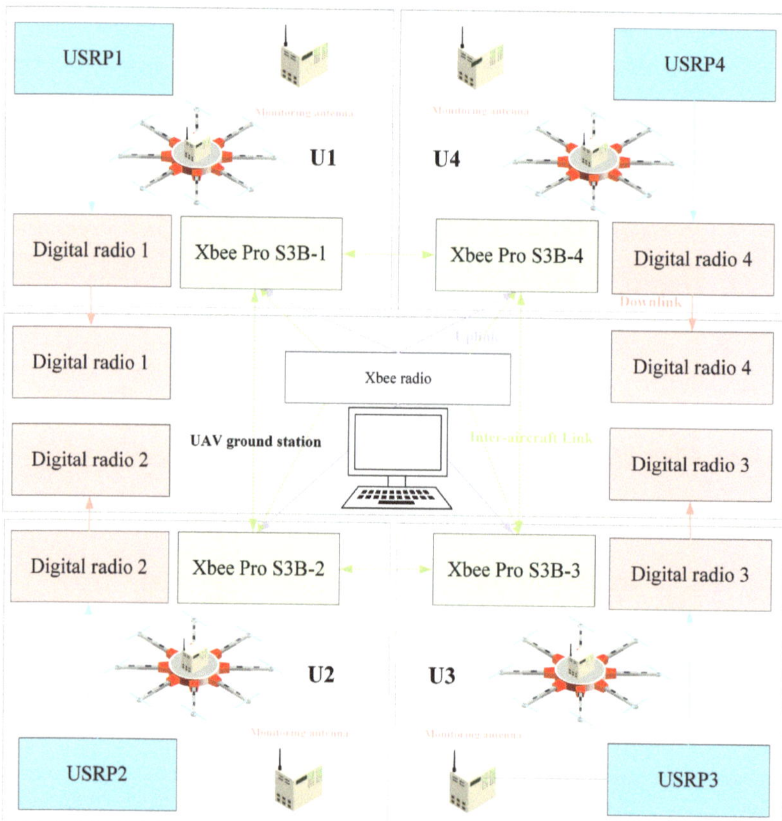

Figure 3. Schematic diagram of four UAVs' data communication design.

The flight control system of UAV is based on the open-source hardware and software Pixhawk 2.4.8 , utilizing the Mavlink communication protocol. The design of the inter-aircraft link and downlink is as follows: For the uplink, the on-board self-assembling device is the Xbee Pro S3B radio, which operates using the DIGI mesh protocol and is configured as a routing model. This setup allows it to receive control commands from the ground station and facilitates data interaction among the four UAVs. As for the ground device, the Xbee radio is chosen and configured as a coordinator. The Mavlink protocol is nested within the outer mesh protocol to enable the ground station to send flight control commands to the UAVs.

To ensure a smooth communication link, it is essential to appropriately reduce the communication load. As a result, the uplink currently transmits only UAV control command packets and UAV-desired position packets. The inter-aircraft link, on the other hand, solely requires UAV flight position data packets. Lastly, the downlink necessitates only the transmission of UAV flight position packages, UAV flight attitude packages, UAV flight status packages, and radio spectrum packages.

4. Simulation of Four-UAV Time–Frequency Difference Positioning Model for Interference Sources

We have set up simulated interference sources within the university premises, with a frequency set around 442 MHz. The joint positioning method for civil aviation radio interference sources based on four UAVs was implemented using the MATLAB program. The simulation involved analyzing the UAV deployment configurations and different moving speeds of radio interference sources independently.

4.1. Positioning Performance of UAVs with Different Deployment Configurations

According to the star, flat rhombus, inverted triangle, and parallelogram deployment patterns, four UAVs were set up with their respective spatial Cartesian coordinate systems. A GDOP positioning accuracy analysis was conducted for each deployment pattern, and the specified UAV coordinates can be found in Table 1.

Table 1. Cartesian coordinates of UAVs under different deployment configurations.

Deployment Configurations	x_i (km)	y_i (km)	z_i (km)
Star deployment configuration UAV1	0	0	0.1
Star deployment configuration UAV2	−17	10	0.1
Star deployment configuration UAV3	17	10	0.1
Star deployment configuration UAV4	0	−20	0.1
Flat rhombus deployment configuration UAV1	0	0	0.1
Flat rhombus deployment configuration UAV2	−17	10	0.1
Flat rhombus deployment configuration UAV3	17	10	0.1
Flat rhombus deployment configuration UAV4	0	20	0.1
Inverted triangle deployment configuration UAV1	0	0	0.1
Inverted triangle deployment configuration UAV2	−20	20	0.1
Inverted triangle deployment configuration UAV3	20	20	0.1
Inverted triangle deployment configuration UAV4	0	20	0.1
Parallelogram deployment configuration UAV1	0	0	0.1
Parallelogram deployment configuration UAV2	−14	14	0.1
Parallelogram deployment configuration UAV3	14	14	0.1
Parallelogram deployment configuration UAV4	28	0	0.1

The yellow markers represent the star deployment configuration, the red markers represent the flat rhombus deployment configuration, the green markers represent the inverted triangle deployment configuration, and the blue markers represent the parallelogram deployment configuration, as shown in Figure 4.

The measurement time error is set to 10 ns, and the UAV station error is set to 5 m. The interference source is considered a fixed source, and the observation ranges are denoted as $x = -200$ km∼200 km, $y = -200$ km∼200 km, with the target height as $z = 10$ km. In the spatial Cartesian coordinate system, interference source positions are randomly generated. Each interference source location is traversed, and the GDOP is calculated for UAV positioning.

Based on Figure 5a, the positioning errors of the star deployment configuration in the vertical dimension can be observed. From Figure 5b, it can be seen that when using star deployment configuration, the time–frequency difference positioning method has a balanced performance in three-dimensional space, with UAV1 as the center, and the farther the interference source is from UAV1, the larger the positioning error is. The position of the interference source is about 75 km from the position of UAV1, and the positioning error is about 360 m; the position of the interference source is about 100 km from the position of UAV1, and the positioning error is about 700 m; and the position of the interference source is about 120 km from the position of UAV1, and the positioning error is more than 1 km.

Figure 4. Diagram of UAV deployment configurations.

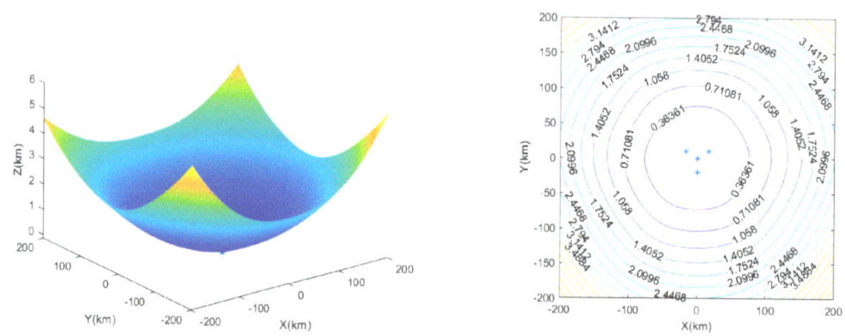

(**a**) (**b**)

Figure 5. The GDOP of UAV star deployment configuration. (**a**) Three-dimensional plot of positioning error for star deployment configuration. (**b**) The GDOP of star deployment configuration.

Based on Figure 6a, the positioning errors of the flat rhombus deployment configuration in the vertical dimension can be observed. Figure 6b illustrates the performance of the time–frequency difference positioning method in three-dimensional space using a flat rhombus deployment configuration with UAV1 as the center. The results show a relatively balanced performance, but there is a notable monitoring blind area. As the interference source moves farther away from UAV1, the range of this blind area increases. Moreover, the positioning performance of the interference source in the vertical direction of UAV1 is better than that in the horizontal direction. At approximately 40 km from the horizontal direction of UAV1, the positioning error of the interference source location is approximately 930 m. At the same positioning error, the distance from the vertical direction of UAV1 is approximately 140–160 km.

Based on Figure 7a, the positioning errors of the inverted triangle deployment configuration in the vertical dimension can be observed. Figure 7b presents the performance analysis of the time–frequency difference positioning method in three-dimensional space, employing an inverted triangle deployment configuration with UAV1 positioned at the center. The results demonstrate a relatively balanced performance; however, there is a significant monitoring blind area when the interference source is located 100 km away from the horizontal direction of UAV1. Furthermore, as the interference source moves farther away from UAV1, the range of this monitoring blind area expands. Moreover, the positioning performance of the interference source in the vertical direction of UAV1 surpasses that in the horizontal direction. At a distance of approximately 50 km from the horizontal direction of UAV1, the positioning error for the interference source location is approximately 930 m. At the same positioning error, the distance from the vertical direction of UAV1 is approximately 140–170 km.

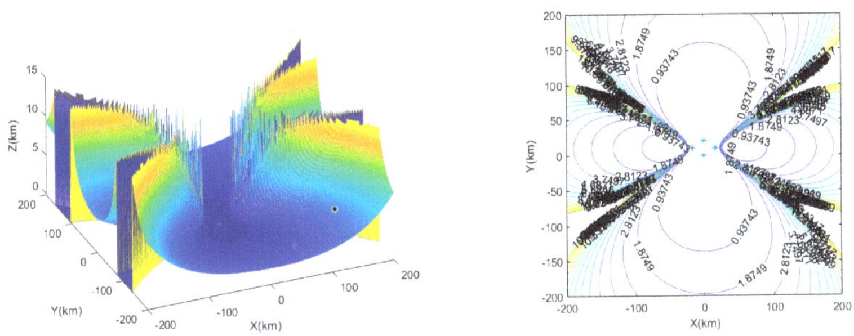

(a) (b)

Figure 6. The GDOP of UAV flat rhombus deployment configuration. (**a**) Three-dimensional plot of positioning error for flat rhombus deployment configuration. (**b**) The GDOP of flat rhombus deployment configuration.

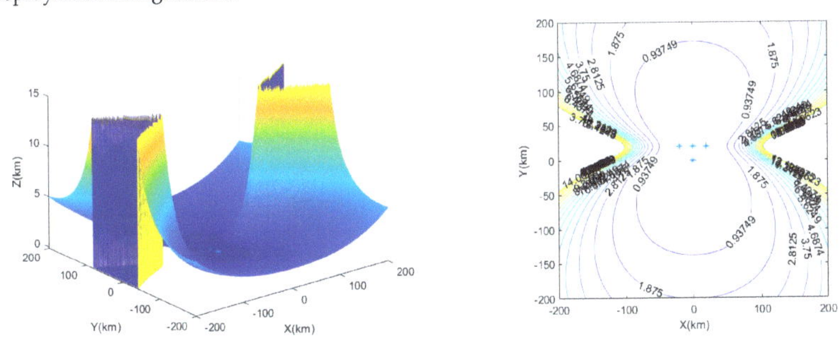

(a) (b)

Figure 7. The GDOP of UAV inverted triangle deployment configuration. (**a**) Three-dimensional plot of positioning error for inverted triangle deployment configuration. (**b**) The GDOP of inverted triangle deployment configuration.

Based on Figure 8a, the positioning errors of the parallelogram deployment configuration in the vertical dimension can be observed. Figure 8b illustrates the performance of the time–frequency difference positioning method in three-dimensional space using a parallelogram deployment configuration with UAV1 as the center. The results indicate a relatively balanced performance; however, a monitoring blind area is present. Notably, as the interference source moves farther away from UAV1, the range of this monitoring blind area expands. Moreover, the positioning performance of the interference source in the vertical direction of UAV1 outperforms that in the horizontal direction. The positioning error of the interference source location is approximately 930 m at distances ranging from about 40 km to 55 km from the horizontal direction of UAV1. For the same positioning error, the distance from the vertical direction of UAV1 is approximately 150 km to 160 km.

Through the GDOP analysis of the four UAV deployment configurations, it is evident that the star-based deployment configuration exhibits a balanced positioning performance with smaller errors compared to other deployment configurations. When employing the UAV platform for monitoring, the star-based formation flight yields comprehensive positioning coverage and high positioning accuracy.

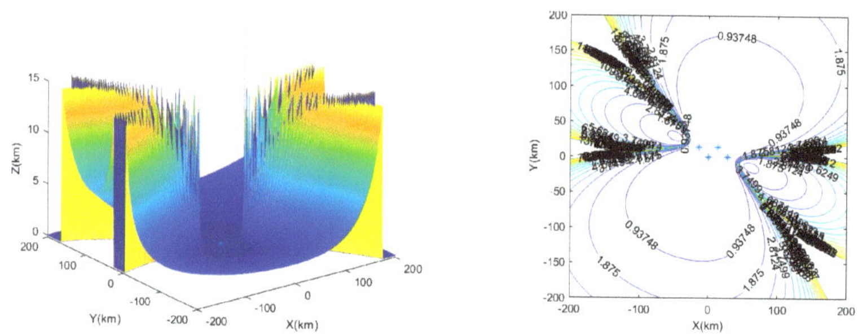

(a) (b)

Figure 8. The GDOP of UAV parallelogram deployment configuration. (**a**) Three-dimensional plot of positioning error for parallelogram deployment configuration. (**b**) The GDOP of parallelogram deployment configuration.

4.2. Interference Source Positioning Performance at Different Moving Speeds

Assuming fixed initial coordinates and speeds for the four UAVs, Table 2 presents their respective coordinates and speed information. The initial position of interference source is $R(280, 320, 270)$, and its speed is set at $\dot{R}_1(0,0,0)$, $\dot{R}_1(20,0,0)$, $\dot{R}_1(40,0,0)$, and $\dot{R}_1(80,0,0)$. To analyze the positioning performance, we utilize the time–frequency difference positioning algorithm through 5000 Monte Carlo simulations. The TDOA measurement value is subject to noise variance levels $10\log(c^2\sigma_d^2)$ ranging from -20 to 20, while the FDOA measurement value experiences noise variance levels set at 0.1 times that of TDOA. We evaluate the positioning performance using both bias and RMSE.

Table 2. Coordinates and speed information of four UAVs.

UAV	x_i (m)	y_i (m)	z_i (m)	$\dot{x}_i$ (m/s)	$\dot{y}_i$ (m/s)	$\dot{z}_i$ (m/s)
U_1	290	100	150	20	-20	20
U_2	380	150	100	-20	10	20
U_3	300	490	200	10	-20	10
U_4	340	200	90	10	20	30

When the moving speed of the interference source is set to $\dot{R}_1(0,0,0)$, indicating the interference source is stationary, the four UAVs conduct aerial monitoring at the speeds specified in Table 2. The obtained simulation data were visualized to analyze the positioning bias and RMSE variation trends, as shown in Figure 9.

The speed of the interference source is $\dot{R}_1(0,0,0))$. The position bias remains below 1 m when the noise variance level is not greater than 4. For noise variance levels ranging from -20 to 4, the position bias shows minimal changes. However, the position bias is greater than 1 m when the noise variance level is greater than 4. The position bias changes more when the noise variance level is from 6 to 20. At a noise variance level of 20, the position bias reaches approximately 39 m. Similarly, for speed bias, when the noise variance level is not greater than 0, the velocity bias is less than 1 m/s. The velocity bias changes less when the noise variance level is from -20 to 0. The velocity bias is greater than 1 m/s when the noise variance level is greater than 0. The position bias varies more when the noise variance level is from 0 to 20. At a noise variance level of 20, the velocity bias reaches about 15 m/s.

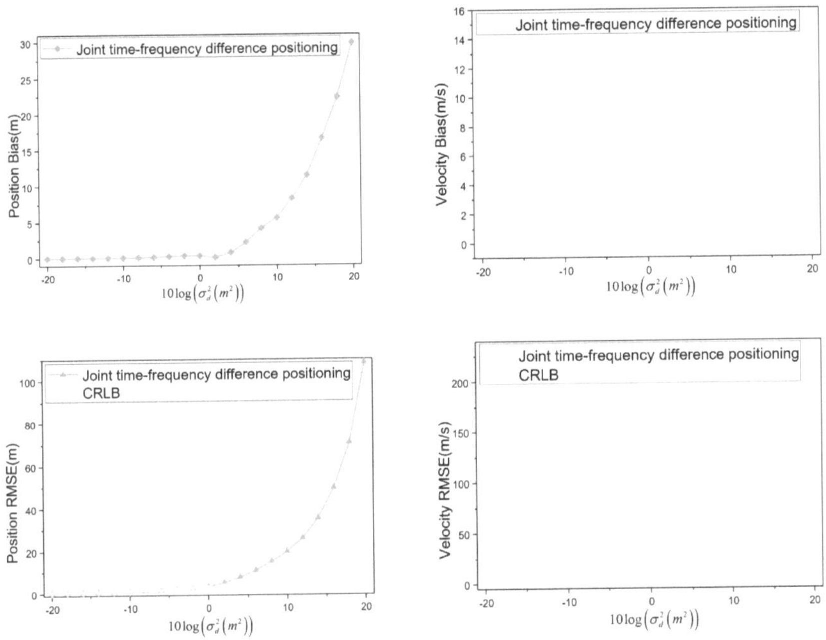

Figure 9. Positioning bias and RMSE ($\dot{R}_1(0,0,0)$).

When the noise variance level is not greater than 2, the position RMSE remains below 5 m. In the range of −20 to 2, the position RMSE is comparable to the CRLB. However, for noise variance levels greater than 2 and from 2 to 20, the position RMSE diverges significantly from the CRLB. At a noise variance level of 20, the position RMSE reaches approximately 108 m, whereas the CRLB is around 27 m. For the velocity RMSE, when the noise variance level is not greater than −4, it remains below 1 m/s. From −20 to −4, the velocity RMSE shows minimal deviations from the CRLB. However, when the noise variance level exceeds −4 and ranges from −4 to 20, the velocity RMSE diverges more significantly from the CRLB. At a noise variance level of 20, the velocity RMSE reaches about 230 m/s, while the CRLB is approximately 13 m/s.

When the moving speed of the interference source is set to $\dot{R}_1(20,0,0)$, indicating the interference source moves horizontally at a speed of 20 m/s, the four UAVs conduct aerial monitoring at the speeds specified in Table 2. The obtained simulation data was visualized to analyze the positioning bias and RMSE variation trends, as shown in Figure 10.

The speed of the interference source is $\dot{R}_1(20,0,0)$). The position bias remains below 1 m when the noise variance level is not greater than 4. For noise variance levels ranging from −20 to 4, the position bias shows minimal changes. However, the position bias is greater than 1 m when the noise variance level is greater than 4. The position bias changes more when the noise variance level is from 6 to 20. At a noise variance level of 20, the position bias reaches approximately 33 m. Similarly, for speed bias, when the noise variance level is not greater than 12, the velocity bias is less than 1 m/s. The velocity bias changes less when the noise variance level is from −20 to 12. The velocity bias is greater than 1 m/s when the noise variance level is greater than 12. The position bias varies more when the noise variance level is from 12 to 20. At a noise variance level of 20, the velocity bias reaches about 25 m/s.

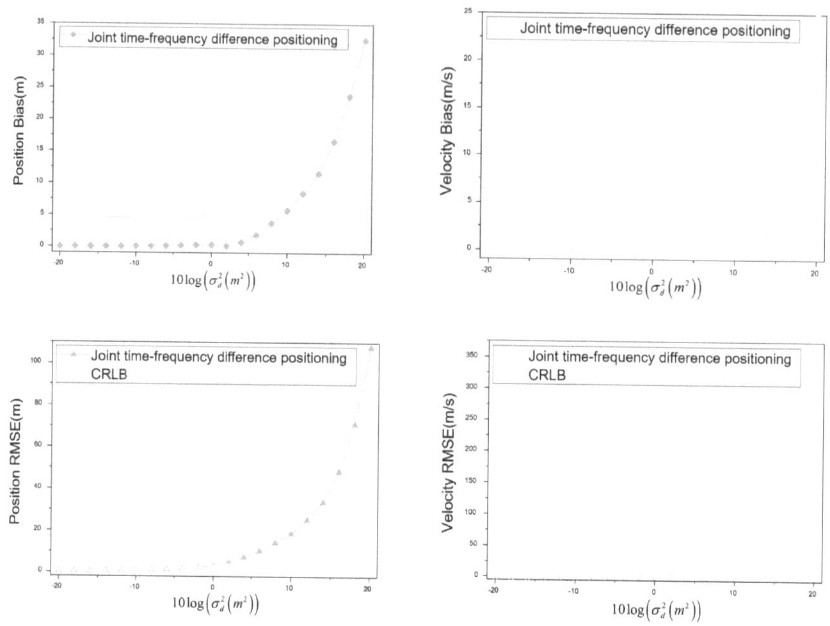

Figure 10. Positioning bias and RMSE ($\dot{R}_1(20, 0, 0)$).

When the noise variance level is not greater than 2, the position RMSE remains below 5 m. In the range of −20 to 2, the position RMSE is comparable to the CRLB. However, for noise variance levels greater than 2 and from 2 to 20, the position RMSE diverges significantly from the CRLB. At a noise variance level of 20, the position RMSE reaches approximately 108 m, whereas the CRLB is around 28 m. For the velocity RMSE, when the noise variance level is not greater than −2, it remains below 1.5 m/s. From −20 to −2, the velocity RMSE shows minimal deviations from the CRLB. However, when the noise variance level exceeds −2 and ranges from −2 to 20, the velocity RMSE diverges more significantly from the CRLB. At a noise variance level of 20, the velocity RMSE reaches about 363 m/s, while the CRLB is approximately 13 m/s.

When the moving speed of the interference source is set to $\dot{R}_1(40, 0, 0)$, indicating that the interference source moves horizontally at a speed of 40 m/s, the four UAVs conduct aerial monitoring at the speeds specified in Table 2. The obtained simulation data were visualized to analyze the positioning bias and RMSE variation trends, as shown in Figure 11.

The speed of the interference source is $\dot{R}_1(40, 0, 0)$). The position bias remains below 1 m when the noise variance level is not greater than 4. For noise variance levels ranging from −20 to 4, the position bias shows minimal changes. However, the position bias is greater than 1 m when the noise variance level is greater than 4. The position bias changes more when the noise variance level is from 6 to 20. At a noise variance level of 20, the position bias reaches approximately 33 m. Similarly, for speed bias, when the noise variance level is not greater than 6, the velocity bias is less than 1 m/s. The velocity bias changes less when the noise variance level is from −20 to 6. The velocity bias is greater than 2 m/s when the noise variance level is greater than 8. The position bias varies more when the noise variance level is from 8 to 20. At a noise variance level of 20, the velocity bias reaches about 36 m/s.

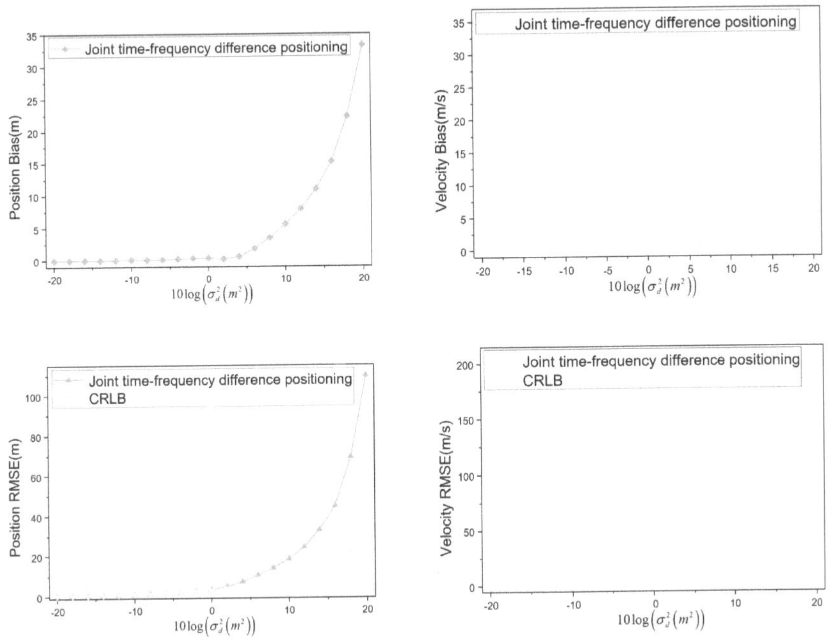

Figure 11. Positioning bias and RMSE ($\dot{R}_1(40,0,0)$).

When the noise variance level is not greater than 2, the position RMSE remains below 5 m. In the range of −20 to 2, the position RMSE is comparable to the CRLB. However, for noise variance levels greater than 4 and from 4 to 20, the position RMSE diverges significantly from the CRLB. At a noise variance level of 20, the position RMSE reaches approximately 110m, whereas the CRLB is around 28 m. For the velocity RMSE, when the noise variance level is not greater than −6, it remains below 1 m/s. From −20 to −6, the velocity RMSE shows minimal deviations from the CRLB. However, when the noise variance level exceeds −2 and ranges from −2 to 20, the velocity RMSE diverges more significantly from the CRLB. At a noise variance level of 20, the velocity RMSE reaches about 204 m/s, while the CRLB is approximately 13 m/s.

When the moving speed of the interference source is set to $\dot{R}_1(80,0,0)$, indicating the interference source moves horizontally at a speed of 80 m/s, the four UAVs conduct aerial monitoring at the speeds specified in Table 2. The obtained simulation data were visualized to analyze the positioning bias and RMSE variation trends, as shown in Figure 12.

The speed of the interference source is $\dot{R}_1(80,0,0)$). The position bias remains below 1 m when the noise variance level is not greater than 4. For noise variance levels ranging from −20 to 4, the position bias shows minimal changes. However, the position bias is greater than 1 m when the noise variance level is greater than 4. The position bias changes more when the noise variance level is from 6 to 20. At a noise variance level of 20, the position bias reaches approximately 28 m. Similarly, for speed bias, when the noise variance level is not greater than 6, the velocity bias is less than 3 m/s. The velocity bias changes less when the noise variance level is from −20 to 6. The velocity bias is greater than 7 m/s when the noise variance level is greater than 8. The position bias varies more when the noise variance level is from 8 to 20. At a noise variance level of 20, the velocity bias reaches about 65 m/s.

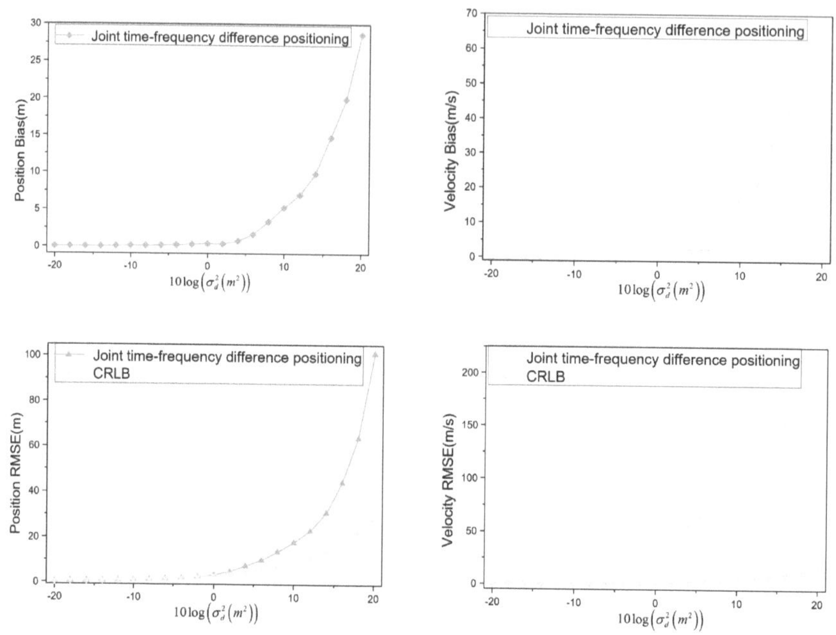

Figure 12. Positioning bias and RMSE ($\dot{R}_1(80,0,0)$).

When the noise variance level is not greater than 2, the position RMSE remains below 5 m. In the range of −20 to 2, the position RMSE is comparable to the CRLB. However, for noise variance levels greater than 4 and from 4 to 20, the position RMSE diverges significantly from the CRLB. At a noise variance level of 20, the position RMSE reaches approximately 101 m, whereas the CRLB is around 28 m. For the velocity RMSE, when the noise variance level is not greater than −4, it remains below 2.5 m/s. From −20 to −4, the velocity RMSE shows minimal deviations from the CRLB. However, when the noise variance level exceeds −2 and ranges from −2 to 20, the velocity RMSE diverges more significantly from the CRLB. At a noise variance level of 20, the velocity RMSE reaches about 220 m/s, while the CRLB is approximately 15 m/s.

Based on the comparative analysis above, the following conclusions can be drawn: The velocity of the interference source in the horizontal direction ranges from 0 m/s to 80 m/s when considering certain initial positions and velocities for the UAV platform and the initial position of the interference source. For position estimation, when the noise variance level is below 4, the position bias remains below 1m. At a noise variance level of 20, the position bias ranges from 29 m to 39 m. For the RMSE of position, at noise variance levels below 2, it remains below 5 m. At a noise variance level of 20, the RMSE of the position ranges from 101 m to 110 m. These findings indicate that the movement velocity of the interference source has little effect on the positioning location results. On the other hand, the velocity of the interference source has a more significant impact on the positioning velocity results. The velocity bias remains below 1 m/s for different noise variance levels. At a noise variance level of 20, the velocity bias ranges from 15 m/s to 65 m/s. For the RMSE of velocity, when the noise variance level is below −6, it remains below 1 m/s and shows similarity to the CRLB. At a noise variance level of 20, the RMSE of velocity ranges from 204 m/s to 363 m/s, while the CRLB is approximately 13 m/s.

5. Conclusions

In this paper, through a combination of theoretical analysis, simulations, and other technical means, we conducted a study of the civil aviation radio interference source positioning method using four UAVs. Our study encompassed the design of time synchronization and communication for the four UAVs, the model of locating civil aviation radio interference sources by four UAVs, and the simulation of the positioning performance of this model. The simulation results show that the positioning performance of the four UAVs' star-based deployment configuration is balanced and the positioning error is small, and the interference source movement velocity has a small impact on the positioning location accuracy and a large impact on the positioning velocity accuracy. The simulation results provide data support for the next experiments.

6. Outlook

The application of civil aviation radio interference source positioning based on multiple UAVs is still in the development stage, and limited by time, experimental equipment, experimental conditions, personal ability, and other factors, this paper is not perfect. Therefore, the following points are now proposed to carry out in-depth research in the following work.

(1) In order to be able to apply multi-UAV localized radio interference source equipment in practice, it is also necessary to integrate multi-UAV collaboration techniques, including UAV formation, UAV obstacle avoidance, and integration of all ground-based software into a single system.

(2) This paper does not consider the atmospheric refractive index error of the radio signals received by UAVs for the time being, and in order to improve the positioning accuracy, the empirical model of atmospheric refractive index can be incorporated into the multi-UAV positioning algorithm.

Author Contributions: Conceptualization, C.Z., X.Z. and R.X.; methodology, X.Z.; software, X.Z. and R.X.; validation, X.Z. and K.H.; formal analysis, C.Z., X.Z. and K.H.; investigation, F.O., C.H. and T.H.; resources, C.Z.; data curation, X.Z.; writing—original draft preparation, X.Z. and K.H.; writing—review and editing, C.Z., X.Z. and R.X.; visualization, X.Z. and R.X.; supervision, C.Z.; project administration, C.Z., C.H. and T.H.; funding acquisition, C.Z., X.Z. and F.O. All authors have read and agreed to the published version of the manuscript.

Funding: This research was funded by the Civil Aviation Security Capacity Project in 2022(MHAQ2022028), the Scientific Research Project of Civil Aviation Flight University of China (ZX2021-03), and the Independent Research Project of Key Laboratory of Civil Aviation Flight Technology and Flight Safety (FZ2021ZZ03).

Institutional Review Board Statement: Not applicable.

Informed Consent Statement: Not applicable.

Data Availability Statement: The data that support the findings of this research are available from the author X.Z. upon reasonable request.

Conflicts of Interest: The authors declare no conflicts of interest.

References

1. Yozevitch, R.; Marbel, R.; Flysher, N.; Ben-Moshe, B. Save Our Roads from GNSS Jamming: A Crowdsource Framework for Threat Evaluation. *Sensors* **2021**, *21*, 4840. [CrossRef] [PubMed]
2. Nicola, M.; Falco, G.; Morales Ferre, R.; Lohan, E.S.; De La Fuente, A.; Falletti, E. Collaborative solutions for interference management in GNSS-based aircraft navigation. *Sensors* **2020**, *20*, 4085. [CrossRef] [PubMed]
3. Cheng, J.; Ren, P.; Deng, T. A Novel Ranging and IMU-Based Method for Relative Positioning of Two-MAV Formation in GNSS-Denied Environments. *Sensors* **2023**, *23*, 4366. [CrossRef]
4. Jeong, S.; Kim, T.; Kim, J. Spoofing detection test of GPS signal interference mitigation equipment. In Proceedings of the 2014 International Conference on Information and Communication Technology Convergence (ICTC), Busan, Republic of Korea, 22–24 October 2014; pp. 651–652.

5. Jeong, S.; Sin, C.S. GNSS interference signal generation scenario for GNSS interference verification platform. In Proceedings of the 2015 15th International Conference on Control, Automation and Systems (ICCAS), Busan, Republic of Korea, 13–16 October 2015; pp. 1363–1365.
6. Mariappan, P.M.; Raghavan, D.R.; Aleem, S.H.A.; Zobaa, A.F. Effects of electromagnetic interference on the functional usage of medical equipment by 2G/3G/4G cellular phones: A review. *J. Adv. Res.* **2016**, *7*, 727–738. [CrossRef]
7. Crespo, G.; Glez-de Rivera, G.; Garrido, J.; Ponticelli, R. Setup of a communication and control systems of a quadrotor type Unmanned Aerial Vehicle. In Proceedings of the Design of Circuits and Integrated Systems, Madrid, Spain, 26–28 November 2014; pp. 1–6.
8. Sorbi, L.; De Capua, G.P.; Fontaine, J.G.; Toni, L. A behavior-based mission planner for cooperative autonomous underwater vehicles. *Mar. Technol. Soc. J.* **2012**, *46*, 32–44. [CrossRef]
9. Ramesh, R.; John-Sabu, A.; Ramesh, S.; Arunachalam, M.; Amrutur, B.; Navada, V.B.; Harshitha, S. Robust and Scalable Techniques for TWR and TDoA based localization using Ultra Wide Band Radios. *arXiv* **2020**, arXiv:2008.04248.
10. Kilari, A. Indoor Location for Emergency Responders Using Asynchronous Differential TDOA in LTE D2D Communications Waveform: Principle and Implementation Using USRP. Ph.D. Thesis, University of Cincinnati, Cincinnati, OH, USA, 2020.
11. Biswas, S.K.; Cetin, E. GNSS interference source tracking using kalman filters. In Proceedings of the 2020 IEEE/ION Position, Location and Navigation Symposium (PLANS), Portland, OR, USA, 20–23 April 2020; pp. 877–882.
12. Perkins, A.; Chen, Y.H.; Lo, S.; Enge, P. Vision Based UAS Navigation for RFI Localization. In Proceedings of the 31st International Technical Meeting of the Satellite Division of The Institute of Navigation (ION GNSS+ 2018), Miami, FL, USA, 24–28 September 2018; pp. 2726–2736.
13. Hao, C.; Liu, H. High Accuracy Satellite Interference Geolocation Using UAV Assist. *J. Signal Process.* **2016**, *32*, 1412–1417.
14. Xu, B.; Li, Y.; Wei, G.; Cao, R. The Research on Fast Positioning based on Analysis of Radio Frequency Parameter. *Digit. Commun. World* **2013**, 83–85. [CrossRef]
15. Jin, P.; Xu, C. Location of interfering signal source in cognitive radio. *J. Yanshan Univ.* **2010**, *34*, 542–545.
16. Wang, G. Location Algorithm for and Error Analysis of MSSR Radio Interference Source. *Electron. Sci. Technol.* **2014**, *27*, 89–92.
17. Li, J.; Shao, Y.; Long, H. Research on Wireless Interference Source Localization Based on Grid Spectrum Monitoring. *Comput. Sci.* **2017**, *44*, 274–275. [CrossRef]
18. Stefanski, J. Asynchronous time difference of arrival (ATDOA) method. *Pervasive Mob. Comput.* **2015**, *23*, 80–88. [CrossRef]
19. Sadowski, J.; Stefanski, J. Asynchronous phase-location system. *J. Mar. Eng. Technol.* **2017**, *16*, 400–408. [CrossRef]
20. Ziółkowski, C.; Kelner, J.M. Doppler-based navigation for mobile protection system of strategic maritime facilities in GNSS jamming and spoofing conditions. *IET Radar Sonar Navig.* **2020**, *14*, 643–651. [CrossRef]
21. Alotaibi, E.T.; Alqefari, S.S.; Koubaa, A. Lsar: Multi-uav collaboration for search and rescue missions. *IEEE Access* **2019**, *7*, 55817–55832. [CrossRef]
22. Kelner, J.M.; Ziółkowski, C. Portable beacon system for emergency mountain landing pad. In Proceedings of the 2019 European Navigation Conference (ENC), Warsaw, Poland, 9–12 April 2019; pp. 1–5.
23. Li, B.; Fei, Z.; Zhang, Y. UAV communications for 5G and beyond: Recent advances and future trends. *IEEE Internet Things J.* **2018**, *6*, 2241–2263. [CrossRef]
24. Wang, D.; Zhang, P.; Yang, Z.; Wei, F.; Wang, C. A novel estimator for TDOA and FDOA positioning of multiple disjoint sources in the presence of calibration emitters. *IEEE Access* **2019**, *8*, 1613–1643. [CrossRef]
25. Musicki, D.; Kaune, R.; Koch, W. Mobile emitter geolocation and tracking using TDOA and FDOA measurements. *IEEE Trans. Signal Process.* **2009**, *58*, 1863–1874. [CrossRef]
26. Yu, H.; Huang, G.; Gao, J.; Liu, B. An efficient constrained weighted least squares algorithm for moving source location using TDOA and FDOA measurements. *IEEE Trans. Wirel. Commun.* **2011**, *11*, 44–47. [CrossRef]
27. Yu, Z.; Wei, J.; Liu, H. An energy-efficient target tracking framework in wireless sensor networks. *EURASIP J. Adv. Signal Process.* **2009**, *2009*, 524145. [CrossRef]
28. Sathyan, T.; Sinha, A.; Kirubarajan, T. Passive geolocation and tracking of an unknown number of emitters. *IEEE Trans. Aerosp. Electron. Syst.* **2006**, *42*, 740–750. [CrossRef]
29. Kim, W.C.; Song, T.L.; Mušicki, D. Mobile emitter geolocation and tracking using correlated time difference of arrival measurements. In Proceedings of the 2012 15th International Conference on Information Fusion, Singapore, 9–12 July 2012; pp. 700–706.
30. Kelner, J.M.; Ziółkowski, C. Effectiveness of mobile emitter location by cooperative swarm of unmanned aerial vehicles in various environmental conditions. *Sensors* **2020**, *20*, 2575. [CrossRef] [PubMed]
31. Zhou, C.; Xiong, R.; Zeng, H.; Xiao, J.; Wang, Y.; Jia, P.; Ye, J.; Zhao, T.; Hu, K. Aerial locating method design for civil aviation RFI: UAV monitoring platform and ground terminal system. *J. Intell. Robot. Syst.* **2021**, *103*, 1–13. [CrossRef]
32. Zhou, C.; He, S.; Ye, J.; Jia, P. Design and Implementation of Ground Terminal for Aerial Radio Monitoring System Based on UAV. *J. Phys. Conf. Ser.* **2020**, *1626*, 012084. [CrossRef]
33. He, S.; Zhou, C.; Ye, J.; Jia, P. Analysis and Research on the Development of UAV Ground Control Station. *Digit. Technol. Appl.* **2019**. [CrossRef]
34. Zhou, C.; Xiong, R.; Wang, Y.; Xiao, J.; Ye, J. Research on Improving Innovation Ability of Professional Master in Communication, Navigation and Surveillance. *Digit. Technol. Appl.* **2021**. [CrossRef]

35. Zhou, C.; Ye, J.; Jia, P. Research on optimal speed of radio interference for UAV autonomous localization. *Inf. Technol.* **2021**. [CrossRef]
36. Zhou, C.; Jia, P.; Ye, J.; He, S.; Li, S. Development of UAV Ground Station System for Civil Aviation Radio Monitoring. *Electron. Opt. Control* **2021**, *28*, 107–110.
37. Zhou, C.; Xiao, J.; Xiong, R.; Wang, Y.; Zhao, T. Design of 3D Location System of Civil Aviation Radio Interference Source Based on GIS. *Aeronaut. Comput. Tech.* **2021**. [CrossRef]
38. Zhou, C.; Wang, Y.; Xiong, R.; Xiao, J. Research on Improvement of Efficiency of Radio Monitoring and Positioning Based on UAV. *Comput. Simul.* **2022**, *39*, 62–66.
39. Zhou, C.; Xiao, J.; Xiong, R.; Wang, Y.; Jia, P. Research on the location of civil aviation radio interference source based on visualization. *Inf. Technol.* **2022**. [CrossRef]
40. Zhang, Y.; Hu, B.; Li, J.; Zhang, J. UAV Multi-Mission Reconnaissance Decision-Making under Uncertainty Environment. *J. Northwestern Polytech. Univ.* **2016**, *34*, 1028–1034.
41. Zhu, G.; Feng, D.; Zhou, Y.; Nie, W. TOA localization algorithm using the linear-correction technique. *J. Xidian Univ.* **2015**, *42*, 22–25.
42. Li, C.; Li, X.; Zhang, J.; Cao, C. Analysis of Airborne Passive Location Precision Based on Multi-static Cooperation. *Mod. Radar* **2017**, *39*, 11–14.

sensors

Article

SiaN-VO: Siamese Network for Visual Odometry †

Bruno S. Faiçal *, Cesar A. C. Marcondes and Filipe A. N. Verri *

Computer Science Division, Aeronautics Institute of Technology—ITA, Sao Jose dos Campos 12228-900, SP, Brazil; cmarcondes@ita.br

* Correspondence: bruno.faical@ga.ita.br (B.S.F.); filipe.verri@gp.ita.br (F.A.N.V.)

† This paper is an extended version of our paper published in the 2022 Latin American Robotics Symposium (LARS), 2022 Brazilian Symposium on Robotics (SBR), and 2022 Workshop on Robotics in Education (WRE), São Bernardo do Campo, Brazil, 18–21 October 2022.

Abstract: Despite the significant advancements in drone sensory device reliability, data integrity from these devices remains critical in securing successful flight plans. A notable issue is the vulnerability of GNSS to jamming attacks or signal loss from satellites, potentially leading to incomplete drone flight plans. To address this, we introduce SiaN-VO, a Siamese neural network designed for visual odometry prediction in such challenging scenarios. Our preliminary studies have shown promising results, particularly for flights under static conditions (constant speed and altitude); while these findings are encouraging, they do not fully represent the complexities of real-world flight conditions. Therefore, in this paper, we have furthered our research to enhance SiaN-VO, improving data integration from multiple sensors and enabling more accurate displacement predictions in dynamic flight conditions, thereby marking a significant step forward in drone navigation technology.

Keywords: visual odometry; drone; autonomous flight

1. Introduction

Unmanned Aerial Vehicles (UAVs) use Global Navigation Satellite System (GNSS) signals as their primary location tool. Knowing the global position (latitude and longitude coordinates) allows the flight control system to perform missions in outdoor environments. However, this system is susceptible to various types of attacks and interruptions in signal reception [1,2].

During the aircraft's flights, particularly during search and rescue (SAR) missions, an accurate position estimation of the aircraft is crucial, even in the absence of Global Navigation Satellite System (GNSS) data [3]. Research in this domain has explored various methodologies, including sensor fusion, the use of Inertial Measurement Units (IMUs), and image-based inference techniques. Notably, image-based navigation employs multiple strategies for location estimation, such as recognizing environmental landmarks and analyzing optical flow [4–7]. However, these image-based methods typically demand extensive computational resources due to their reliance on intricate image processing pipelines.

In the existing literature, various studies have proposed image-based methods for estimating displacement from a reference point, such as the initial position, to facilitate collision avoidance [8–10]. However, there needs to be more research regarding the exclusive use of Convolutional Neural Networks (CNNs) and dataflow for global position estimation of Unmanned Aerial Vehicles (UAVs). Hence, using lightweight Artificial Neural Network (ANN) models in image-based approaches could reduce computational costs during the inference phase.

Moreover, CNNs have become a cornerstone in computer vision, often serving as a fundamental component in image processing within machine learning frameworks [11,12]. Compared to traditional fully connected ANN layers, CNNs are more efficient in processing local spatial patterns in images and require significantly fewer parameters [13].

Building on this, we introduce the Siamese Neural Network for Visual Odometry (SiaN-VO), a method capable of determining displacement between a pair of sequential images captured by a drone. SiaN-VO represents an advanced version of a method previously discussed in another study [14]. It uses inferred displacement and the haversine formula to calculate the UAV's new geographic coordinates (latitude and longitude). We assume that flight direction and vehicle altitude are acquired from other sensors, such as a compass and a radio altimeter. This novel approach enables displacement inference during flights with varying altitudes, addressing the limitations of earlier methods that required constant altitude flights. We also keep the number of parameters in the network low, so that we can enable the future implementation of the solution in an embedded system on the aircraft.

The remainder of this paper is organized as follows: In Section 2, we review the relevant scientific literature that relates to our proposal. Section 3 provides a comprehensive description of the system we propose. Section 4 elaborates on the methodology employed in our experiments, including details about the dataset, the training phase of our proposed method, and the testing procedures. Section 5 presents the results we obtained. The paper concludes with Section 6, wherein we discuss our conclusions and directions for future research.

2. Scientific Literature

Numerous works in the literature have highlighted the importance of investigating methods and approaches for different vehicles (ground or air) to have location information without the dependency on external sources [8–10]. In general, these works investigate approaches that allow vehicles to move safely and reach their goal.

Visual approaches, such as visual odometry, have attracted attention for several reasons, among them, price, accessibility, accuracy, and the independence of external signals, as in the case of GNSS-based methods [15,16]. Visual odometry is the process of estimating the motion of an agent (e.g., vehicle, human, and robot) using only the input from a single or multiple cameras [17].

There are works describing good results in merging visual odometry information with other information to achieve better accuracy in location inference [18,19]. The results show that merging information from visual odometry with other sensors can increase the accuracy of the positioning and movement information. Visual odometry commonly uses an important concept called optical flow.

Optical flow has been used to detect the motion of objects and scenery to help autonomously drive vehicles and avoid collisions [20]. An example of this scenario can be seen in [21], where the proposed method (named LiteFlowNet2) is evaluated on datasets from different contexts. The MPI Sintel dataset is a dataset derived from the open-source 3D animation short film Sintel. In this setting, the method receives a pair of sequential images, and its output is a segmented image of the regions occupied by the characters' movements in the time interval between the images received as input. Another dataset used is KITTI [20,22], which is a set of images captured by a car on urban routes. In this dataset, the method is evaluated for the goal of detecting the surrounding scenery in motion.

Works that use CNN to estimate the position and displacement of unmanned aerial vehicles are found in the scientific literature [14,23,24]. Olawoye and Gross [23] propose applying a deep learning approach to 3D object detection to calculate the relative position of an Unmanned Aerial Vehicle (UAV). However, this approach requires an Unmanned Ground Vehicle (UGV) equipped with a LiDAR sensor in order to operate in GPS-denied environments. In a study by Araveti et al. [24], another method is proposed to estimate drone displacement, but this method is not designed to deal with altitude variation. This weakness is also observed in the paper originally published by the authors of the current work [14], where the proposed method was unable to deal with altitude variation. In view of this, it is possible to note that the current work proposes a method capable of overcoming the obstacles of altitude variation and the requirement for additional vehicles.

One can observe in the literature works that use the concepts above to estimate the movement of unmanned aerial vehicles [25,26]. The proposed methods can be used in outdoor and indoor environments.

However, it is common for navigation systems to use geographic coordinate information to manage UAV flights. Considering this context, we were not able to find works with the objective of inferring the geographic coordinates of UAVs during flight.

3. Siamese Network for Visual Odometry (SiaN-VO)

In this section, we describe the proposed new ANN model and its training procedure. It is important to emphasize that the proposed method presents the evolution of previously published work and the overcoming of the limitation on flight dynamics.

Network Architecture

The proposed neural network model must be able to receive two images and two other matrices as input. One of them must contain the altitude value, while the other matrix must contain data from a sensor that captures the aircraft yaw variable (the angle at which the front of the drone is facing, based on magnetic north.). The matrices have the same dimensions as the images and their values are repeated in all the cells. The transformation of the unit values into matrices with the value repeated in all the cells aims to provide stimulation similar to that of the images. This prevents the altitude and yaw data stimuli from being ignored because they are weak stimuli compared to the number of values in the image pair. The detailed architecture can be seen in Figure 1. The total number of trainable params is 2,004,801, i.e., around 7.65 MB of data.

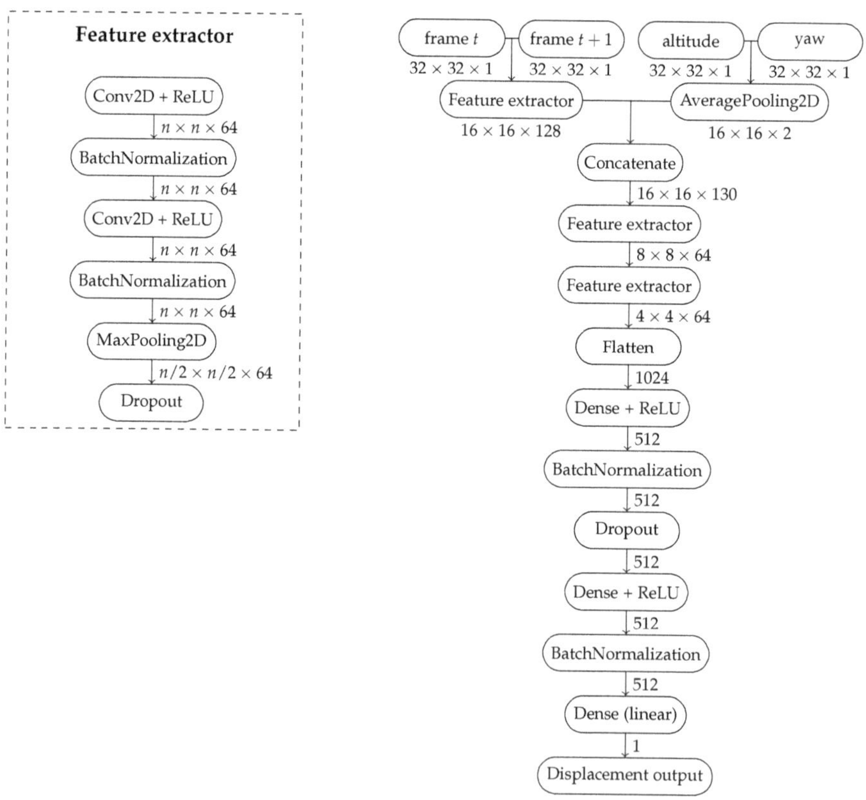

Figure 1. Architecture of the Siamese Network for Visual Odometry (SiaN-VO).

Pixel values are normalized between 0 and 1 for the image pairs. Yaw is also normalized between 0 and 1, where 0.5 denotes magnetic north, 0 denotes −180 degrees, and 1 denotes 180 degrees. Altitude input that the aircraft flies with maximum altitude h from the ground. Thus, we divide the input by h so its maximum value is also 1.

Our network can be divided into two parts:

1. In the first part, we have a Siamese network, where the model receives a pair of images taken at consecutive time steps, and these images are passed through identical neural network layers and weights. We chose such a design assuming that the images could be processed independently. The neural network can find useful coarse patterns in them, like edges, before joining both images. Parallel to the Siamese network, there is an AveragePooling step that will adjust the dimensions of the altitude and yaw matrices to the same dimensions as the feature maps resulting from the Siamese network.
2. The results of the Siamese network and AveragePooling are concatenated, effectively overlaying the image maps and the two complementary data matrices. The concatenated feature maps then pass through a normal CNN pipeline, ending in a prediction head containing three fully connected layers, the last of which has only a single output—the displacement output.

During the search for the best hyperparameters for the neural network, we find that larger filter sizes work best for our problem, and we ultimately employ 7×7 filters throughout the network. Apart from the filter size, our design choices took inspiration from the VGGNet architecture [27]: at every new set of convolutional layers, we double the number of filters, as well as reduce, by half, the spatial dimensions of the image maps by means of applying MaxPooling layers with a kernel size of 2×2. In summary, our model is composed of six convolutional layers (two layers in the Siamese network and four sequential convolutional layers) and three fully connected layers. Dropout layers are set at 20%. In total, the neural network is nine layers deep. It is important to note that the Siamese network is encapsulated in the figure within the "Feature extractor" element.

Once the model is to be employed in a small unmanned aircraft, computational power and storage are important restrictions. With this in mind, we designed our neural network to receive gray-scale images of dimensions 32×32. The use of gray-scale images should suffice for the task of displacement prediction, as the color information transmitted by the red, green, and blue channels should not give any significant insights into the movement of the vehicle. We highlight that we have reduced the size of the images used compared to the method proposed previously, allowing for a less computationally expensive execution.

4. Methodology for the Experiments

In this section, we will describe the dataset used and the methodology employed in the training and testing stages to evaluate SiaN-VO.

4.1. Dataset

In order to develop a large-scale simulated dataset for the task of UAV displacement estimation, we leverage the capabilities of AirSim [28], which is an open-source simulator for autonomous vehicles, including self-driving cars and drones.

We simulated drone flying in three different scenarios: a mountainous arctic region (mountains), a tropical forest (forest), and a city with a green area (downtown). These scenarios include artifacts like lakes and rivers, as well as different sizes of streets, buildings, and vegetation. Besides the inherent heterogeneity of these choices of maps, we also varied the weather conditions, adding dust, mist, and rain, creating a diversified range of settings. This variability in scenarios is important if we want the machine learning model trained on this data to be able to generalize well to scenes never seen before, which is paramount once we employ these models in real-world applications. We also vary the flight dynamics and characteristics such as altitude, acceleration, direction, and environment. Figure 2 shows an example of the images in the dataset.

Figure 2. Examples of images considering the environments in which the flights were simulated.

Table 1 shows details of the flights simulated in AirSim. The length of the flights varies, as do their maximum and minimum altitudes. It is important to note that the "map" column refers to the region in which the flight took place. Thus, the environment may be forest, but in regions other than forest. It is important to note that, in some cases, the number of flights is less than the standard 50 flights. This is because the map we used was smaller, and we did not want to use it to exhaustion.

Table 1. Details of the flights that make up the simulated dataset.

Environment	Map	Number of Routes
Downtown	A	50
Downtown	B	50
Downtown	C	50
Forest	A	12
Forest	B	10
Forest	C	50
Mountains	A	50
Mountains	B	50
Mountains	C	50

The images taken from the drone have a resolution of 720×480 pixels. Even though most applications may need a smaller resolution, the operation of resizing in the data pipeline is very cheap, and on the plus side, the choice of a large resolution gives the user a choice to apply operations of data augmentation, like random-cropping, which need an image larger than the input of the network. The use of data augmentation operations in the preprocessing step can greatly increase the generalization ability of the models trained, as well as virtually increasing the dataset size.

The images are taken from a camera mounted under the vehicle, pointing vertically to the ground. In this configuration, the captured images should obtain the maximum information about the drone movement in the horizontal plane. Additionally, the images are taken with approximately 3 frames per second, which is a slow rate but sufficient to obtain enough superposition between each pair of images, as well as easy to replicate.

Unlike the previous work, the current study shows variation in the drone's horizontal position and also in its altitude. In addition, the speed sampled for each flight segment comes from a uniform distribution.

Furthermore, each image has an associated file with the ground truth information of the complete status of the drone at the instant the picture was taken, which includes linear and angular speeds, linear and angular accelerations, position, latitude and longitude, and attitude of the vehicle. We explicitly annotated the images with this set of information for the prediction of the vehicle displacement.

4.2. *Training*

For training our model, we employed ReLU activation functions throughout the network, except in the last layer, where we used a linear activation function. Additionally, we used batch normalization layers after every weight layer and dropout layers with a probability of 0.2.

We trained the network on the Tensorflow framework and applied the Adam optimizer with a learning rate of 0.001. For our loss function, we used mean squared error, and we trained the network for 30 epochs. At the end of each epoch during the training stage, the model was evaluated against a validation group. The model that showed an improvement in the value of the metric (considering the validation group) was saved as the best model found. This was the model used for the testing stage.

The dataset was divided into three sets for training: training, validation, and testing. To make up the test set, one flight made on each map of each environment was preserved with the complete sequence of images. In this way, the prediction and its impact during the flight could be evaluated. For the other sets, 80% was set aside for the training set and 20% for the validation set. In the training stage, the sets used were the training and validation sets. Thus, the training set was presented to the model, and at the end of each epoch, the model was evaluated on the validation set. If it showed an improvement in the value of the metric, the model was saved as the best model found so far.

Additionally, during training, the image pairs were shuffled before being presented to the neural network: This ensures that we do not feed correlated data to the model, as the image pairs seen in a single batch will not all be from the same flight or will not have been taken in a sequence by the camera mounted on the drone.

4.3. Test

The model generated in the training stage was evaluated in the test stage. In this stage, we used one route from each map (and from each environment) to predict displacement, considering execution during a flight. Table 2 shows the size of each route used in this stage. The name of the route was composed of the name of the environment and the conversion of the map into a numerical value (i.e., A is 1 and B is 2).

Table 2. Details of the name and size of each route used in the test stage.

Name	Flight Size (m)
Downtown1	346.03
Downtown2	674.80
Downtown3	458.49
Forest1	783.77
Forest2	321.01
Forest3	1001.24
Mountains1	906.62
Mountains2	784.24
Mountains3	569.84

Three levels of prediction were made for the final parts of each route. This approach allowed us to better represent the negative impact of prediction on the completion of the mission. The three levels of prediction corresponded to 20%, 40%, and 60%. Finally, the flight was also predicted for the entire route, assuming that only the initial displacement coordinate was known.

It is important to note that although the drone is subject to unknown influences (such as weather conditions) during flight, causing it to move in an undesired direction, we used the yaw (these data can be provided by sensors such as a gyroscope) data to calculate the drone's new geographical position. This approach allowed us to use the geographical position reported by the drone's flight system as the "expected position". The geographical position indicated in our proposal is the "inferred position", which is used by the drone if there is no GNSS signal. By using data from different sources, we aimed to maintain the integrity of the evaluation stage.

5. Results

We emphasize that we conducted experiments to assess the proposed method's ability to estimate the displacement based on the received inputs. Therefore, the method was used to infer the aircraft's displacement, and we calculated the new geographical coordinates (lat and long). In this case study, we considered that heading and altitude are obtained by other variables and sensors, such as yaw and radio altimeter.

Firstly, Figure 3 shows the routes used in this test stage. The green dot on each route indicates the starting point of the flight. The blue line outlines the expected route, while the red line represents the route taken using the displacement estimation model. In the flights shown in this picture, doom was suffered in the final 20% of the flight. It is important to note that the flights used speed variations, meaning that the distance covered during the prediction varied in size.

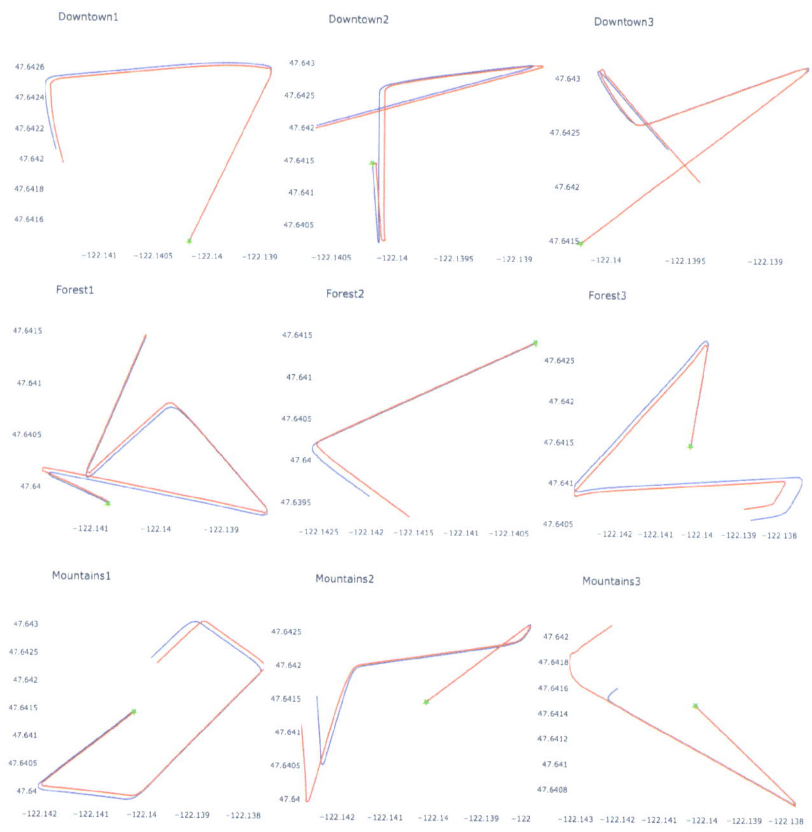

Figure 3. Example of expected routes (blue line) compared with routes that used prediction (red line) for 20% of the flight. The green dot signals the starting position of the flight.

More details of the flights can be seen in Table 3, which summarizes the size of the route, the number of predictions made, the distance of the drone's final position from the expected position, and the average distance between each inferred and expected geographical positions of the drone. It should be noted that this last piece of information is calculated based only on the period in which the displacement was inferred. Tables 4–6 present similar information to Table 3. This makes it possible to compare the changes that occurred at different points in the prediction.

Table 3. Summarization of results for flights with 20% predictions.

Route	Number of Predictions	Distance between Endpoints (m)	Average Distance between Expected and Predicted Position (m)
Downtown1	136	11.15	5.03
Downtown2	193	5.07	3.99
Downtown3	111	36.71	17.74
Forest1	72	1.17	6.28
Forest2	72	42.66	27.74
Forest3	83	19.14	21.17
Mountains1	197	13.32	21.17
Mountains2	275	52.84	43.68
Mountains3	195	55.71	48.33

Table 4. Summarization of results for flights with 40% predictions.

Route	Number of Predictions	Distance between Endpoints (m)	Average Distance between Expected and Predicted Position (m)
Downtown1	271	8.64	7.28
Downtown2	386	28.55	19.48
Downtown3	221	16.87	27.93
Forest1	143	1.00	5.94
Forest2	143	65.17	36.05
Forest3	166	36.05	23.79
Mountains1	393	22.46	23.33
Mountains2	550	94.10	56.95
Mountains3	390	58.78	34.87

Table 5. Summarization of results for flights with 60% predictions.

Route	Number of Predictions	Distance between Endpoints (m)	Average Distance between Expected and Predicted Position (m)
Downtown1	406	7.44	12.18
Downtown2	579	7.22	18.19
Downtown3	331	24.75	36.70
Forest1	214	15.96	16.82
Forest2	214	91.74	48.61
Forest3	249	34.53	19.67
Mountains1	589	28.07	32.06
Mountains2	825	129.05	69.73
Mountains3	585	49.05	49.85

Table 6. Summarization of results for flights with 100% predictions.

Route	Number of Predictions	Distance between Endpoints (m)	Average Distance between Expected and Predicted Position (m)
Downtown1	677	17.73	19.28
Downtown2	985	16.96	27.06
Downtown3	551	30.40	29.33
Forest1	357	15.58	12.23
Forest2	357	81.69	25.93
Forest3	415	48.94	28.81
Mountains1	981	22.82	15.31
Mountains2	1375	127.61	54.67
Mountains3	975	146.51	115.04

In these results, it was impossible to find a clear link between the number of predictions and the metrics relating to comparing the final position and the average positional error during in-flight predictions. In other words, even if the number of predictions is higher on one flight than on another, it does not mean the errors will also be higher. This growth characteristic between the metrics mentioned could not be affirmed when we analyzed the increase in the prediction period considering isolated flights. This behavior occurs because, presumedly, the errors are generated by a normal distribution with the center close to the expected value, causing predictions with positive (greater than expected) and negative (less than expected) error polarities. Figure 4 makes it possible to compare the predicted values with the expected values, preserving the order in which the predictions were made during the flight.

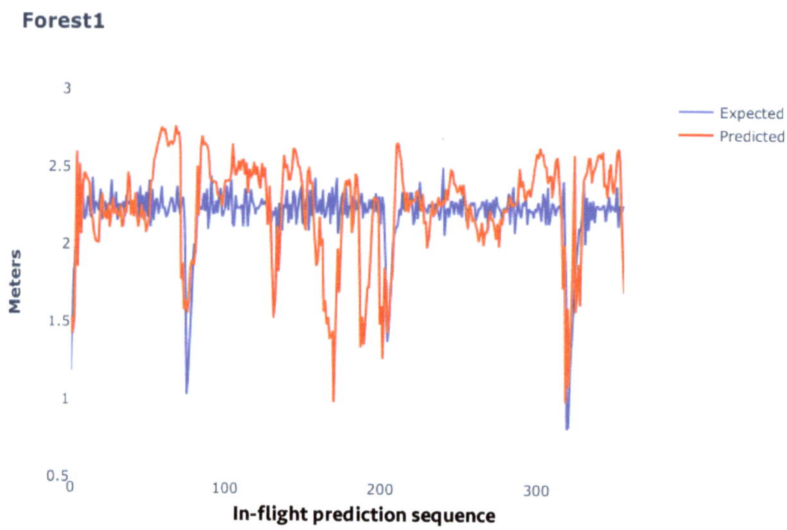

Figure 4. Predicted and expected displacement values along the route Forest1.

With the variation in environments and routes used in the experiments presented, the SiaN-VO method is capable of predicting displacement in a variety of conditions (even with a change in altitude). Another interesting feature presented by the method in the data set is its ability not to accumulate noise in its prediction.

6. Conclusions

Inferring the position of a UAV with high accuracy without the use of a GNSS is an obstacle to several studies in the scientific literature. The increasing evolution of UAVs and the high range of possible contexts in which they can be applied further highlights the need for independence from the GNSS signals for safe navigation.

The evolution provided by the proposed new architecture, giving rise to the SiaN-VO method, could infer the drone's displacement and allow the value to be used to calculate the new geographical position. SiaN-VO predicts based on a sequential pair of ground images, altitude and yaw data.

The results suggest that noise does not accumulate (or is imperceptible) in the predictions during flights on the routes described. Although we bear in mind that it is possible that noise accumulation becomes noticeable on long journeys and that this characteristic needs to be investigated further, we would like to emphasize that this characteristic is exciting and makes us believe that the SiaN-VO method is a promising approach in the field of visual odometry.

Furthermore, it is important to emphasize that the SiaN-VO method overcame the obstacle of predicting drone displacement on routes with altitude variation. The successful displacement prediction in these flight characteristics means that SiaN-VO surpasses the previous proposal, which exclusively uses images. Another point that surpasses the current approach is the possibility of using images with smaller dimensions (1/4 of the previous size), which allows for less computational processing.

Given the results obtained in this work, we aim to advance our studies in the following directions:

- Evaluate the performance of SiaN-VO with real flight data;
- Evaluate SiaN-VO on long routes and with more variations in flight dynamics;
- Measure the performance of SiaN-VO in the face of data failures and inconsistencies;
- Test our proposed model during a UAV flight, effectively embedding the network in the aircraft.

Author Contributions: Conceptualization, B.S.F. and F.A.N.V.; methodology, B.S.F. and F.A.N.V.; software, B.S.F. and F.A.N.V.; validation, B.S.F., C.A.C.M. and F.A.N.V.; formal analysis, B.S.F., C.A.C.M. and F.A.N.V.; investigation, B.S.F. and F.A.N.V.; resources, B.S.F. and F.A.N.V.; data curation, B.S.F.; writing—original draft preparation, B.S.F., C.A.C.M. and F.A.N.V.; writing—review and editing, B.S.F., C.A.C.M. and F.A.N.V.; visualization, B.S.F. and F.A.N.V.; supervision, F.A.N.V.; project administration, F.A.N.V.; funding acquisition, F.A.N.V. All authors have read and agreed to the published version of the manuscript.

Funding: This research was funded by Embraer and FAPESP (grant number: 2021/06872-6).

Institutional Review Board Statement: Not applicable.

Informed Consent Statement: Not applicable.

Data Availability Statement: The source code for this work can be accessed at https://github.com/verri/sian-vo (accessed on 1 January 2024), and the dataset for training, validation, and testing is available at http://doi.org/10.5281/zenodo.10140028.

Acknowledgments: The authors would like to thank Leonardo Silveira for their contribution to generating the dataset.

Conflicts of Interest: The authors declare no conflicts of interest.

References

1. Oruc, A. Potential cyber threats, vulnerabilities, and protections of unmanned vehicles. *Drone Syst. Appl.* **2022**, *10*, 51–58. [CrossRef]
2. Moore, A.B.; Johnson, M. Geospatial Support for Agroecological Transition through Geodesign. In *Drones and Geographical Information Technologies in Agroecology and Organic Farming Contributions to Technological Sovereignty*; CRC Press: Boca Raton, FL, USA, 2022; pp. 174–203.
3. Mohsan, S.A.H.; Khan, M.A.; Noor, F.; Ullah, I.; Alsharif, M.H. Towards the Unmanned Aerial Vehicles (UAVs): A Comprehensive Review. *Drones* **2022**, *6*, 147. [CrossRef]
4. Braga, J.R.G.; Velho, H.F.C.; Conte, G.; Doherty, P.; Shiguemori, É.H. An image matching system for autonomous UAV navigation based on neural network. In Proceedings of the 2016 14th International Conference on Control, Automation, Robotics and Vision (ICARCV), Phuket, Thailand, 13–15 November 2016; pp. 1–6. [CrossRef]
5. da Penha Neto, G.; de Campos Velho, H.F.; Shiguemori, E.H. UAV Autonomous Navigation by Image Processing with Uncertainty Trajectory Estimation. In Proceedings of the 5th International Symposium on Uncertainty Quantification and Stochastic Modelling, Rouen, France, 29 June–3 July 2021; De Cursi, J.E.S., Ed.; Springer: Cham, Switzerland, 2021; pp. 211–221.
6. Gupta, A.; Fernando, X. Simultaneous Localization and Mapping (SLAM) and Data Fusion in Unmanned Aerial Vehicles: Recent Advances and Challenges. *Drones* **2022**, *6*, 85. [CrossRef]
7. Noviello, C.; Gennarelli, G.; Esposito, G.; Ludeno, G.; Fasano, G.; Capozzoli, L.; Soldovieri, F.; Catapano, I. An Overview on Down-Looking UAV-Based GPR Systems. *Remote Sens.* **2022**, *14*, 3245. [CrossRef]
8. Rathnayake, B.S.S.; Ranathunga, L. Lane Detection and Prediction under Hazy Situations for Autonomous Vehicle Navigation. In Proceedings of the 2018 18th International Conference on Advances in ICT for Emerging Regions (ICTer), Colombo, Sri Lanka, 6–7 September 2018; pp. 99–106. [CrossRef]
9. Bergman, N.; Ljung, L.; Gustafsson, F. Terrain Navigation Using Bayesian Statistics. *IEEE Control Syst. Mag.* **1999**, *19*, 33–40. [CrossRef]
10. Titterton, D.H.; Weston, J.L. *Strapdown Inertial Navigation Technology*, 2nd ed.; The Institution of Engineering and Technology: Stevenage, UK, 2004.

11. Krizhevsky, A.; Sutskever, I.; Hinton, G.E. ImageNet Classification with Deep Convolutional Neural Networks. In *Advances in Neural Information Processing Systems*; Pereira, F., Burges, C., Bottou, L., Weinberger, K., Eds.; Curran Associates, Inc.: Nice, France, 2012; Volume 25.
12. He, K.; Zhang, X.; Ren, S.; Sun, J. Deep Residual Learning for Image Recognition. In Proceedings of the IEEE Conference on Computer Vision and Pattern Recognition, Las Vegas, NV, USA, 27 June–30 June 2016.
13. Goodfellow, I.; Bengio, Y.; Courville, A. *Deep Learning*; MIT Press: Cambridge, MA, USA, 2016. Available online: http://www.deeplearningbook.org (accessed on 31 January 2020).
14. Silveira, L.; Rodrigues, M.; Faiçal, B.S.; da Silva, A.S.Q.; Marcondes, C.; Maximo, M.R.O.A.; Verri, F.A.N. Navigation Aids Based on Optical Flow and Convolutional Neural Network. In Proceedings of the 2022 Latin American Robotics Symposium (LARS), 2022 Brazilian Symposium on Robotics (SBR), and 2022 Workshop on Robotics in Education (WRE), São Bernardo do Campo, Brazil, 18–21 October 2022; pp. 318–323. [CrossRef]
15. Aqel, M.O.; Marhaban, M.H.; Saripan, M.I.; Ismail, N.B. Review of visual odometry: Types, approaches, challenges, and applications. *SpringerPlus* **2016**, *5*, 1897. [CrossRef] [PubMed]
16. Yu, Y.; Hua, C.; Li, R.; Li, H. Pose Estimation Method Based on Bidirectional Recurrent Neural Network in Visual Odometry. Available online: https://papers.ssrn.com/sol3/papers.cfm?abstract_id=4155315 (accessed on 31 January 2020).
17. Scaramuzza, D.; Fraundorfer, F. Visual Odometry [Tutorial]. *IEEE Robot. Autom. Mag.* **2011**, *18*, 80–92. [CrossRef]
18. Rehder, J.; Gupta, K.; Nuske, S.; Singh, S. Global pose estimation with limited GPS and long range visual odometry. In Proceedings of the 2012 IEEE International Conference on Robotics and Automation, Saint Paul, MI, USA, 14–18 May 2012; IEEE: Toulouse, France, 2012; pp. 627–633.
19. Cai, G.S.; Lin, H.Y.; Kao, S.F. Mobile robot localization using gps, imu and visual odometry. In Proceedings of the 2019 International Automatic Control Conference (CACS), Keelung, Taiwan, 13–16 November 2019; IEEE: Toulouse, France, 2019; pp. 1–6.
20. Geiger, A.; Lenz, P.; Urtasun, R. Are we ready for autonomous driving? the kitti vision benchmark suite. In Proceedings of the 2012 IEEE Conference on Computer Vision and Pattern Recognition, Saint Paul, MI, USA, 16–21 June 2012; IEEE: Toulouse, France, 2012; pp. 3354–3361.
21. Hui, T.-W.; Tang, X.; Loy, C.C. A lightweight optical flow CNN—Revisiting data fidelity and regularization. *IEEE Trans. Pattern Anal. Mach. Intell.* **2020**, *43*, 2555–2569. [CrossRef] [PubMed]
22. Menze, M.; Geiger, A. Object scene flow for autonomous vehicles. In Proceedings of the IEEE Conference on Computer Vision and Pattern Recognition, Boston, MA, USA, 7–12 June 2015; pp. 3061–3070.
23. Olawoye, U.; Gross, J.N. UAV Position Estimation Using a LiDAR-based 3D Object Detection Method. In Proceedings of the 2023 IEEE/ION Position, Location and Navigation Symposium (PLANS), Monterey, CA, USA, 24–27 April 2023; pp. 46–51. [CrossRef]
24. Araveti, V.; Vats, U.; Chiddarwar, S.S. Fully End-to-End Visual Odometry of a Minidrone. In *Advances in Mechanism and Machine Science*; Okada, M., Ed.; Springer: Cham, Switzerland, 2023; pp. 775–788.
25. Goppert, J.; Yantek, S.; Hwang, I. Invariant Kalman filter application to optical flow based visual odometry for UAVs. In Proceedings of the 2017 Ninth International Conference on Ubiquitous and Future Networks (ICUFN), Milan, Italy, 4–7 July 2017; IEEE: Toulouse, France, 2017; pp. 99–104.
26. Romero, H.; Salazar, S.; Santos, O.; Lozano, R. Visual odometry for autonomous outdoor flight of a quadrotor UAV. In Proceedings of the 2013 International Conference on Unmanned Aircraft Systems (ICUAS), Atlanta, GA, USA, 28–31 May 2013; IEEE: Toulouse, France, 2013; pp. 678–684.
27. Simonyan, K.; Zisserman, A. Very Deep Convolutional Networks for Large-Scale Image Recognition. *arXiv* **2014**, arXiv:1409.1556. [CrossRef]
28. Shah, S.; Dey, D.; Lovett, C.; Kapoor, A. AirSim: High-Fidelity Visual and Physical Simulation for Autonomous Vehicles. *arXiv* **2017**, arXiv:1705.05065. [CrossRef]

sensors

Article

Attention Mechanism and LSTM Network for Fingerprint-Based Indoor Location System

Zhen Wu *, Peng Hu, Shuangyue Liu and Tao Pang

Department of Mobile Communications and Terminal Research, China Telecom Research Institute, Guangzhou 510000, China; hup11@chinatelecom.cn (P.H.); liusy24@chinatelecom.cn (S.L.); pangtao@chinatelecom.cn (T.P.)
* Correspondence: wuz16@chinatelecom.cn

Abstract: The demand for precise indoor localization services is steadily increasing. Among various methods, fingerprint-based indoor localization has become a popular choice due to its exceptional accuracy, cost-effectiveness, and ease of implementation. However, its performance degrades significantly as a result of multipath signal attenuation and environmental changes. In this paper, we propose an indoor localization method based on fingerprints using self-attention and long short-term memory (LSTM). By integrating a self-attention mechanism and LSTM network, the proposed method exhibits outstanding positioning accuracy and robustness in diverse experimental environments. The performance of the proposed method is evaluated under two different experimental scenarios, which involve 2D and 3D moving trajectories, respectively. The experimental results demonstrate that our approach achieves an average localization error of 1.76 m and 2.83 m in the respective scenarios, outperforming the existing state-of-the-art methods by 42.67% and 31.64%.

Keywords: fingerprinting; indoor localization system; long short-term memory (LSTM); self-attention mechanism

1. Introduction

The rapid development of global digitization has created a high demand for location-based services (LBS) in many industries [1]. These services have become essential for various systems and applications, including transportation [2], logistics [3,4], emergency response [5], etc. [6,7]. In outdoor environments, mobile users already have access to established outdoor positioning technologies such as the Global Positioning System (GPS) [8] and the BeiDou Satellite Navigation System (BDS) [9] to obtain accurate location information. However, the effectiveness of these technologies is often limited in indoor environments due to the scattering and attenuation effects of satellite signals.

In the field of indoor localization, various wireless signals have been proposed and utilized, including Wi-Fi [10–13], Bluetooth [14,15], ultra-wide bandwidth (UWB) [16,17], radio frequency identification (RFID) [18], and custom radios [19]. Typical ranging-based methods for processing wireless signals in indoor localization involve using information such as angle of arrival (AOA) or time of arrival (TOA) to estimate the specific positions of the user equipment (UE) [20]. However, these methods require prior knowledge of the locations of access points (APs) and are susceptible to errors in the distance measurement between the UE and APs, which can negatively impact the accuracy of the positioning. In contrast to these methods, the fingerprint-based indoor localization method is characterized by simplicity and efficiency [21]. This technique relies on the unique characteristics of wireless signals in indoor environments to create a map or "fingerprint" of the received signal strength indicator (RSSI) at different locations. The fingerprint can then be used to estimate the position of the UE based on the signal strengths measured at that location. Fingerprint-based methods are highly accurate and can offer sub-meter-level positioning accuracy in many cases, making them a promising alternative to ranging-based methods.

However, in the context of fingerprint-based methods, the radio propagation environment introduces multi-path effects, shadowing, signal fading, and other forms of signal degradation and distortion, leading to significant fluctuations in RSSI values. In the experiments described in this paper, the observed RSSI values for different APs at a fixed location exhibit a wide range of fluctuations, as illustrated in Figure 1. The fluctuation in RSSI makes it challenging to discern the pattern of RSSI between the test points (TPs) and reference points (RPs), thereby significantly impacting the accuracy of positioning.

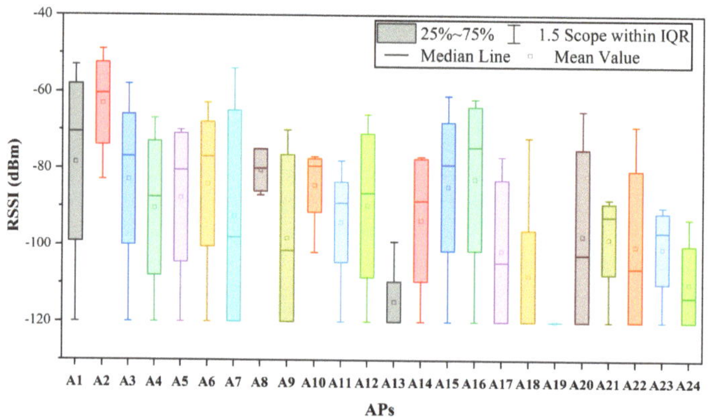

Figure 1. The range of variation in RSSI for APs observed at a fixed location.

With the development of machine learning algorithms in recent decades, numerous machine learning algorithms have been proven to be effective in recognizing the RSSI pattern [22]. M. Brunato et al. proposed applying support vector machines (SVMs) in location fingerprint positioning systems [23]. Hoang et al. introduced a soft range-limited k-nearest neighbors (KNN) fingerprinting algorithm that addresses spatial ambiguity in localization by scaling the fingerprint distance with a range factor based on the physical distance between the previous position of users and the reference location in the database [24]. Fang et al. utilized feedforward neural networks (FNNs) to extract fingerprint features from the RSSI, enabling the accurate localization of the actual position [25]. However, the performance of these algorithms can easily be limited when learning features in complex indoor environments. To achieve superior performance, some research studies have suggested using long short-term memory (LSTM) for handling sequential trajectory prediction in indoor localization systems [10,26,27], which has been experimentally demonstrated to be more effective than the conventional KNN method. Meanwhile, self-attention has been proposed as a promising technique for enhancing the performance of sequence processing tasks [28–31]. By enabling the model to attend to various regions of the input sequence, self-attention improves its capacity to capture the connections between various features in a sequence.

This paper introduces a novel method named self-attention and LSTM (SA-LSTM) that effectively improves the positioning accuracy and robustness. We conducted experiments in two different scenarios to validate the effectiveness and robustness of the proposed approach. The experimental results demonstrate that SA-LSTM exhibits greater robustness and higher accuracy in indoor localization compared to some of the most advanced algorithms.

The main contributions of this paper are as follows:

1. We propose a novel deep neural network that integrates the self-attention mechanism and LSTM networks for indoor localization. The proposed SA-LSTM method processes the RSSI values of consecutive time instances and predicts the position at the final moment in the input sequence. To the best of our knowledge, this is the

first time that the self-attention mechanism and LSTM networks have been fused for RSSI-based fingerprinting localization.

2. Based on LSTM, the SA-LSTM introduces the self-attention mechanism, which enables the LSTM to effectively capture the interdependencies between the RSSI values at different time instances, thereby facilitating the improved extraction of location information and reducing the localization error.
3. We conducted a comparative analysis between our proposed model and several state-of-the-art methods. The experimental results reveal that our proposed SA-LSTM model achieves the highest localization accuracy in both experimental scenarios, demonstrating its robustness and precision.

The rest of this paper is structured as follows. Section 2 provides an overview of related works in the area of fingerprint indoor localization systems. Section 3 presents the technical details of our proposed model. Section 4 outlines the experimental setup utilized in our study. Section 5 presents and analyzes the experimental results obtained from various datasets. Finally, Section 6 offers concluding remarks and outlines our future research plans.

2. Related Work

In this section, we present an overview of the existing research on fingerprint-based indoor localization and the application of self-attention mechanisms.

In the current landscape, numerous clustering-based and ensemble-based models have been applied in the field of fingerprint-based indoor localization. In terms of clustering-based models, Bahl et al. [32] were the first to propose the utilization of the KNN algorithm in fingerprint-based indoor localization. By evaluating the Euclidean distance of the RSSI vector from multiple base stations, the KNN algorithm assigns the nearest reference points for target points and computes the average position as their predicted positions. Expanding on KNN, Brunato et al. [23] introduced weighted KNN (WKNN), which calculates the weighted average of reference point positions and enhances the overall positioning accuracy. Within the realm of ensemble-based models, Jedari E. et al. [33] employed a random forest classifier for RSSI-based indoor positioning. Experimental outcomes demonstrate that the random forest classifier outperforms KNN in terms of positioning accuracy. Furthermore, the effectiveness of the AdaBoost method is validated in [34], where AdaBoost is utilized to leverage the channel state information (CSI) from Wi-Fi signals for localization. In another study, Singh N. et al. [35] presented an indoor localization scheme based on XGBoost, capable of accurately classifying the positions of mobile devices in indoor environments, achieving an average positioning error of 4.93 m, 7.02 m, and 1.5 m in three different environments. Moreover, Tekler Z. D. et al. [36] proposed a supervised ensemble model and a semi-supervised clustering model and evaluation revealed that the supervised ensemble model outperforms in terms of positioning accuracy.

Except for ensemble-based models, Yerbolat Khassanov et al. explored the use of end-to-end sequence models for Wi-Fi-based indoor localization at a finer level [10]. The study showed that the localization task can be effectively formulated as a sequence learning problem using recurrent neural networks (RNNs) with regression output. The use of regression output allows for estimating three-dimensional positions and enables scalability to larger areas. The experiments conducted on the Wi-Fi dataset reveal that RNN models outperform non-sequential models such as KNN and FNN, achieving an average positioning error of 3.05 m for finer-level localization tasks. Furthermore, Zhenghua Chen et al. proposed a deep LSTM network for indoor localization using Wi-Fi fingerprinting [37]. The network incorporates a local feature extractor that enables the encoding of temporal dependencies and the learning of high-level representations based on the extracted sequential local features. The experimental results demonstrate that the proposed approach achieves state-of-the-art localization performance, with mean localization errors of 1.48 m and 1.75 m in research lab and office environments, respectively.

In the field of neural machine translation tasks, Bahdanau et al. introduced the self-attention mechanism to the encoder–decoder model. This enables the model to learn alignment and translation simultaneously, allowing for the adaptive selection of encoded vectors [38]. Building on the effectiveness of the self-attention mechanism, several other deep learning architectures have been redesigned to incorporate self-attention for performance enhancement. Yang C. H. et al. [39] integrated self-attention into DNN to effectively improve the adversarial speech signals. Additionally, Mittag G. et al. [40] proposed a deep CNN-self-attention model for multidimensional speech quality prediction, which outperformed CNN. Moreover, an LSTM structure based on the self-attention mechanism was introduced in [41], which showed a superior performance in forecasting temporal sequences compared to other benchmark methods.

In general, LSTM has demonstrated exceptional performance in sequence prediction tasks, including fingerprint localization. It has been experimentally verified that it outperforms conventional methods such as KNN and WKNN. Additionally, the self-attention mechanism enables the model to consider the relationship between each element in the sequence. This leads to a better understanding of contextual information and a more precise processing of sequence data. Based upon that, we propose an SA-LSTM model with high accuracy and strong robustness for indoor localization systems based on fingerprinting.

3. Methodology

In this section, we will begin by introducing the framework of the SA-LSTM-based localization algorithm. Subsequently, we will provide detailed introductions to the working principles of its subcomponents.

3.1. SA-LSTM-Based Localization Algorithm

Figure 2 illustrates the framework of the SA-LSTM-based localization algorithm, comprising an offline training stage and an online estimation stage. During the offline training stage, the RSSI values collected at different points and their corresponding coordinates of locations are recorded and stored in the fingerprint database. Subsequently, the collected RSSI data are normalized and used to train the SA-LSTM network. The trainable weights of the SA-LSTM network will be updated to minimize the loss between the output and the ground true locations. The trainable weights of the SA-LSTM network are adjusted to minimize the loss between the output and the actual locations. During the online estimation stage, real-time RSSI data from the device are normalized and input into the trained SA-LSTM model, which then generates real-time location estimates.

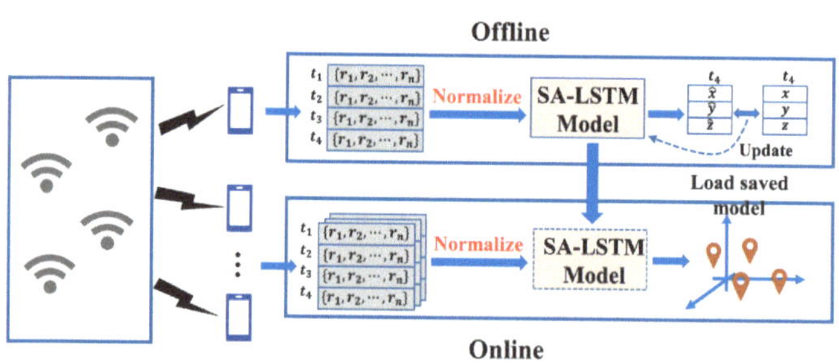

Figure 2. The framework of SA-LSTM-based localization algorithm. t_i: the ith time slice. r_i: the RSSI value from AP i. $(\hat{x}, \hat{y}, \hat{z})$: the coordinates of predicted positions. (x, y, z): the coordinates of real positions.

3.2. LSTM Network

LSTM is a unique form of recurrent neural network that has been extensively researched in deep learning. In contrast to conventional RNN, the LSTM network introduces gated states to modulate the flow of information, thereby enabling it to selectively retain relevant information over extended periods while filtering out irrelevant data, which allows it to effectively analyze the long temporal sequences.

Figure 3 shows the common architecture of LSTM, which is composed of connected memory units. In this context, $\boldsymbol{C_t}$ and $\boldsymbol{H_t}$ represent the unit state and hidden state at time t, respectively. Focus on the time t, the memory unit receives the $\boldsymbol{C_{t-1}}$ and $\boldsymbol{H_{t-1}}$ from the previous memory unit, as well as the current input value x_t. After performing internal arithmetic operations, the unit generates the updated cell state $\boldsymbol{C_t}$ and hidden state $\boldsymbol{H_t}$, which are subsequently passed on to the next memory unit. The hidden state $\boldsymbol{H_t}$ also serves as the output result y_t corresponding to the current time step.

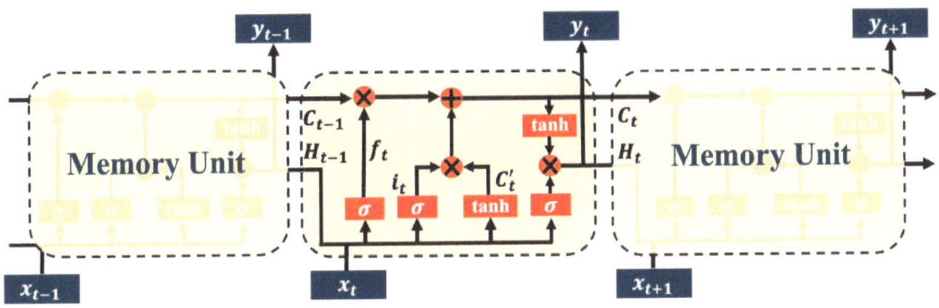

Figure 3. Architecture of LSTM.

Each memory unit in the LSTM architecture comprises three components: a forget gate, an input gate, and an output gate. The forget gate can be mathematically expressed as follows:

$$f_t = \sigma(\boldsymbol{W_f}[\boldsymbol{H_{t-1}}, x_t] + \boldsymbol{b_f}) \tag{1}$$

Here, σ represents the activation function, while $\boldsymbol{W_f}$ and $\boldsymbol{b_f}$ denote the weights and bias of the forget gate, respectively. By multiplying with $\boldsymbol{C_{t-1}}$, the forget gate aims to decide what information should be forgotten in it. For the implementation of the input gate, the sigmoid activation function [42] is initially employed to determine the values that require updating, as illustrated in (2), where $\boldsymbol{W_i}$ and $\boldsymbol{b_i}$ are the weight matrices and the bias. Subsequently, the tanh activation function generates a new candidate value, denoted by $\boldsymbol{C_t'}$. The mathematical expression is shown in (3), where $\boldsymbol{W_c}$ and $\boldsymbol{b_c}$ represent the weight matrices and the bias, respectively.

$$\boldsymbol{i_t} = \sigma(\boldsymbol{W_i}[\boldsymbol{H_{t-1}}, x_t] + \boldsymbol{b_i}) \tag{2}$$

$$\boldsymbol{C_t'} = tanh(\boldsymbol{W_c}[\boldsymbol{H_{t-1}}, x_t] + \boldsymbol{b_c}) \tag{3}$$

These two stages are subsequently combined to generate an updated state value, which is then added to the unit state to update the long-term memory of LSTM (i.e., $\boldsymbol{C_t}$), as indicated by the following equation:

$$\boldsymbol{C_t} = f_t \odot \boldsymbol{C_{t-1}} + \boldsymbol{i_t} \odot \boldsymbol{C_t'} \tag{4}$$

$\odot$ represents the Hadamard product operation. The output gate is responsible for generating the hidden state, which can be calculated as:

$$\boldsymbol{H_t} = \sigma(\boldsymbol{W_o}[\boldsymbol{H_{t-1}}, x_t] + \boldsymbol{b_o}) \odot tanh(\boldsymbol{C_t}) \tag{5}$$

where $\boldsymbol{W_o}$ and $\boldsymbol{b_o}$ are the weight matrix and the bias of the output gate. LSTM is capable of selectively memorizing and forgetting features via the regulation of three gates, thereby mitigating the issue of long-term dependency. Additionally, LSTM addresses the issue of vanishing gradients that often occurs in RNN. As a result, LSTM has gained widespread adoption in time series prediction tasks.

3.3. Self-Attention Mechanism

The attention mechanism is inspired by the human visual attention mechanism, which selectively focuses on specific regions of interest and allocates more attentional resources to extract relevant information while suppressing irrelevant information. Self-attention is a type of attention mechanism, which enables the model to capture the degree of association between each position in a sequence and all other positions. By computing the attention weight of each position with respect to all other positions, the model is able to selectively focus on the most relevant parts of the input sequence and generate more precise predictions or representations.

The self-attention mechanism is based on the query matrix $\boldsymbol{Q}$, the key matrix $\boldsymbol{K}$, and the value matrix $\boldsymbol{V}$, the generation of which is depicted in Figure 4. Given an input sequence $\boldsymbol{X}$, the attention mechanism employs three trainable weight matrices (corresponding to $\boldsymbol{W_Q}$, $\boldsymbol{W_K}$, and $\boldsymbol{W_V}$ in Figure 4) to compute the query matrix, the key matrix, and the value matrix $\boldsymbol{V}$, respectively. By computing the dot product between $\boldsymbol{Q}$ and $\boldsymbol{K}$, and normalizing the resulting scores using a softmax function, the attention weight coefficients can be obtained, which can be expressed as:

$$AW(\boldsymbol{Q},\boldsymbol{K}) = softmax\left(\frac{\boldsymbol{QK}^T}{\sqrt{d}}\right) \tag{6}$$

where d refers to the dimension of the hidden layer in the key and query matrices. Due to the potentially large dot product of the query matrix $\boldsymbol{Q}$ and the key matrix $\boldsymbol{K}$ when their dimensions are high, numerical instability may occur during training. To address this issue, dividing the dot product by $\sqrt{d}$ normalizes the scale of the product across all dimensions, enhancing the stability and performance of the model. Furthermore, based on the attention weight $AW(\boldsymbol{Q},\boldsymbol{K})$, the attention value can be expressed as:

$$A(\boldsymbol{Q},\boldsymbol{K},\boldsymbol{V}) = softmax\left(\frac{\boldsymbol{QK}^T}{\sqrt{d}}\right)\boldsymbol{V} \tag{7}$$

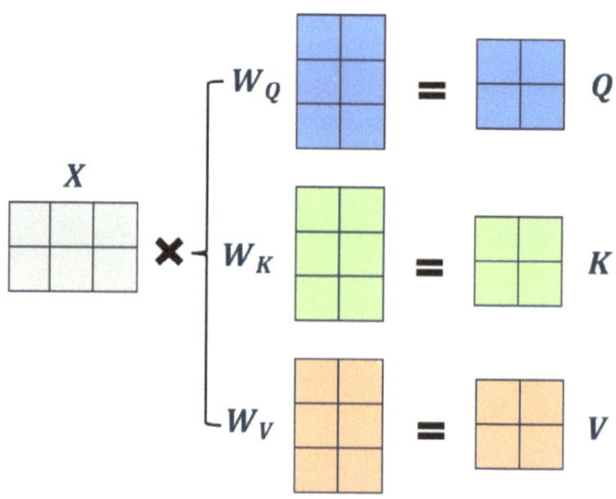

Figure 4. The generation of the $\boldsymbol{Q}$, $\boldsymbol{K}$, and $\boldsymbol{V}$ matrices.

Specifically, for each position in the sequence, the corresponding value vector is multiplied by its attention weight coefficient. The resulting products are then summed to obtain the attention value, allowing the model to place greater emphasis on the most relevant positions. This process is illustrated in Figure 5, where $\{\alpha_{i,1}, \alpha_{i,2}, \cdots, \alpha_{i,d}\}$ represents the attention weight coefficients.

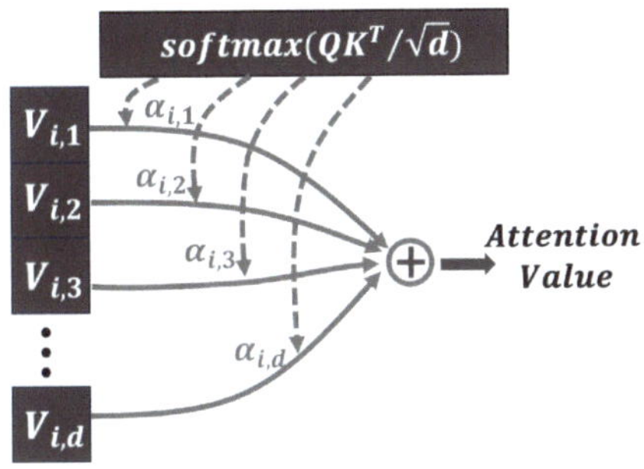

Figure 5. The generation of the attention value.

3.4. Proposed SA-LSTM Network

Based on the LSTM model and the self-attention mechanism, this paper proposes an SA-LSTM model for indoor localization enhancement. The framework of the SA-LSTM model is depicted in Figure 6. The input data for SA-LSTM are constructed using the collected RSSI data.

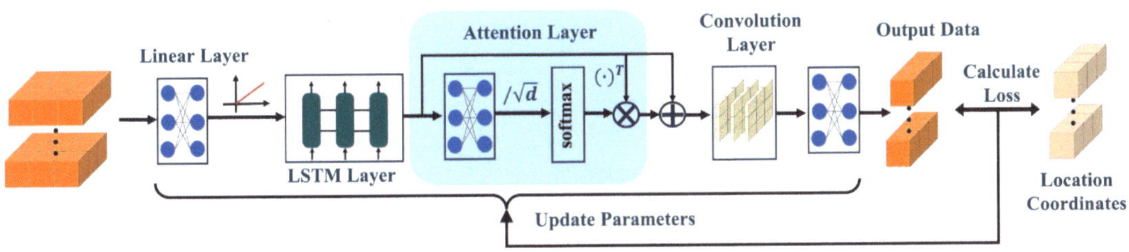

Figure 6. The framework of proposed SA-LSTM model.

3.4.1. Input Sequence Data

At first, a recorded trajectory can be expressed as a matrix:

$$R = \begin{bmatrix} r_{1,1} & r_{1,2} & \cdots & r_{1,N} \\ r_{2,1} & r_{2,2} & \cdots & r_{2,N} \\ \vdots & \vdots & \ddots & \vdots \\ r_{T,1} & r_{T,2} & \cdots & r_{T,N} \end{bmatrix} \tag{8}$$

In this context, N refers to the total number of APs, while T represents the length of a trajectory. Each element in the matrix $\boldsymbol{R}$ corresponds to the received RSSI. To prepare the data for analysis, we apply the normalization method described in [43]. This involves using the following expression:

$$r'_{i,j} = \left(\frac{r_{i,j} - c}{-c}\right)^e \tag{9}$$

where e represents the Euler's number [44]. The constant value c should be set to a number less than or equal to the minimum value of RSSI. This ensures that all RSSI values can be scaled between 0 and 1 through normalization. Once the normalization is complete, trajectory segmentation will be performed on all the collected trajectories. Considering trajectories as $[(\widetilde{\boldsymbol{r}}_1, \boldsymbol{l}_1), (\widetilde{\boldsymbol{r}}_2, \boldsymbol{l}_2), \cdots, (\widetilde{\boldsymbol{r}}_T, \boldsymbol{l}_T)]$, where $\widetilde{\boldsymbol{r}}_i = [r_{1,1}, r_{1,2}, \cdots, r_{1,N}]$ represents the RSSI from all APs in a given position, while $\boldsymbol{l}_i = [x_i, y_i]$ represents the corresponding coordinates of this position. To facilitate the analysis, each trajectory is divided into smaller segments using a sliding window of a fixed length, denoted by L. These segments are then used as inputs for the SA-LSTM model. Mathematically, this process can be expressed as follows:

$$l_{i+L} = \mathcal{F}(\widetilde{r}_i, \widetilde{r}_{i+1}, \cdots, \widetilde{r}_{i+L-1}) \tag{10}$$

where $\mathcal{F}(\cdot)$ is the mathematical expression of SA-LSTM, and the l_{i+L} represents the position of the last time step for the input data.

3.4.2. The Layers of Network

After preprocessing the data, the prepared dataset will be fed into the SA-LSTM model. The input layer of the SA-LSTM model employs a fully connected layer with a rectified linear (ReLU) activation function to increase the dimension of the feature space. Mathematically, this can be expressed as follows:

$$\boldsymbol{X}' = ReLU(\boldsymbol{W}_1\boldsymbol{X} + \boldsymbol{b}_1) \tag{11}$$

The resulting output will then be passed through an LSTM layer to generate the corresponding output for each time step. This output will serve as the input for the subsequent self-attention layer. Within the self-attention layer, several enhancements are implemented to decrease the number of network parameters. As shown in Equation (6), the attention weights are computed using the query matrix $\boldsymbol{Q}$ and the key matrix $\boldsymbol{K}$. This computation can be further simplified as follows:

$$\begin{aligned} AW(\boldsymbol{X}, \boldsymbol{W}_a) &= softmax\left(\frac{\boldsymbol{Q}\boldsymbol{K}^T}{\sqrt{d}}\right) \\ &= softmax\left(\frac{(\boldsymbol{X}\boldsymbol{W}_Q)(\boldsymbol{X}\boldsymbol{W}_K)^T}{\sqrt{d}}\right) \\ &= softmax\left(\frac{\boldsymbol{X}(\boldsymbol{W}_Q\boldsymbol{W}_K^T)\boldsymbol{X}^T}{\sqrt{d}}\right) \end{aligned} \tag{12}$$

Given the relationship $\boldsymbol{W}_A = \boldsymbol{W}_Q\boldsymbol{W}_K^T$, it follows that a fully connected layer with trainable weights $\boldsymbol{W}_A$ can be utilized in the attention layer to facilitate the computation of attention weights. Afterward, the output of the fully connected layer will be divided by $\sqrt{d}$ and normalized using the softmax function to obtain the attention weights. It is noteworthy that the output of the LSTM layer contains the information required for SA-LSTM, which means it can be directly considered as the key matrix $\boldsymbol{K}$. After calculating the attention weights, the next step involves performing a dot product operation between the attention weights and the transposed output from the LSTM layer.

SA-LSTM utilizes a shortcut connection [45] to propagate the attention values obtained from the attention layer, which enhances the backpropagation of gradients and mitigates gradient vanishing. A convolutional layer is then applied to modify the data channels before moving on to the final layer. In the final layer, a fully connected layer is employed to convert the input into location coordinates. The model then calculates the mean square error (MSE) between the predicted output $\tilde{\boldsymbol{Y}}$ and the practical location coordinates $\boldsymbol{Y}$. The loss is calculated as:

$$\mathcal{L}_{MSE}(\tilde{\boldsymbol{Y}}, \boldsymbol{Y}) = \frac{\sum_{i=1}^{n}(||\boldsymbol{Y} - \tilde{\boldsymbol{Y}}||^2)}{n} \tag{13}$$

where n denotes the number of samples in a batch. According to the loss value, the gradients of the trainable parameters in the model will be computed through backpropagation. Simultaneously, the trainable parameters will be updated in the direction of the negative gradient to minimize the loss value.

4. Experimental Setup

To verify the performance of the proposed SA-LSTM method, Bluetooth, and Wi-Fi fingerprint data are applied, which are collected from 2D and 3D moving scenarios, respectively.

4.1. Two-Dimensional-Moving Experiment Setup

The experimental location for the 2D-moving scenarios is located in an office room on the 28th floor of the Guangdong Telecom Science and Technology Building in China. In this experiment, we deployed 24 Bluetooth beacons at various locations within an office room. These beacons are used to track the movement and location of individuals. Figure 7 shows the layout of the office room, which has an area of 9.6 m × 20.4 m. The solid red dot in Figure 7 represents the origin point in a customized absolute coordinate system. The trajectories used for feature analysis are based on the coordinates of an absolute coordinate system, which serves as a reference point for all position measurements. Additionally, the green cross marks in Figure 7 represent the positions of the Bluetooth beacons, while the blue dashed line indicates the trajectories followed during data collection. The E5 Pilot Positioning Beacon version V006 is applied as the Bluetooth signal transmitter. The specific product parameters are shown in Table 1.

Table 1. The product parameters of the Bluetooth beacon.

Parameters	Values
Bluetooth version	BLE 5.0
Bluetooth protocol	iBeacon
Working temperature	−30 ∼ 25 °C
Maximum transmission distance	120 m
Transmitted power	−30 ∼ +4 dBm (default: 0 dBm)
Broadcast interval	100 ms ∼ 10 s (default: 500 ms)

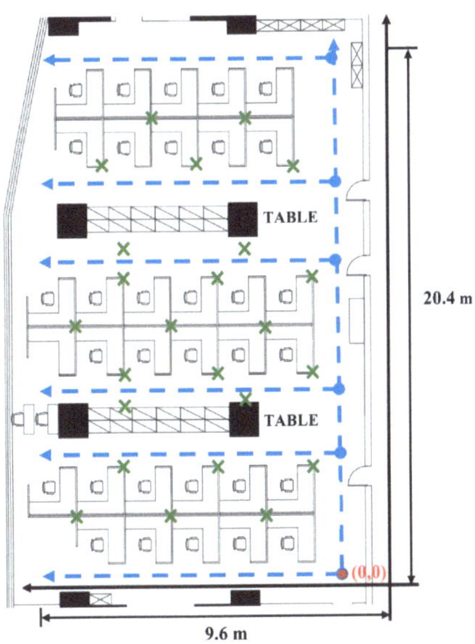

Figure 7. The layout of the office room.

During the experiment, we employed a Xiaomi 10 Pro mobile phone (Xiaomi, Beijing, China) and a ZTE Axon 40 mobile phone (ZTE, Shenzhen, China), both equipped with cameras. To facilitate the data collection task, we developed a mobile phone data collection application capable of capturing Bluetooth signals and logging user positions. In Figure 8, we depict the page of the application. This application leverages the visual simultaneous localization and mapping (VSLAM) framework to acquire real-time coordinates, which were then logged onto files for further analysis. The working principle of VSLAM involves analyzing the visual data captured by the camera to track the movement of the camera and identify features in the environment. By comparing these features with those from previous frames, VSLAM can estimate the motion of the camera and update its position in real time.

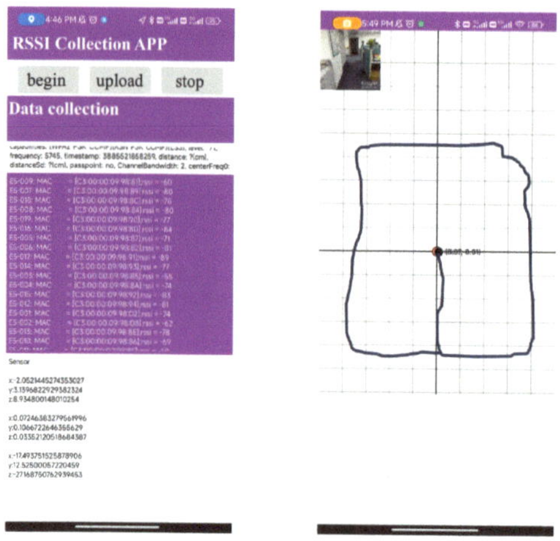

(a) (b)

Figure 8. The application used for the (**a**) RSSI collection and (**b**) position recording.

To ensure the accuracy of the collected position coordinates, we conducted data acquisition by moving the acquisition device at a constant speed along the predetermined trajectories. The trajectory data for RSSI collection were obtained by following the blue-dashed lines shown in Figure 7. Specifically, we followed each dashed line from the starting point to the end and then retraced our steps from the end back to the exit point, creating two distinct trajectories. The two mobile phones used for data acquisition were programmed to perform signal acquisition and collect corresponding addresses at different times. Overall, these measures ensured that the collected data were of sufficient quality to support our research objectives. The sampling frequency of the collecting devices was set to 1 Hz while moving along the trajectories. In total, we collected 28 trajectories, which were subsequently partitioned into three sets: training, validation, and test sets, in a ratio of 3:1:1. The test and validation datasets mainly contain two categories of trajectories. The first category consists of trajectories that were not included in the training set. The second category includes trajectories that are identical to those in the training set but were collected using different devices.

4.2. Three-Dimensional-Moving Experiment Setup

The 3D-moving experiment dataset is publicly available as an open source dataset [10]. In contrast to the 2D-moving experiment, the 3D-moving experiment dataset is based on Wi-Fi fingerprints and covers trajectories across the fourth, fifth, and sixth floors of the C4 building at Nazarbayev University. This dataset provides a comprehensive and representative set of data, enabling a thorough evaluation of the performance of indoor localization systems in complex, multi-floor environments. This Wi-Fi dataset comprises 290 trajectories that were sequentially collected with a fine spatiotemporal resolution. The dataset covers a total area of over 9564 m^2 across three floors. The experimental environment is equipped with 439 wireless access points. During the experiment, the validation and test trajectories were collected a few days after obtaining the training set.

These trajectories were uniquely designed to be dissimilar from the training trajectories. Moreover, the users were authorized to switch floors using the four elevators installed in the building while collecting the data, which helps to evaluate the performance of the model in 3D-moving scenarios. A total of 170 unique trajectories were collected, with an even distribution between the validation and test sets.

4.3. SA-LSTM Training Setup

In the two experimental scenarios, the hyperparameters of the SA-LSTM model were adjusted differently. The details of these hyperparameters are presented in Table 2. For each L of consecutive input RSSI vectors at a given moment, the network predicts the exact location of the last recorded time point. The initial learning rate is set to 0.001 for both scenarios. During the training process, we reduce the learning rate to one-tenth of the previous rate after a fixed number of training epochs. In the 2D scenario, the learning rate was adjusted every 30 epochs, while in the 3D scenario, the learning rate was adjusted every 20 epochs. All models were trained using an NVIDIA GeForce RTX 2080 Ti GPU, manufactured by NVIDIA, based in Santa Clara, CA, United States.

Table 2. The hyperparameters of SA-LSTM.

Layer	2D Experiment	3D Experiment
Linear layer 1	(24 × 64)	(436 × 128)
LSTM layer	(64 × 64)	(128 × 128)
Linear layer 2	(64 × 4)	(128 × 4)
Convolution layer	3 × 3 kernels, 1 filter	3 × 3 kernels, 1 filter
Linear layer 3	(62 × 2)	(126 × 3)
Batch size	2	2
Initial learning rate	0.001	0.001
Optimizer	Adam	Adam
Loss function	MSE	MSE
Training epochs	200	100

5. Results and Discussion

Before comparing the performance of various methods, the sliding window length L for the SA-LSTM method needs to be determined. Figure 9 illustrates the mean positioning error as a function of the window size. As shown in the figure, SA-LSTM performs poorly when L is set to 1 or 2. As L increases, the average localization error of SA-LSTM shows a significant decrease. This occurs because when L is set to a smaller value, the network model obtains less information, resulting in lower positioning accuracy. When L is taken to 5 or 6, the average localization error fluctuates within a small range. To avoid additional computational complexity, L is determined to be set to 4.

To compare our indoor localization approach, we implemented an indoor localization system network based on LSTM, as described in [37]. Additionally, we implemented other methods such as RNN [10], KNN, WKNN, FNN, and linear regression. We adjusted the parameters of these models within a certain range to optimize their performance. During the training process, all the model was validated using the validation set after each training epoch, and the model with the minimum average position error was saved for further evaluation.

The average and maximum positioning errors of all these methods are presented in Table 3. The SA-LSTM method outperforms other methods in terms of average positioning accuracy. Among these methods, the LSTM approach achieves the second-best performance in mean positioning accuracy, following the proposed SA-LSTM method. On the test set, the LSTM method results in a maximum error of 13.73 m and an average error of 3.07 m, which is 0.98 m and 1.31 m higher than the proposed SA-LSTM method. Compared to the RNN method, which has a mean positioning error of 4.16 m and a maximum error of 12.64 m, SA-LSTM improves the positioning accuracy by 2.4 m and 0.29 m. Moreover,

SA-LSTM achieves a maximum improvement of 66.85% in average positioning accuracy compared to the linear regression method.

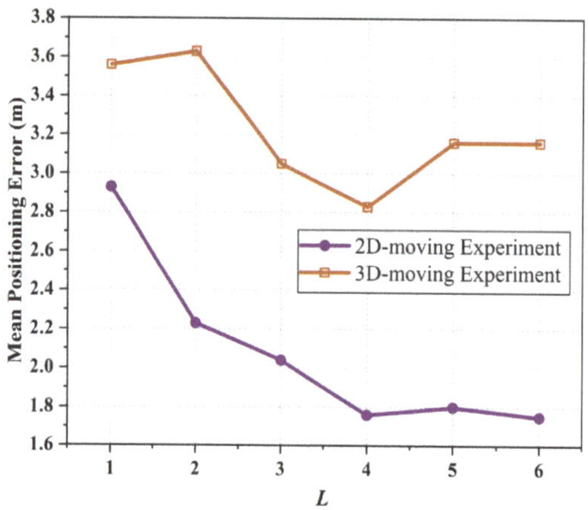

Figure 9. The length of the sliding window against the mean positioning error.

Table 3. The positioning error for 2D-moving experiment.

Method	Average Error (m)		Maximum Error (m)	
	Validation Set	Test Set	Validation Set	Test Set
KNN	2.53	3.36	18.39	15.22
WKNN	2.53	3.33	18.41	15.42
FNN	3.49	5.28	12.44	12.54
Linear regression	3.64	5.31	13.06	**12.34**
RNN	3.37	4.16	12.67	12.64
LSTM	2.57	3.07	13.73	13.73
SA-LSTM	**1.67**	**1.76**	**12.35**	12.35

Figure 10 illustrates the MSE loss curve of the SA-LSTM and LSTM methods during the training process with 2D-moving trajectories. Our results indicate that exhibits a faster convergence rate in terms of training loss compared to the LSTM model. Moreover, after 200 epochs of training, the training loss of SA-LSTM converges to around 0, while the training loss of LSTM converges to around 0.5. The validation loss of SA-LSTM converges faster to near-stabilization values compared to LSTM, as demonstrated in the black-dotted box in Figure 10. Throughout the entire training process, we observed that the SA-LSTM model achieved a slightly lower minimum validation loss than the LSTM model. These results suggest that the SA-LSTM model is more effective in terms of training efficiency with the help of a self-attention mechanism and shortcut connection.

Figure 11 illustrates the cumulative distribution function (CDF) of localization errors for the 2D-moving experiment. In total, a maximum localization error of 12.35 m is recorded for SA-LSTM, 15.22 m is recorded for KNN, and the largest maximum localization error of 15.42 m is recorded for WKNN. Compared to the KNN and WKNN methods, the SA-LSTM method showed a decrease in the maximum localization error by 2.87 m and 3.07 m, respectively. Meanwhile, the maximum localization error of LSTM is 12.47 m, which is also higher than that of SA-LSTM. When considering the 90% percentile of the CDF, the proposed SA-LSTM model demonstrates a 90% location error of approximately under 3.86 m. In comparison, the LSTM, RNN, and KNN models exhibit location errors of around 4.36 m, 5.74 m, and 6.31 m, respectively. This suggests that the proposed SA-LSTM can achieve an improvement of 11.47%, 32.75%, and 63.47% in the 90% CDF compared to LSTM, RNN, and KNN, respectively.

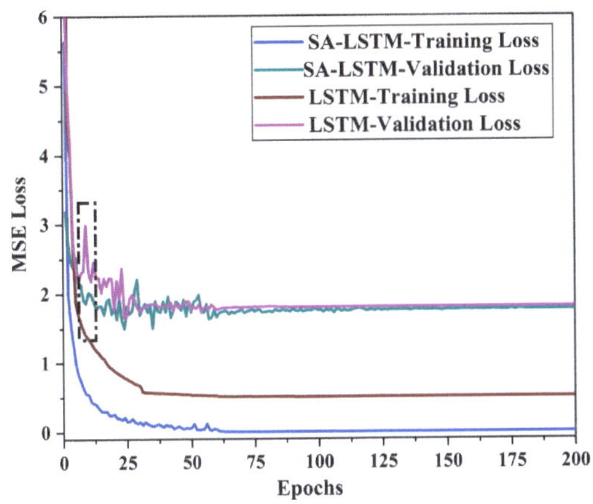

Figure 10. The MSE loss curve of SA-LSTM and LSTM methods in 2D-moving experiment.

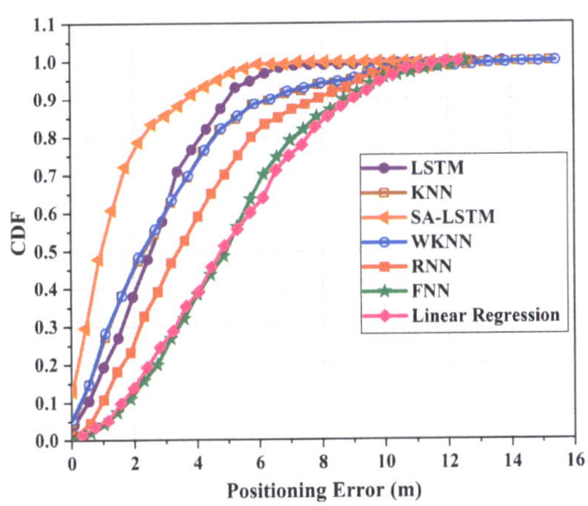

Figure 11. The CDF of localization errors for 2D-moving experiment.

Regarding the 3D-moving experiment, the proposed SA-LSTM model continues to exhibit superior performance in the localization system. Similarly, we compare the average and maximum positioning error of KNN, WKNN, FNN, linear regression, RNN, LSTM, and SA-LSTM. As shown in Table 4, the proposed SA-LSTM achieves an average positioning error of 2.83 m and a maximum positioning error of 57.64 m in the 3D-moving experiment. Compared to LSTM, SA-LSTM improves the average positioning accuracy by 31.64%. In addition, SA-LSTM reduces the average positioning errors by 2.1 m and the maximum localization errors by 3.32 m compared to RNN. Compared to KNN and WKNN, the SA-LSTM has an average positioning error that is 0.62 m and 0.61 m lower, respectively. The SA-LSTM has achieved the lowest average positioning error and the maximum positioning error in scenes involving 3D motion.

The loss curves for SA-LSTM and LSTM in the 3D-moving experiment are depicted in Figure 12. The training loss of SA-LSTM and LSTM converge at a similar rate. As shown in the zoomed-in image in Figure 12, the final convergence value of SA-LSTM is a bit lower. In terms of the validation loss, the SA-LSTM model exhibited a better performance than the LSTM model. Specifically, the validation loss of SA-LSTM could eventually converge to 3, while that of LSTM remained above 4. Based on these findings, we can conclude that our proposed SA-LSTM model is significantly more efficient in terms of training efficiency compared to the conventional LSTM model.

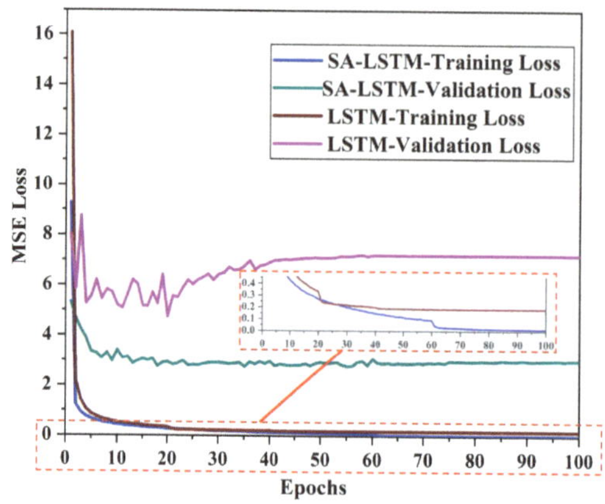

Figure 12. The MSE loss curve of SA-LSTM and LSTM methods in 3D-moving experiment.

Table 4. The positioning error for 3D-moving experiment.

Method	Average Error (m)		Maximum Error (m)	
	Validation Set	Test Set	Validation Set	Test Set
KNN	3.42	3.45	68.95	69.99
WKNN	3.41	3.44	68.95	69.99
FNN	6.41	6.81	68.79	58.70
Linear regression	7.06	7.56	100.44	89.74
RNN	3.73	4.93	40.71	60.96
LSTM	3.91	4.14	66.91	69.29
SA-LSTM	**2.56**	**2.83**	**28.46**	**57.64**

Figure 13 illustrates the CDF of localization errors for the 3D-moving experiment. Overall, the proposed SA-LSTM still outperforms the other classical algorithms. The LSTM network performs the second best, which achieves a 90% location error below 6 m, while RNN achieves a 90% location error below 8.45 m. Compared to LSTM and RNN, SA-LSTM decreased the 90% CDF by 1.99 m and 4.44 m.

Furthermore, a couple of estimated trajectories are drawn in a 3D-moving experiment using the SA-LSTM model. Figure 14a,b depict the moving trajectories, which involve transitions between two and three different floors, respectively. The red lines correspond to the reference trajectory, whereas the blue lines depict the estimated trajectories generated by SA-LSTM. The experimental results indicate that the measured position points in the referenced trajectories exhibit anomalous behavior during pedestrian transitions between different floors. This behavior is attributed to the reliance on elevators for inter-floor movement, which leads to abnormal fluctuations in the measurement signal, resulting in anomalous measured positions. From the trajectories shown in Figure 14a,b, it can be demonstrated that the proposed SA-LSTM model exhibits a satisfactory performance when the pedestrians under test move within a single floor. However, when pedestrians move between floors, the estimated position points generated by the SA-LSTM model may exhibit some fluctuations within a narrow range. Nevertheless, once the pedestrians reach a specific floor, the SA-LSTM model can promptly resume its effective operation.

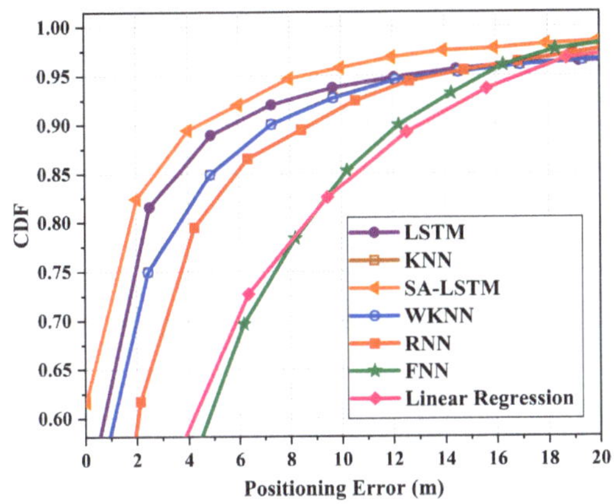

Figure 13. The CDF of localization errors for 3D-moving experiment.

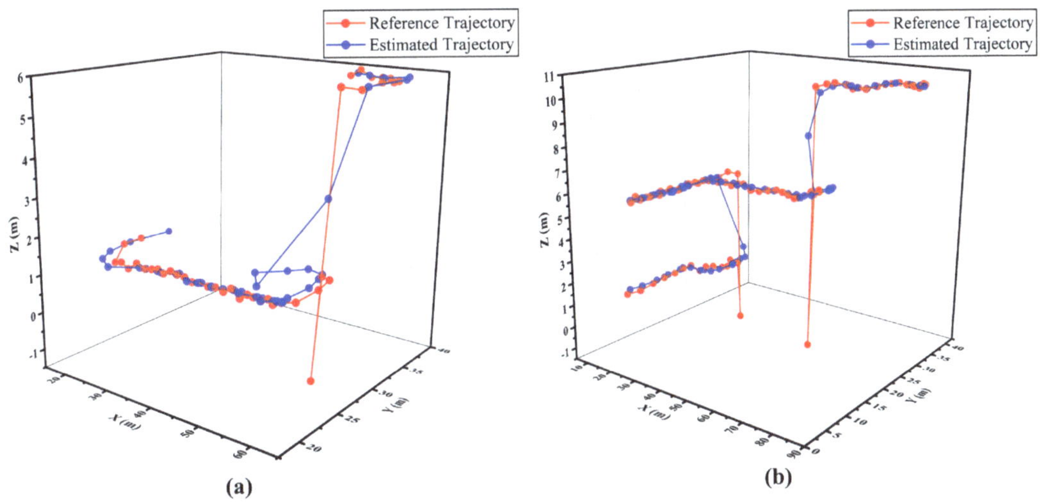

Figure 14. Schematic diagram of referenced and estimated trajectories with a range of movement involving (**a**) two floors and (**b**) three floors.

The 90% quantile of CDF is an important performance evaluation metric in location systems, as highlighted in 3GPP Rel.18 [46]. To comprehensively evaluate the performance of each algorithm in both 2D-moving and 3D-moving experiments, we calculate the 90% error for each algorithm and present the results in Figure 15.

In both experimental scenarios, SA-LSTM demonstrates the highest localization accuracy compared to the other algorithms, as indicated by its remarkably low 90% positioning error. Under the 3D-moving experimental environment, SA-LSTM achieves a 90% localization error under 3.86 m, which is 0.5 m and 1.88 m lower than that of LSTM and RNN, respectively. Compared to classical KNN algorithms, the SA-LSTM model consistently exhibits a lower 90% positioning error under both experimental environments. These results suggest that SA-LSTM demonstrates high accuracy and stability in the field of indoor positioning, highlighting its potential to outperform traditional methods and pave the way for more advanced and reliable indoor positioning systems.

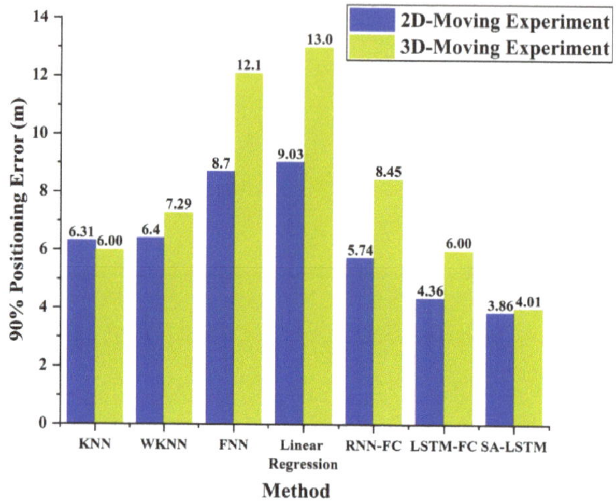

Figure 15. The histogram of a 90% positioning error for two experiments.

Furthermore, we implemented several ensemble-based algorithms in the mainstream and compared their performance to that of the proposed SA-LSTM. As depicted in Figure 16, the random forest and AdaBoost exhibited a similar positioning accuracy in the 2D-moving experiment, with an average positioning error of 3.96 m. In the 3D-moving experiment, random forest and AdaBoost demonstrate average positioning errors of 5.69 m and 4.37 m, respectively. Additionally, the SA-LSTM model shows lower positioning errors regarding the 90% CDF in both experimental environments. When compared to the SA-LSTM and LSTM algorithms, the ensemble-based models only focus on the wireless fingerprint signal characteristics at the current moment and do not consider the temporal characteristics of the signal. Moreover, the fluctuation of RSSI can lead to changes in the RSSI pattern at a particular location. These factors seriously impair the performance of these ensemble-based models.

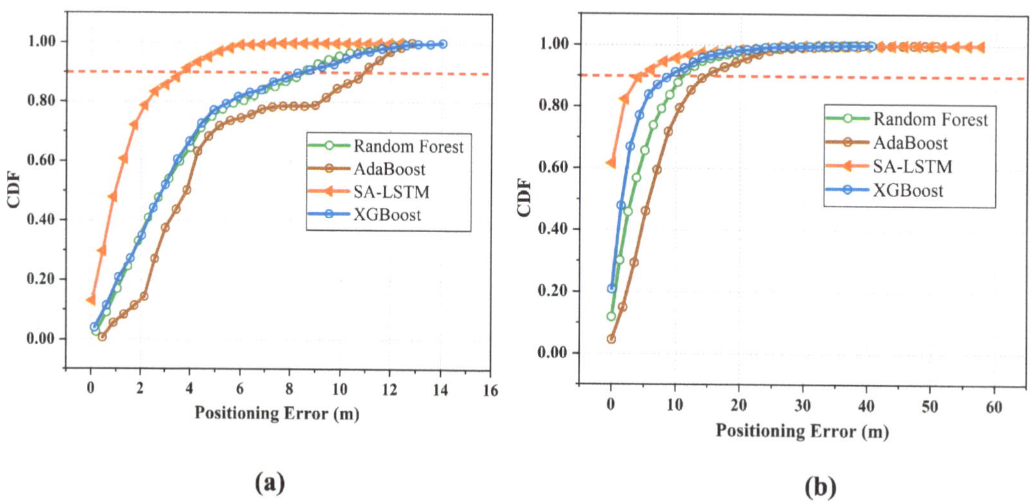

Figure 16. The CDF of SA-LSTM and implemented ensemble-based models in (**a**) 2D-moving experiment and (**b**) 3D-moving experiment.

Based on our experimental results, the proposed SA-LSTM shows an outstanding performance in RSSI-based fingerprinting indoor positioning. However, there are still a number of limitations that need to be addressed in our future work. However, there are still several limitations that need to be addressed in our future work. We identified that the density of deployed beacons has a significant impact on the performance of SA-LSTM. To achieve a high positioning accuracy, we tried to have a Bluetooth beacon within every 8 m^2 based on our beacon configuration. Nevertheless, this strategy necessitates a great number of beacons for large areas. Our future plan involves developing a positioning method that integrates Bluetooth signal data fusion with Wi-Fi and cellular signals. By leveraging these existing wireless signals, we aim to reduce the number of required Bluetooth beacons. Furthermore, we observed that the performance of SA-LSTM is influenced by the movement trajectory. While the training and testing trajectories do not necessarily need to align in the experiments discussed in this paper, it is essential for the training trajectory to comprehensively cover the entire experimental area to ensure localization accuracy. In our future research, we will focus on enhancing the fingerprint acquisition method to mitigate the challenges and costs associated with RSSI acquisition. Finally, due to resource constraints, the performance of SA-LSTM was only validated in two specific environments. As illustrated in the experimental results, SA-LSTM demonstrated a superior performance in an office room compared to the C4 building. This discrepancy can be attributed to the larger size of the C4 building and the increased obstruction by objects within it. Theoretically, the localization accuracy of SA-LSTM is anticipated to be higher in less obstructed environments. For future research endeavors, we aim to validate our approach in a more diverse array of environments.

6. Conclusions

This paper introduces a novel SA-LSTM method for fingerprint-based indoor localization systems. The proposed model utilizes the self-attention mechanism to calculate attention scores between each element and all other elements in the output sequence of the LSTM. This enables the SA-LSTM model to focus on the relationship between the position features at different time steps, thereby improving the accuracy of real-time position estimation. The performance of SA-LSTM has been evaluated under various experimental environments that involve 2D and 3D moving trajectories. The experimental results show that SA-LSTM achieves an average localization error of 1.76 m and 2.83 m in the respective scenarios, with 90% of the positioning errors being under 3.86 m and 4.01 m, respectively. Furthermore, when compared with existing state-of-the-art methods in the same test environment, SA-LSTM exhibits a significant improvement in positioning accuracy by 42.67% to 31.64% under the same test environment.

Our study has successfully showcased the potential of the self-attention mechanism in enhancing the accuracy and efficiency of indoor localization systems. In our future work, we plan to conduct further research to explore the applicability and effectiveness of this mechanism in improving the accuracy of indoor localization.

Author Contributions: Conceptualization, Z.W.; Data curation, P.H. and S.L.; Formal analysis, Z.W.; Methodology, Z.W.; Project administration, T.P.; Software, Z.W.; Writing—review and editing, Z.W. All authors have read and agreed to the published version of the manuscript.

Funding: This research was funded by the National Key Research and Development Program of China under Grant 2022YFB3904700.

Institutional Review Board Statement: Not applicable.

Informed Consent Statement: Not applicable.

Data Availability Statement: Data are contained within the article.

Conflicts of Interest: The authors declare no conflicts of interest.

Abbreviations

The following abbreviations are used in this manuscript:

LBS	Location-based services
GPS	Global Positioning System
BDS	BeiDou Satellite Navigation System
UWB	Ultra-wide bandwidth
RFID	Radio frequency identification
AOA	Angle of arrival
TOA	Time of arrival
APs	Access points
UE	User equipment
LSTM	Long short-term memory
RSSI	Received signal strength indicator
TPs	Test points
RPs	Reference points
KNN	K-Nearest neighbors
WKNN	Weighted K-nearest neighbors
SVM	Support vector machines
FNN	Feedforward neural networks
SA-LSTM	Self-attention and LSTM
VSLAM	Simultaneous localization and mapping
RNN	Recurrent neural networks

References

1. Zhong, S.; Li, L.; Liu, Y.G.; Yang, Y.R. *Privacy-Preserving Location-Based Services for Mobile Users in Wireless Networks*; Technical Report ALEU/DCS/TR-1297; Department of Computer Science, Yale University: New Haven, CT, USA, 2004; Volume 26.
2. Gao, L.; Xiong, L.; Xia, X.; Lu, Y.; Yu, Z.; Khajepour, A. Improved vehicle localization using on-board sensors and vehicle lateral velocity. *IEEE Sens. J.* **2022**, *22*, 6818–6831. [CrossRef]
3. Balaji, B.; Xu, J.; Nwokafor, A.; Gupta, R.; Agarwal, Y. Sentinel: Occupancy based HVAC actuation using existing WiFi infrastructure within commercial buildings. In Proceedings of the 11th ACM Conference on Embedded Networked Sensor Systems, Roma, Italy, 11–15 November 2013; pp. 1–14.
4. Tekler, Z.D.; Chong, A. Occupancy prediction using deep learning approaches across multiple space types: A minimum sensing strategy. *Build. Environ.* **2022**, *226*, 109689. [CrossRef]
5. Filippoupolitis, A.; Oliff, W.; Loukas, G. Bluetooth low energy based occupancy detection for emergency management. In Proceedings of the 2016 15th International Conference on Ubiquitous Computing and Communications and 2016 International Symposium on Cyberspace and Security (IUCC-CSS), Rovaniemi, Finland, 12–15 December 2016; pp. 31–38.
6. Nemra, A.; Aouf, N. Robust INS/GPS sensor fusion for UAV localization using SDRE nonlinear filtering. *IEEE Sens. J.* **2010**, *10*, 789–798. [CrossRef]
7. Mulloni, A.; Seichter, H.; Schmalstieg, D. Handheld augmented reality indoor navigation with activity-based instructions. In Proceedings of the 13th International Conference on Human Computer Interaction with Mobile Devices and Services, Stockholm, Sweden, 30 August–2 September 2011; pp. 211–220.
8. Spilker, J.J., Jr.; Axelrad, P.; Parkinson, B.W.; Enge, P. *Global Positioning System: Theory and Applications*; American Institute of Aeronautics and Astronautics: Reston, VA, USA, 1996; Volume I.
9. Yang, Y.; Gao, W.; Guo, S.; Mao, Y.; Yang, Y. Introduction to BeiDou-3 navigation satellite system. *Navigation* **2019**, *66*, 7–18. [CrossRef]
10. Khassanov, Y.; Nurpeiissov, M.; Sarkytbayev, A.; Kuzdeuov, A.; Varol, H.A. Finer-level sequential wifi-based indoor localization. In Proceedings of the 2021 IEEE/SICE International Symposium on System Integration (SII), Iwaki, Japan, 11–14 January 2021; pp. 163–169.
11. Salamah, A.H.; Tamazin, M.; Sharkas, M.A.; Khedr, M. An enhanced WiFi indoor localization system based on machine learning. In Proceedings of the 2016 International Conference on Indoor Positioning and Indoor Navigation (IPIN), Alcala de Henares, Spain, 4–7 October 2016; pp. 1–8.
12. Abbas, M.; Elhamshary, M.; Rizk, H.; Torki, M.; Youssef, M. WiDeep: WiFi-based accurate and robust indoor localization system using deep learning. In Proceedings of the 2019 IEEE International Conference on Pervasive Computing and Communications (PerCom), Kyoto, Japan, 11–15 March 2019; pp. 1–10.
13. Chen, C.; Chen, Y.; Lai, H.Q.; Han, Y.; Liu, K.R. High accuracy indoor localization: A WiFi-based approach. In Proceedings of the 2016 IEEE International Conference on Acoustics, Speech and Signal Processing (ICASSP), Shanghai, China, 20–25 March 2016; pp. 6245–6249.

14. Altini, M.; Brunelli, D.; Farella, E.; Benini, L. Bluetooth indoor localization with multiple neural networks. In Proceedings of the IEEE 5th International Symposium on Wireless Pervasive Computing 2010, Modena, Italy, 5–7 May 2010; pp. 295–300.
15. Wang, Y.; Ye, Q.; Cheng, J.; Wang, L. RSSI-based bluetooth indoor localization. In Proceedings of the 2015 11th International Conference on Mobile Ad-Hoc and Sensor Networks (MSN), Shenzhen, China, 16–18 December 2015; pp. 165–171.
16. Zhang, C.; Kuhn, M.; Merkl, B.; Fathy, A.E.; Mahfouz, M. Accurate UWB indoor localization system utilizing time difference of arrival approach. In Proceedings of the 2006 IEEE Radio and Wireless Symposium, San Diego, CA, USA, 17–19 October 2006; pp. 515–518.
17. Poulose, A.; Han, D.S. UWB indoor localization using deep learning LSTM networks. *Appl. Sci.* **2020**, *10*, 6290. [CrossRef]
18. Montaser, A.; Moselhi, O. RFID indoor location identification for construction projects. *Autom. Constr.* **2014**, *39*, 167–179. [CrossRef]
19. Chen, Y.; Lymberopoulos, D.; Liu, J.; Priyantha, B. Indoor localization using FM signals. *IEEE Trans. Mob. Comput.* **2013**, *12*, 1502–1517. [CrossRef]
20. Lan, T.; Wang, X.; Chen, Z.; Zhu, J.; Zhang, S. Fingerprint augment based on super-resolution for WiFi fingerprint based indoor localization. *IEEE Sens. J.* **2022**, *22*, 12152–12162. [CrossRef]
21. Torres-Sospedra, J.; Montoliu, R.; Martínez-Usó, A.; Avariento, J.P.; Arnau, T.J.; Benedito-Bordonau, M.; Huerta, J. UJIIndoorLoc: A new multi-building and multi-floor database for WLAN fingerprint-based indoor localization problems. In Proceedings of the 2014 International Conference on Indoor Positioning and Indoor Navigation (IPIN), Busan, Republic of Korea, 27–30 October 2014; pp. 261–270.
22. Roy, P.; Chowdhury, C. A survey of machine learning techniques for indoor localization and navigation systems. *J. Intell. Robot. Syst.* **2021**, *101*, 63. [CrossRef]
23. Brunato, M.; Battiti, R. Statistical learning theory for location fingerprinting in wireless LANs. *Comput. Netw.* **2005**, *47*, 825–845. [CrossRef]
24. Hoang, M.T.; Zhu, Y.; Yuen, B.; Reese, T.; Dong, X.; Lu, T.; Westendorp, R.; Xie, M. A soft range limited K-nearest neighbors algorithm for indoor localization enhancement. *IEEE Sens. J.* **2018**, *18*, 10208–10216. [CrossRef]
25. Fang, S.H.; Lin, T.N. Indoor location system based on discriminant-adaptive neural network in IEEE 802.11 environments. *IEEE Trans. Neural Netw.* **2008**, *19*, 1973–1978. [CrossRef] [PubMed]
26. Nurpeiissov, M.; Kuzdeuov, A.; Assylkhanov, A.; Khassanov, Y.; Varol, H.A. End-to-end sequential indoor localization using smartphone inertial sensors and WiFi. In Proceedings of the 2022 IEEE/SICE International Symposium on System Integration (SII), Narvik, Norway, 9–12 January 2022; pp. 566–571.
27. Zhang, Y.; Qu, C.; Wang, Y. An indoor positioning method based on CSI by using features optimization mechanism with LSTM. *IEEE Sens. J.* **2020**, *20*, 4868–4878. [CrossRef]
28. Gibbons, F.X. Self-attention and behavior: A review and theoretical update. *Adv. Exp. Soc. Psychol.* **1990**, *23*, 249–303.
29. Zhao, H.; Jia, J.; Koltun, V. Exploring self-attention for image recognition. In Proceedings of the IEEE/CVF Conference on Computer Vision and Pattern Recognition, Seattle, WA, USA, 13–19 June 2020; pp. 10076–10085.
30. Humphreys, G.W.; Sui, J. Attentional control and the self: The self-attention network (SAN). *Cogn. Neurosci.* **2016**, *7*, 5–17. [CrossRef] [PubMed]
31. Vaswani, A.; Shazeer, N.; Parmar, N.; Uszkoreit, J.; Jones, L.; Gomez, A.N.; Kaiser, Ł.; Polosukhin, I. Attention is all you need. In Proceedings of the Advances in Neural Information Processing Systems 30 (NIPS 2017), Long Beach, CA, USA, 4–9 December 2017.
32. Bahl, P.; Padmanabhan, V.N. RADAR: An in-building RF-based user location and tracking system. In Proceedings of the Proceedings IEEE INFOCOM 2000, Conference on Computer Communications, Nineteenth Annual Joint Conference of the IEEE Computer and Communications Societies (Cat. No. 00CH37064), Tel Aviv, Israel, 26–30 March 2000; Volume 2, pp. 775–784.
33. Jedari, E.; Wu, Z.; Rashidzadeh, R.; Saif, M. Wi-Fi based indoor location positioning employing random forest classifier. In Proceedings of the 2015 International Conference on Indoor Positioning and Indoor Navigation (IPIN), Banff, AB, Canada, 13–16 October 2015; pp. 1–5.
34. Ding, J.; Wang, Y.; Fu, S.; Si, H.; Zhang, J.; Gao, S. Multiview features fusion and Adaboost based indoor localization on Wifi platform. *IEEE Sens. J.* **2022**, *22*, 16607–16616. [CrossRef]
35. Singh, N.; Choe, S.; Punmiya, R.; Kaur, N. XGBLoc: XGBoost-Based Indoor Localization in Multi-Building Multi-Floor Environments. *Sensors* **2022**, *22*, 6629. [CrossRef]
36. Tekler, Z.D.; Low, R.; Gunay, B.; Andersen, R.K.; Blessing, L. A scalable Bluetooth Low Energy approach to identify occupancy patterns and profiles in office spaces. *Build. Environ.* **2020**, *171*, 106681. [CrossRef]
37. Chen, Z.; Zou, H.; Yang, J.; Jiang, H.; Xie, L. WiFi fingerprinting indoor localization using local feature-based deep LSTM. *IEEE Syst. J.* **2019**, *14*, 3001–3010. [CrossRef]
38. Chorowski, J.K.; Bahdanau, D.; Serdyuk, D.; Cho, K.; Bengio, Y. Attention-based models for speech recognition. In *Advances in Neural Information Processing Systems*; MIT Press: St. Cambridge, MA, USA, 2015; pp. 577–585.
39. Yang, C.H.; Qi, J.; Chen, P.Y.; Ma, X.; Lee, C.H. Characterizing speech adversarial examples using self-attention u-net enhancement. In Proceedings of the ICASSP 2020—2020 IEEE International Conference on Acoustics, Speech and Signal Processing (ICASSP), Barcelona, Spain, 4–8 May 2020; pp. 3107–3111.
40. Mittag, G.; Naderi, B.; Chehadi, A.; Möller, S. Nisqa: A deep cnn-self-attention model for multidimensional speech quality prediction with crowdsourced datasets. *arXiv* **2021**, arXiv:2104.09494.

41. Zang, H.; Xu, R.; Cheng, L.; Ding, T.; Liu, L.; Wei, Z.; Sun, G. Residential load forecasting based on LSTM fusing self-attention mechanism with pooling. *Energy* **2021**, *229*, 120682. [CrossRef]
42. Finney, D.J. Probit analysis: A statistical treatment of the sigmoid response curve. *J. R. Stat. Soc.* **1947**, *110*, 263–266.
43. Torres-Sospedra, J.; Montoliu, R.; Trilles, S.; Belmonte, Ó.; Huerta, J. Comprehensive analysis of distance and similarity measures for Wi-Fi fingerprinting indoor positioning systems. *Expert Syst. Appl.* **2015**, *42*, 9263–9278. [CrossRef]
44. Song, X.; Fan, X.; Xiang, C.; Ye, Q.; Liu, L.; Wang, Z.; He, X.; Yang, N.; Fang, G. A novel convolutional neural network based indoor localization framework with WiFi fingerprinting. *IEEE Access* **2019**, *7*, 110698–110709. [CrossRef]
45. He, K.; Zhang, X.; Ren, S.; Sun, J. Deep residual learning for image recognition. In Proceedings of the IEEE Conference on Computer Vision and Pattern Recognition, Las Vegas, NV, USA, 27–30 June 2016; pp. 770–778.
46. Morselli, F.; Razavi, S.M.; Win, M.Z.; Conti, A. Soft information based localization for 5G networks and beyond. *IEEE Trans. Wirel. Commun.* **2023**, *22*, 9923–9938. [CrossRef]

sensors

MDPI

Article

Deep Compressed Communication and Application in Multi-Robot 2D-Lidar SLAM: An Intelligent Huffman Algorithm

Liang Zhang * and Jinghui Deng

School of Electrical Engineering and Automation, Anhui University, Hefei 230093, China
* Correspondence: liangzhang@ahu.edu.cn

Abstract: Multi-robot Simultaneous Localization and Mapping (SLAM) systems employing 2D lidar scans are effective for exploration and navigation within GNSS-limited environments. However, scalability concerns arise with larger environments and increased robot numbers, as 2D mapping necessitates substantial processor memory and inter-robot communication bandwidth. Thus, data compression prior to transmission becomes imperative. This study investigates the problem of communication-efficient multi-robot SLAM based on 2D maps and introduces an architecture that enables compressed communication, facilitating the transmission of full maps with significantly reduced bandwidth. We propose a framework employing a lightweight feature extraction Convolutional Neural Network (CNN) for a full map, followed by an encoder combining Huffman and Run-Length Encoding (RLE) algorithms to further compress a full map. Subsequently, a lightweight recovery CNN was designed to restore map features. Experimental validation involves applying our compressed communication framework to a two-robot SLAM system. The results demonstrate that our approach reduces communication overhead by 99% while maintaining map quality. This compressed communication strategy effectively addresses bandwidth constraints in multi-robot SLAM scenarios, offering a practical solution for collaborative SLAM applications.

Keywords: deep compressed network; multi-robot system; Huffman encoder; 2D-lidar SLAM; communication-limited application

1. Introduction

Recent years have witnessed significant advancements in multi-robot SLAM algorithms due to their potential applications in search and rescue operations, environmental monitoring, and the exploration of unknown or hazardous environments [1–5]. Single-robot SLAM, where a solitary robot independently navigates and maps its surroundings, is often constrained by time-consuming exploration and limited observation capabilities in large areas. In contrast, multi-robot SLAM has the advantage of efficiently exploring large-scale environments, enhancing mapping accuracy, and improving robustness through the collaborative efforts of multiple robots in various challenging scenarios.

Traditional multi-robot SLAM is developed directly from expanding the single SLAM into multiple robots under a centralized scheme. For example, the CCM-SLAM [6] stands out as a well-established centralized system for visual-inertial multi-robot SLAM. In this system, a central server takes charge of multi-robot map management, fusion, and optimization. By building upon [6], COVINS [7,8] is an extension that has been demonstrated to scale effectively to 12 robots. Another recent addition to centralized CSLAM systems is CVIDS [9], which has achieved the collaborative localization and dense reconstruction of multiple agents in a unified co-ordinate system by using a monocular visual-inertial sensor suite and a centralized loosely coupled framework. LAMP [10,11] is a computationally efficient and outlier-resilient centralized multi-robot SLAM system that consists of a scalable loop closure detection module and GNC (graduated nonconvexity)-based pose map optimization module. Though centralized systems are simply realized and with high precision, they require stable communications and are not robust to the failure of the central node.

Distributed systems can alleviate these problems by reducing the reliance on a central server. Ref. [12] proposed a metric-semantic 3D mesh model for a distributed multi-robot SLAM system, which includes an outlier-resistant fully distributed trajectory estimation method based on GNC and a novel outlier-resistant distributed PGO algorithm. In [13], Kimera-Multi [12,14] is improved to adapt to large-scale real-world deployments, with special emphasis on handling intermittent and unreliable communications. DOOR SLAM [15] uses a distributed frontend to detect inter-robot loop closures without exchanging raw sensor data, successfully rejecting outliers and obtaining accurate trajectory estimates while requiring low communication bandwidth. At the same time, lidar-based distributed multi-robot SLAM is also underway. In [16], the Scan-Context algorithm is proposed to project a 3D point cloud onto a low-resolution 2D plane, thus converting it into a compact feature vector for retrieving the created map and enabling localization, greatly reducing the amount of data and computational complexity. In [17], a distributed backend is used to remove inter-robot loop closure redundancy and perform position map optimization to optimize the global robot cluster position.

The selection of appropriate map representations of multi-robot SLAM directly impacts its performance on computation efficiency, memory usage, and communication burden. For instance, in environments where ground robots navigate indoor spaces, employing a 2D map is sufficient [18]. Studies have demonstrated that occupancy grid maps provide a compact and accurate solution compared to feature-based maps [19,20]. However, certain applications necessitate 3D representations despite the associated computational and storage complexities. This complexity presents challenges, particularly for resource-constrained robotic platforms. Given the communication limitations in multi-robot SLAM systems, there is a preference for compact or sparse map representations, such as the topological maps used in [21,22]. Additionally, efforts are underway to develop semantic-based representations, such as sparse maps annotated with labeled regions [23]. The choice of map representation is critical for long-term operations due to increasing memory requirements, posing a persistent challenge in multi-robot SLAM [24]. Communication bottlenecks in multi-robot SLAM systems typically arise from the exchange of sensor data or representations used for computing inter-robot loop closures [25]. Robots must exchange sufficient data to determine if other robots have explored the same areas and subsequently estimate map alignment using overlapping map sections. Therefore, advancements in the frontend of multi-robot SLAM systems often involve efficient methods for searching loop closure candidates across a team while considering communication constraints. The increasing challenge of balancing efficientinformation transmission with limited communication bandwidth is a pressing issue in multi-robot SLAM systems.

Distributed multi-robot SLAM highly relies on the exchange of data among the network to realize co-ordination processes, such as place recognition, relative pose computation, loop closure detection, and so on. There is increasing work attempting to relieve the communication burden such that the multi-robot system can be deployed to critical environments with less communication capacity. Instead of communicating the merged full map, only the key points and descriptors from parts of the keyframes (using the camera) or submaps (using lidar) are exchanged to realize these co-ordination processes. As depicted in Table 1, segments are extracted in [26] from the source point cloud to generate a source map that contains lists of low-dimensional segment descriptors. In [9], BRIEF descriptors are extracted from each frame and exchanged with another robot. In [27], the keyframes and map points of each robot are transmitted to the server and then distributed by the server to other robots. In [28], two communication modes are designed in different situations: compact and greedy. In compact mode, the communication overhead is minimized by sending compact descriptors. In greedy mode, each drone shares as much information as possible, which is suitable for good network conditions. In [17], NetVlAD descriptors are transmitted between the robots to perform place recognition. In [12], the data flow between different robots are key points and feature descriptors. Compact representations have been explored, incorporating semantic features [29] that rely on objects as landmarks. This approach requires communicating only object labels and poses to other robots, presenting a condensed object-based descriptor that depends on the configuration of objects in a scene for place recognition. In addition to compacting representations, ensuring the sharing of only pertinent information is valuable. However, the data types transmitted by these methods are partial or compact representations of the data, which will lead to the loss of detail. In a wide range of task scenarios, if there will be an accumulation of errors, it is meaningful to study the transmission of the merged full map.

Table 1. Comparison of the types of data transferred with the related methods.

Method	Type of Data Transfer	Map Density
Dubé et al. [26]	Segmentation description	Sparse
CVIDS [9]	Descriptor extraction	Sparse
CVI-SLAM [27]	Keyframe and map point	Sparse
D^2SLAM [28]	Landmarks and descriptors	Sparse
Zhang et al. [24]	Descriptor extraction	Sparse
DCL-SLAM [17]	Descriptor	Sparse
Door-SLAM [15]	NetVLAD descriptor	Sparse
Kimera-Multi [12]	Keypoints and feature descriptors	Dense
Our methoed	2D occupancy full map	Dense

We present a multi-robot SLAM system utilizing 2D occupancy grid maps as a representation of the environment, enabling merged, full-map transmission. Due to the specific format characteristics of raster maps, we have devised intelligent compression algorithms aimed at significantly reducing redundancy in the information contained within these maps while preserving their overall integrity. These intelligent compression algorithms consist of a lightweight network for both map feature extraction and recovery, along with a map feature coder and decoder.

The main innovations and contributions of this paper are as follows:

1. We present a compressed communication framework for multi-robot SLAM, enabling merged, full-map transmission among robots, which can reduce the duration of exploration and produce a map of the whole environment quickly. Such a whole map plays a key role in many robotic tasks, such as path planning, collision avoidance et al.
2. According to the characteristics of the occupancy grid map, we designed an intelligent compression algorithm by combining a convolutional neural network, Huffman coding, and RLE coding that compress the transmitted full map by both image downsampling and stream encoding.
3. We use Ultra-Wide Band (UWB) as a communication medium to validate the multi-robot SLAM system proposed in this paper. The results show that our method is able to reduce the communication burden by up to 99 percent and the localization error is less than 5 cm compared to when no compression is employed.

The remainder of the paper is organized as follows: In Section 2, we describe the proposed multi-robot SLAM method based on an intelligent compression algorithm. In Section 3, the experimental results are provided to validate the performance of the intelligent compression algorithm. Finally, we conclude this paper in Section 4.

2. The Compressed Communication Approach

The multi-robot SLAM framework based on compressed communication consists of the following three main components: full-map creation, full-map compression and transmission, and full-map fusion. In Figure 1, the robots first use Gmapping to build the full map. The map is then compressed by our proposed algorithm and transmitted to other robots using the UWB device. After receiving maps, a robot can decompress maps and merge them with its own created full maps. Eventually, the robot can navigate through the environment using the merged maps.

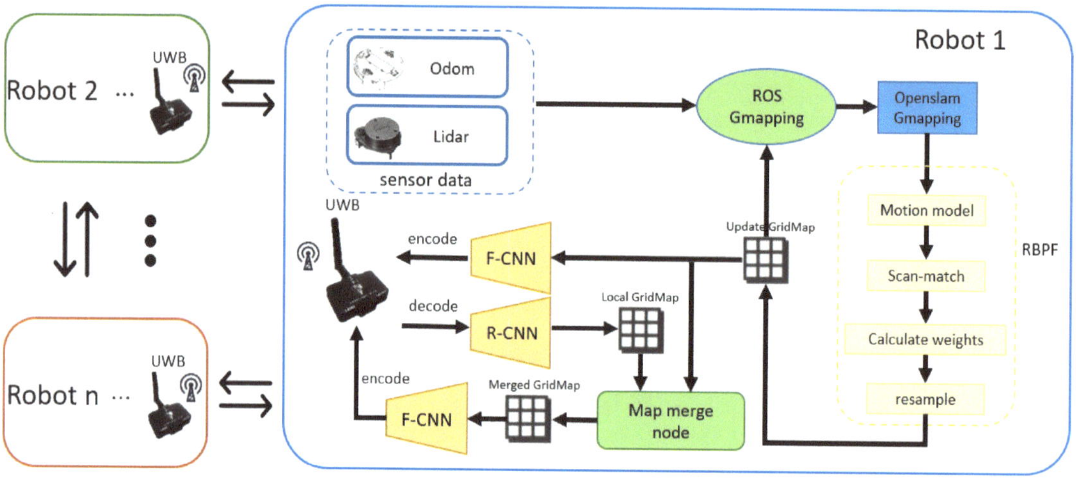

Figure 1. The framework for compressed communication application in multi-robot SLAM.

2.1. Full-Map Creation

Gmapping [30] is a SLAM algorithm that combines particle filtering and FastSLAM. The algorithm predicts the robot's position using a motion model and then uses sensor data for position correction and map updates. As the robot moves through the environment, it gradually builds an accurate occupancy grid map and is able to determine the robot's position on the map. We use Gmapping to create full maps for each robot.

2.2. Full-Map Compression and Decompression

Map compression consists of a map feature extraction CNN $CNN_{feature}$ and an encoder. The map feature $m_{feature}$ that retains the key information of the full map is generated by $CNN_{feature}$. Then, $m_{feature}$ is fed into the encoder to form a string encoding e. Corresponding to the above, map decompression consists of a decoder and a map recovery CNN $CNN_{recovery}$. The decoder recovers e into $m_{feature}$, and then the $m_{feature}$ is processed through $CNN_{recovery}$ to obtain the new full map. Although the new full map is slightly different from the map before compression, it does not affect its ability to provide navigation for the robot, as demonstrated in the experiments in Section 3.

(1) $CNN_{feature}$: As shown in Figure 2, $CNN_{feature}$ consists of three weight layers, which preserve the structural information of the input map while obtaining storage space-saving $m_{feature}$. The first layer consists of convolution (set the convolution kernel to 3×3) and ReLU (enhance the model's expressive power) for the purpose of extracting the features of the map. Considering the computing power of the processing unit on the robot, we set 16 sets of 3×3 filters. The second layer, composed of convolution, Batch Normalization (BN), and ReLU, aims to downsample and enhance features. In order to change the map resolution to a quarter of the original, the parameter stride of the convolution was set to 2, and 16 filters of size $3 \times 3 \times 16$ were used. In the last layer, a filter of size $3 \times 3 \times 16$ is used to get the $m_{feature}$. Compared to downsampling the map directly, the $m_{feature}$ obtained by $CNN_{feature}$ contains more map information.

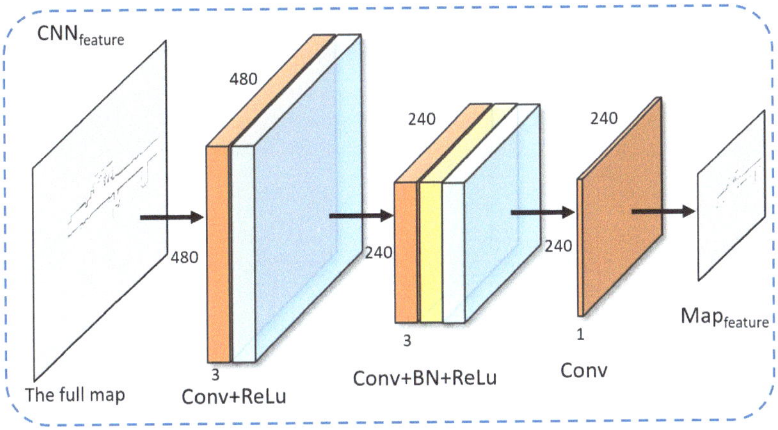

Figure 2. Map feature CNN.

(2) *Encoder for map$_{feature}$*: The encoding process of $m_{feature}$ is shown in Figure 3. The algorithm of $m_{feature}$ encoder based on RLE and Huffman is shown in Algorithm 1. Firstly, the values that appear in the $m_{feature}$ are recorded $\mathcal{V} = \{v_1, v_2, ..., v_n\}$, and the probability of each value in $m_{feature}$ are calculated. We order the probability of occurrence of each value from highest to lowest to obtain the set $\mathcal{P} = \{p_1, p_2, ..., p_n\}$, where p_i represents the probability of the i-th value in $m_{feature}$. $\mathcal{N} = \{n_1, n_2, ..., n_n\}$, which are initialized according to $\mathcal{P} = \{p_1, p_2, ..., p_n\}$, and n_i corresponds to p_i. Then, the two smallest values in the set $\mathcal{N}$ are merged into a new node, so $\mathcal{N}$ and $\mathcal{P}$ are updated to $\{p_1, ..., p_{n-2}, p_{merged}\}$, and $\{n_1, ..., n_{n-2}, n_{merged}\}$. We repeat this step until only one node remains in $\mathcal{N}$. According to the process of node emergence, we can get the Huffman coding tree, e.g., Figure 4, where the grey nodes represent the initial nodes and the white nodes are the new nodes generated by merging the initial nodes. As a result, we can encode each value in $m_{feature}$ to obtain the code table $\mathcal{T} = \{code_1 : v_1, ..., code_n : v_n\}$. Then, according to the code table $\mathcal{T}$, $m_{feature}$ is encoded into the binary code. Finally, the binary code is converted into hexadecimal code h to compress the length of the code. In order to further enhance compression efficiency and reduce communication pressure, consecutive repeated characters in hexadecimal encoding are subsequently processed. We replace the consecutively repeated code segments in h with the repeated characters themselves and the number of times they are repeated to obtain e. Therefore, $m_{feature}$ is compressed into a code table $\mathcal{T}$ and a string e.

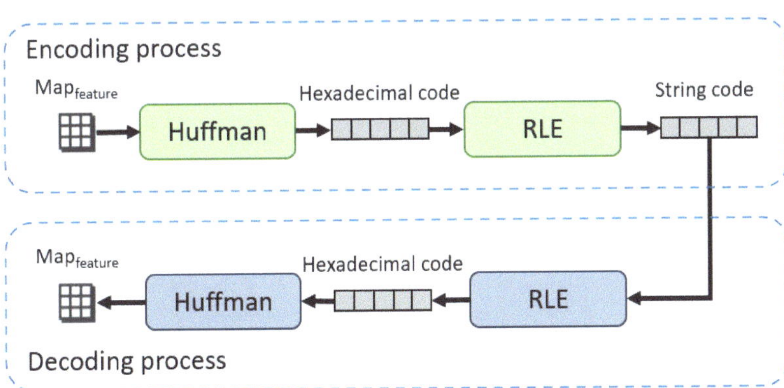

Figure 3. Encoding and decoding process.

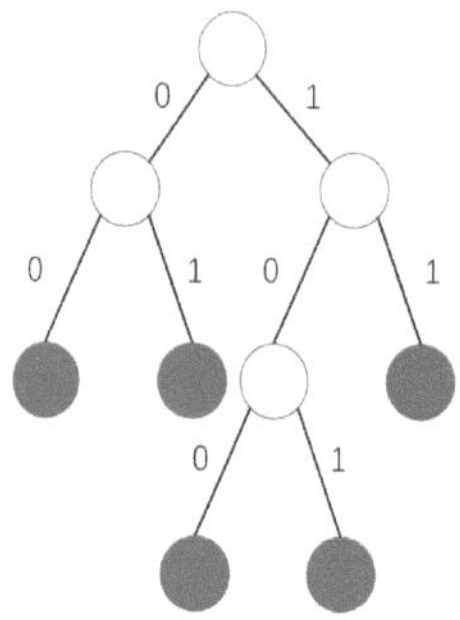

Figure 4. The Huffman coding tree.

Algorithm 1 $m_{feature}$ compression algorithm

Input: $m_{feature}$

1: record the values in the $m_{feature}$ $\mathcal{V} = \{v_1, v_2, ..., v_n\}$, count the probability of different value $\mathcal{P} = \{p_1, p_2, ..., p_n\}$, initialize nodes $\mathcal{N} = \{n_1, n_2, ..., n_n\}$

2: **repeat**

3: merge the two nodes with the lowest probability to get the new node and update $\mathcal{P}, \mathcal{N}$

4: **until** there is only one node left in the set $\mathcal{N}$

5: encode each value according to the Huffman encoding tree to get code table $\mathcal{T} : \{\boldsymbol{code}_1 : v_1, ..., \boldsymbol{code}_n : v_n\}$

6: rewrite $m_{feature}$ data into binary code according to the code table

7: convert binary code to hexadecimal code $\boldsymbol{h}$

8: Initialize $\boldsymbol{e}$ as an empty string and $\boldsymbol{count}$ as 1.

9: For $\boldsymbol{index}$ from 1 to length($\boldsymbol{h}$) - 1:

a. Set currentChar as the character at the current $\boldsymbol{index}$.

b. Set previousChar as the character at the previous $\boldsymbol{index}$.

c. If currentChar equals previousChar, $\boldsymbol{count}$ + 1.

d. If currentChar does not equal previousChar, append $\boldsymbol{count}$ and previousChar to $\boldsymbol{e}$, then reset $\boldsymbol{count}$ to 1.

Output: $\mathcal{T}, \boldsymbol{e}$

---

(3) *Decoder for* $m_{feature}$: The decoding process of $m_{feature}$ is shown in Figure 3. The algorithm of the decoder is shown in Algorithm 2. First, the $\boldsymbol{e}$ obtained by the compression algorithm is restored to hexadecimal code $\boldsymbol{d}$. Then, the hexadecimal code $\boldsymbol{d}$ is converted into the binary code. According to the code table $\mathcal{T}$, we can replace the corresponding code with the corresponding value in $m_{feature}$. Finally, $m_{feature}$ is recovered according to $\mathcal{T}$ and $\boldsymbol{e}$.

Algorithm 2 $m_{feature}$ decompression algorithm

Input: $\mathcal{T}, \boldsymbol{e}$

1: Initialize $\boldsymbol{d}$ as an empty string.

2: While there are characters remaining in $\boldsymbol{e}$:

a. Extract the next $\boldsymbol{count}$-value pair from $\boldsymbol{e}$.

b. Append $\boldsymbol{count}$ occurrences of the associated character to $\boldsymbol{d}$.

3: convert hexadecimal code $\boldsymbol{d}$ to binary code

4: Recover $m_{feature}$ from binary code according to the code table $\mathcal{T}$

Output: $m_{feature}$

---

(4) $CNN_{recovery}$: In Figure 5, $CNN_{recovery}$ consists of six weighting layers, and there are three types, namely Convolution + ReLU, Convolution + BN + ReLU, and Convolution. The purpose of $CNN_{recovery}$ is to recover the full map from $m_{feature}$, and a residual block is

employed in $CNN_{recovery}$. The jump connections of the residual block can help the network learn the difference between the input and the output. This allows $CNN_{recovery}$ to better recover the full map, thus preserving the detailed information of $m_{feature}$. A total of 16 filters of size 3×3 are used to generate 16 feature mappings with ReLU in the first layer. For the second to fifth layers, 16 filters of size $3 \times 3 \times 16$ are utilized, and BN is inserted between Convolution and ReLU. The last layer uses a filter of size $3 \times 3 \times 16$ to generate the final single-channel output, i.e., the full map.

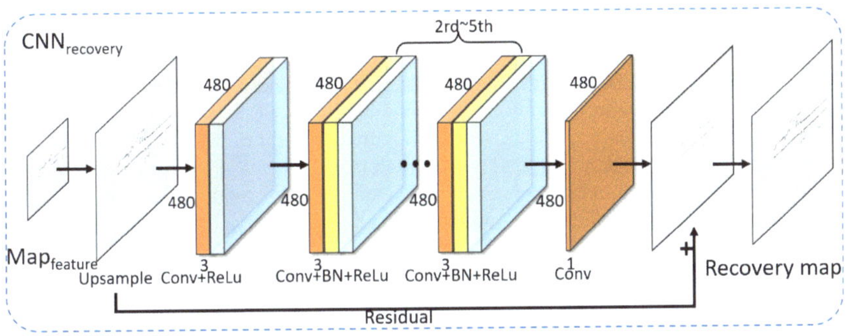

Figure 5. Map recovery CNN.

(5) *Learning algorithm*: The entire network consists of $CNN_{feature}$, the encoder, the decoder, and $CNN_{recovery}$. The optimization function for the network is designed as follows:

$$(\bar{\theta_1}, \bar{\theta_2}) = \arg\min_{\theta_1, \theta_2} \| R(\theta_2, Cod(F(\theta_1, m))) - m \|^2 \tag{1}$$

where $R(\cdot)$ and $F(\cdot)$ represent $CNN_{recovery}$ and $CNN_{feature}$, respectively. $Cod(\cdot)$ represents the encoder and decoder. m is the full map that is the input of the network. θ_1 and θ_2 are the parameters of the optimization function.

Therefore, we calculate the mean square error between the map of network output and the original full map as the loss and the gradient of the whole network parameters concerning loss by backpropagation. Then, we use the Adaptive Moment Estimation (Adam) optimization algorithm to update the parameters θ_1 and θ_2 to minimize the optimization function.

(6) *Code transmission based on UWB:* When $m_{feature}$ is processed by the encoder, it becomes a string. In order to facilitate the transmission, we split the string and transmitted it to other robots using UWB segment by segment. After a string has been transmitted, the integrity of the received string is judged. If any data are missing, the receiver commands the transmitter to resend the string.

2.3. Multi-Occupancy Grid Maps Fusion

We consider two cases of multi-occupancy grid map fusion:

(1) The initial positions of the robots are known: In this case, the rigid transformation (R, t) between the maps can be computed directly from the initial position of the robot (the x, y co-ordinates and the yaw angle) so that accurate fusion results can be obtained.

$$R = \begin{bmatrix} cos\theta & -sin\theta \\ sin\theta & cos\theta \end{bmatrix}, t = \begin{bmatrix} t_x \\ t_y \end{bmatrix} \tag{2}$$

The (R, t) of the two maps are known; then, the two maps are fused as follows:

$$map_{fusion} = map_1 + (map_2 \times R + t) \tag{3}$$

(2) The initial positions of the robots are unknown: We used the approach in [31] to estimate the transformation relationship between maps using feature matching when the initial position is unknown. This requires that the maps have enough overlapping regions for reliable matching. The approach detects the features in each map, computes a match for the map pair, estimates the transform using RANSAC, and computes a confidence score for each match. Then, the matches with sufficient confidence are selected for map fusion. In this way, fusion can be performed in the case of overlapping maps, even if the initial positions of the robots are not known.

3. Experiments

In this section, we build the robot platform in Figure 6 and deploy the trained network model to the robots. The performance of the trained model was verified in Figures 7 and 8a. Figure 7 shows the multi-robot SLAM process with and without the compressed communication method, demonstrating that compressed communication has little effect on map quality. Then, multi-robot co-operative SLAM experiments based on the method in this paper are conducted to validate the method further.

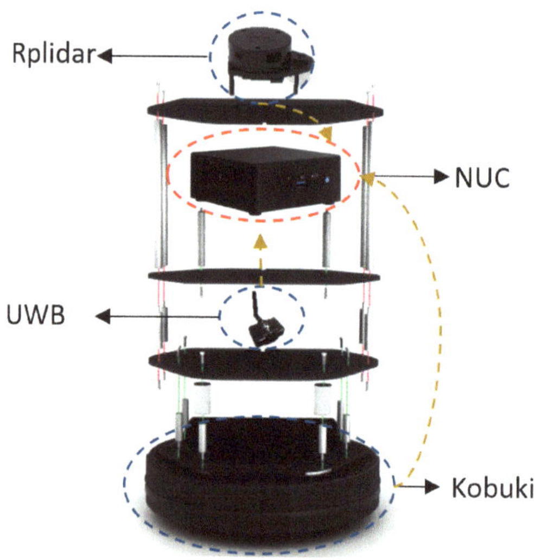

Figure 6. Robot hardware platform.

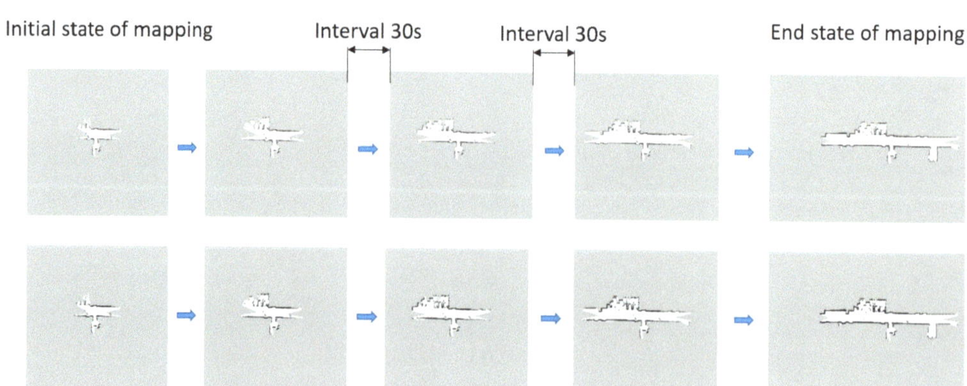

Figure 7. Compressed communication in multi-robot SLAM.

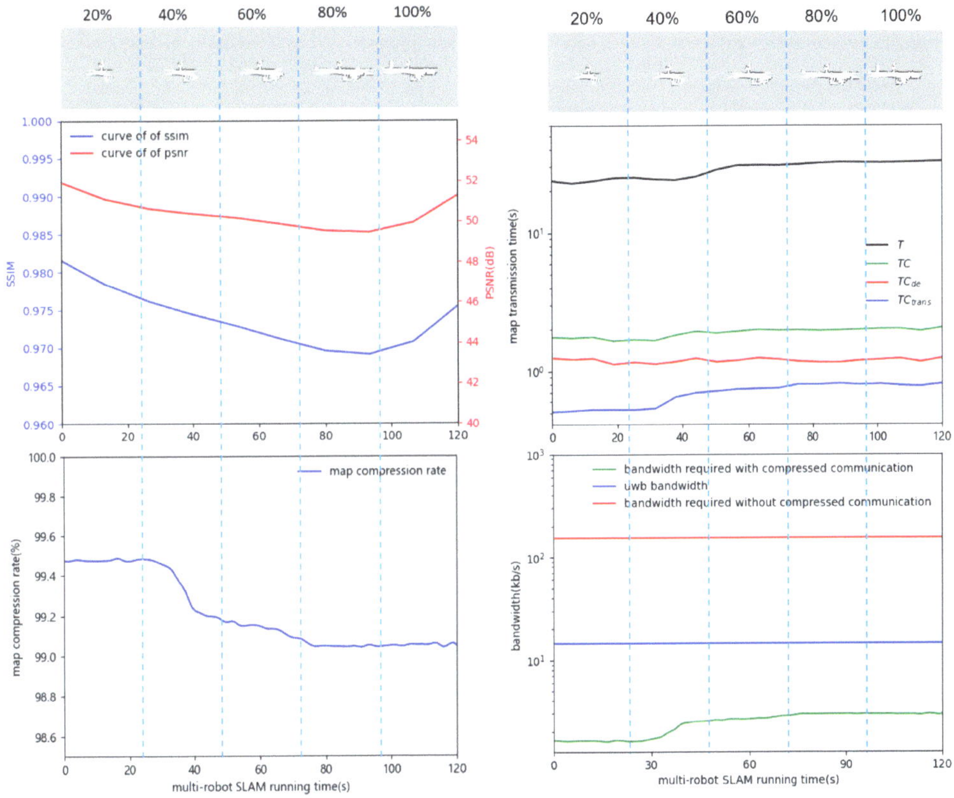

(**a**) The quality of the recovery map and compression rate. (**b**) Transmission time and the required bandwith.

Figure 8. Performance of the proposed methodologies; (**a**) chage in map quality (upper) with change in map compression rate (lower) during the operation of multi-robot SLAM based on compressed communication; (**b**) comparison of map transfer time required with and without compressed communication (upper) versus bandwidth required with and without compressed communication (lower). T and TC denote the time to transmit the map using UWB with compressed communication and without compressed communication, respectively, where TC is the sum of TC_{trans} (time to transmit the compressed map) and TC_{de} (time to decompress the compressed map).

3.1. Datasets and Model Training

We used the Gmapping algorithm to build maps of different environments. A total of 2,800 occupancy grid maps were saved and used for model training. Model training based on the Pytorch framework was implemented on a computer with AMD Ryzen 5 2600X Six-Core Processor 3.60 GHz (AMD, Santa Clara, CA, USA), NVIDIA GeForce GTX 1050 Ti (NVIDIA, Santa Clara, CA, USA), and 16 G RAM. It takes about 10 h to train the network proposed in this paper.

3.2. Robotic Platform

As shown in Figure 6, the robots for collaborative SLAM are mainly composed of a Kobuki drive chassis, an NUC (a microcomputer), Rplidar, and a UWB module. The Gmapping algorithm utilizes information from the chassis' odometer and Rplidar for full-map creation; then, the full maps are processed by the model deployed in the NUC to get the string. Finally, the string is transmitted via the UWB module. Similarly, the robot receives the string through the UWB module, and it is processed by NUC to get the recovery map. The recovered map is fused with the locally created map to obtain the global map.

3.3. Experimental Result

In order to validate the compressed communication frame in Figure 1, we used two robots to collaboratively build maps of a hallway and obtained the map data in Figure 7. In Figure 7, the whole process of the global maps obtained with and without compressed communication for the two robots is shown. The first row of Figure 7 shows the global map obtained without compressed communication; the second row shows the global map obtained with compressed communication; from left to right, the initial state of mapping to the end state of mapping can be seen, and each state of mapping is separated by 30 s.

We evaluated the method of this paper by calculating the metrics for the correspondence maps in Figure 7. The metrics Peak Signal-to-Noise Ratio (PSNR) and Structural Similarity Index Measure (SSIM) were calculated, and the changes in these two metrics with the running time of multi-robot SLAM are plotted in the upper figure of Figure 8a. PSNR is a metric used to quantify the quality of a map, where higher PSNR values typically indicate less distortion introduced by compression and higher map quality. A PSNR above 40 dB indicates excellent map quality (i.e., very close to the original map). On the other hand, SSIM takes values in the range of $[-1, 1]$, with a value closer to 1 indicating more similarity between two maps and a value closer to -1 suggesting greater dissimilarity. The objective is to measure the structural and content similarity between the recovered and original maps. As global map quality depends not only on pixel-level differences but also on factors such as structure and texture, SSIM provides a more accurate assessment of the recovered map's quality. As shown in the upper figure of Figure 8a, the mean values of PSNR and SSIM are 50.649 dB and 0.975. Therefore, the difference between maps obtained with and without compressed communication is very small, and the quality of maps obtained with compressed communication can be guaranteed. As shown in the lower figure of Figure 8a, the map compression rate of up to 99% significantly reduces the communication burden. Based on the data in Figure 8a, we can conclude that the compression method in this paper has a high compression rate while maintaining good recovery quality. The recovered map can be well utilized to construct the global map.

The upper figure of Figure 8b illustrates the time consumed with compressed communication and without compressed communication. The blue and red lines represent the time for map transmission and decompression, respectively. The black line depicts the total time for a map transmission without compressed communication, while the green line shows the time for a map transmission with compression. As shown in the lower figure of Figure 8b, the blue line (12.5 kb/s) is the bandwidth that can be stably achieved by the UWB model in this framework, and the red (124.7 kb/s) and green (Average 720.4 bit/s) lines represent the bandwidth required for transmission of the map without and with compression, respectively. By using compressed communication, the bandwidth required to transmit the maps is greatly reduced. From Figure 8b, we can conclude that the compressed communication in this paper can satisfy the UWB hardware bandwidth and allow for the fast transmission of maps.

In order to verify the effect of the maps obtained from the compressed communication method in robot localization, we used the occupancy grid map obtained with and without compressed communication to provide localization to the robot, respectively. We selected 28 locations in a realistic environment to cover the entire occupancy grid map. Each location was used separately with compressed and uncompressed maps to provide localization to the robot. We used two sets of data and calculated their errors, as in Figure 9. Figure 9a shows the error variation of the robot in the 28 sets of data, and Figure 9b shows the distribution of the error. The mean values of errors $error_x$, $error_y$, and $error_{yaw}$ are 3.59 cm, 3.43 cm, and 0.72°, respectively. This result shows that the compressed communication method in this paper can be applied in a multi-robot SLAM and that the subtle changes produced by compressing the map have little effect on robot localization.

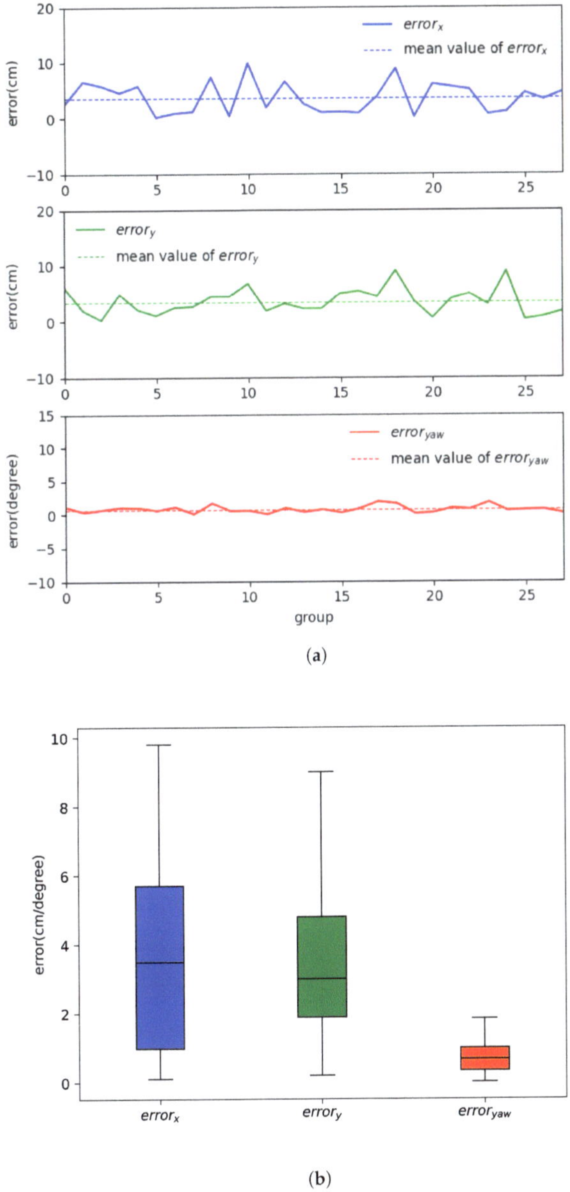

Figure 9. Robot's localization error using the merged map with compressed communication; (**a**) Performance of localization errors regarding the robot's position (x, y) and its yaw angle θ; (**b**) Distribution of the localization error over time.

4. Conclusions

The compressed communication method proposed in this paper consists of the following three main components: (1) full-map creation, (2) full-map compression and transmission, and (3) full-map fusion. We designed a lightweight map-feature extraction CNN and a map recovery CNN that can process occupancy grid maps in real time when processing units with limited arithmetic power. Meanwhile, the encoder and decoder are designed by combining the Huffman and RLE algorithms, which greatly reduce the code length and make the bandwidth required for the transmission code fully satisfy the hardware UWB. The compressed communication framework is validated in a multi-robot system, employing two robots for the collaborative mapping of an unknown environment. One robot compresses and transmits its full map, and the other robot receives and recovers it. Subsequently, the recovery map is fused with its map to generate the global map. The experimental results demonstrate that the method achieves a high compression ratio (99%) and maintains recovered map quality with a PSNR of 50.649 dB and SSIM of 0.975. In summary, the compressed communication method proposed in this paper satisfies the bandwidth-constrained multi-robot SLAM system.

Author Contributions: Conceptualization, L.Z. and J.D.; Data curation, J.D.; Formal analysis, L.Z. and J.D.; Funding acquisition, L.Z.; Investigation, L.Z.; Methodology, L.Z. and J.D.; Project administration, L.Z.; Software, L.Z. and J.D.; Supervision, L.Z.; Validation, J.D.; Visualization, J.D. Writing—original draft, L.Z. and J.D.; Writing—review & editing, L.Z. All authors have read and agreed to the published version of the manuscript.

Funding: This work was supported in part by the National Natural Science Foundation of China under grant No. 62303002 and the Key Program of Natural Science Foundation of Anhui Higher Education Institutions of China under grant No. 2022AH050068.

Institutional Review Board Statement: Not applicable.

Informed Consent Statement: Not applicable.

Data Availability Statement: Data are contained within the article.

Conflicts of Interest: The authors declare no conflict of interest.

Abbreviations

The following abbreviations are used in this manuscript:

SLAM	Simultaneous Localization and Mapping
CNN	Convolutional Neural Network
RLE	Run-Length Encoding
UWB	Ultra-Wide Band
PSNR	Peak Signal-to-Noise Ratio
SSIM	Structural Similarity Index Measure

References

1. Durrant-Whyte, H.; Bailey, T. Simultaneous localization and mapping: Part I. *IEEE Robot. Autom. Mag.* **2006**, *13*, 99–110. [CrossRef]
2. Cadena, C.; Carlone, L.; Carrillo, H.; Latif, Y.; Scaramuzza, D.; Neira, J.; Reid, I.; Leonard, J.J. Past, Present, and Future of Simultaneous Localization and Mapping: Toward the Robust-Perception Age. *IEEE Trans. Robot.* **2016**, *32*, 1309–1332. [CrossRef]
3. Tian, Y.; Liu, K.; Ok, K.; Tran, L.; Allen, D.; Roy, N.; How, J.P. Search and rescue under the forest canopy using multiple UAVs. *Int. J. Robot. Res.* **2020**, *39*, 1201–1221. [CrossRef]
4. Bonin-Font, F.; Burguera, A. Towards multi-robot visual graph-SLAM for autonomous marine vehicles. *J. Mar. Sci. Eng.* **2020**, *8*, 437. [CrossRef]
5. Zhang, L.; Zhang, Z.; Siegwart, R.; Chung, J.J. Probabilistic network topology prediction for active planning: An adaptive algorithm and application. *IEEE Trans. Robot.* **2022**, *39*, 147–164. [CrossRef]
6. Schmuck, P.; Chli, M. CCM-SLAM: Robust and efficient centralized collaborative monocular simultaneous localization and mapping for robotic teams. *J. Field Robot.* **2019**, *36*, 763–781. [CrossRef]
7. Schmuck, P.; Ziegler, T.; Karrer, M.; Perraudin, J.; Chli, M. Covins: Visual-inertial slam for centralized collaboration. In Proceedings of the 2021 IEEE International Symposium on Mixed and Augmented Reality Adjunct (ISMAR-Adjunct), Bari, Italy, 4–8 October 2021; pp. 171–176.
8. Patel, M.; Karrer, M.; Bänninger, P.; Chli, M. COVINS-G: A Generic Back-end for Collaborative Visual-Inertial SLAM. *arXiv* **2023**, arXiv:2301.07147.
9. Zhang, T.; Zhang, L.; Chen, Y.; Zhou, Y. CVIDS: A Collaborative Localization and Dense Mapping Framework for Multi-Agent Based Visual-Inertial SLAM. *IEEE Trans. Image Process.* **2022**, *31*, 6562–6576. [CrossRef] [PubMed]

10. Chang, Y.; Ebadi, K.; Denniston, C.E.; Ginting, M.F.; Rosinol, A.; Reinke, A.; Palieri, M.; Shi, J.; Chatterjee, A.; Morrell, B.; et al. LAMP 2.0: A robust multi-robot SLAM system for operation in challenging large-scale underground environments. *IEEE Robot. Autom. Lett.* **2022**, *7*, 9175–9182. [CrossRef]
11. Ebadi, K.; Chang, Y.; Palieri, M.; Stephens, A.; Hatteland, A.; Heiden, E.; Thakur, A.; Funabiki, N.; Morrell, B.; Wood, S.; et al. LAMP: Large-scale autonomous mapping and positioning for exploration of perceptually-degraded subterranean environments. In Proceedings of the 2020 IEEE International Conference on Robotics and Automation (ICRA), Paris, France, 31 May–31 August 2020; pp. 80–86.
12. Tian, Y.; Chang, Y.; Arias, F.H.; Nieto-Granda, C.; How, J.P.; Carlone, L. Kimera-multi: Robust, distributed, dense metric-semantic slam for multi-robot systems. *IEEE Trans. Robot.* **2022**, *38*, 2022–2038. [CrossRef]
13. Tian, Y.; Chang, Y.; Quang, L.; Schang, A.; Nieto-Granda, C.; How, J.P.; Carlone, L. Resilient and distributed multi-robot visual slam: Datasets, experiments, and lessons learned. *arXiv* **2023**, arXiv:2304.04362.
14. Chang, Y.; Tian, Y.; How, J.P.; Carlone, L. Kimera-multi: A system for distributed multi-robot metric-semantic simultaneous localization and mapping. In Proceedings of the 2021 IEEE International Conference on Robotics and Automation (ICRA), Xi'an, China, 30 May–5 June 2021; pp. 11210–11218.
15. Lajoie, P.Y.; Ramtoula, B.; Chang, Y.; Carlone, L.; Beltrame, G. DOOR-SLAM: Distributed, online, and outlier resilient SLAM for robotic teams. *IEEE Robot. Autom. Lett.* **2020**, *5*, 1656–1663. [CrossRef]
16. Huang, Y.; Shan, T.; Chen, F.; Englot, B. DiSCo-SLAM: Distributed scan context-enabled multi-robot lidar slam with two-stage global-local graph optimization. *IEEE Robot. Autom. Lett.* **2021**, *7*, 1150–1157. [CrossRef]
17. Zhong, S.; Qi, Y.; Chen, Z.; Wu, J.; Chen, H.; Liu, M. Dcl-slam: A distributed collaborative lidar slam framework for a robotic swarm. *arXiv* **2022**, arXiv:2210.11978.
18. Caccavale, A.; Schwager, M. Wireframe mapping for resource-constrained robots. In Proceedings of the 2018 IEEE/RSJ International Conference on Intelligent Robots and Systems (IROS), Madrid, Spain, 1–5 October 2018; pp. 1–9.
19. Saeedi, S.; Paull, L.; Trentini, M.; Li, H. Multiple robot simultaneous localization and mapping. In Proceedings of the 2011 IEEE/RSJ International Conference on Intelligent Robots and Systems, San Francico, CA, USA, 25–30 September 2011; pp. 853–858.
20. Benedettelli, D.; Garulli, A.; Giannitrapani, A. Multi-robot SLAM using M-Space feature representation. In Proceedings of the 49th IEEE Conference on Decision and Control (CDC), Atlanta, GA, USA, 15–17 December 2010; pp. 3826–3831.
21. Chang, H.J.; Lee, C.G.; Hu, Y.C.; Lu, Y.H. Multi-robot SLAM with topological/metric maps. In Proceedings of the 2007 IEEE/RSJ International Conference on Intelligent Robots and Systems, San Diego, CA, USA, 29 October–2 November 2007; pp. 1467–1472.
22. Saeedi, S.; Paull, L.; Trentini, M.; Seto, M.; Li, H. Group mapping: A topological approach to map merging for multiple robots. *IEEE Robot. Autom. Mag.* **2014**, *21*, 60–72. [CrossRef]
23. Choudhary, S.; Carlone, L.; Nieto, C.; Rogers, J.; Liu, Z.; Christensen, H.I.; Dellaert, F. Multi robot object-based slam. In Proceedings of the 2016 International Symposium on Experimental Robotics, Nagasaki, Japan, 3–8 October 2016; pp. 729–741.
24. Zhang, H.; Chen, X.; Lu, H.; Xiao, J. Distributed and collaborative monocular simultaneous localization and mapping for multi-robot systems in large-scale environments. *Int. J. Adv. Robot. Syst.* **2018**, *15*, 1729881418780178. [CrossRef]
25. Tardioli, D.; Montijano, E.; Mosteo, A.R. Visual data association in narrow-bandwidth networks. In Proceedings of the 2015 IEEE/RSJ International Conference on Intelligent Robots and Systems (IROS), Hamburg, Germany, 28 September–2 October 2015; pp. 2572–2577.
26. Dubé, R.; Gawel, A.; Sommer, H.; Nieto, J.; Siegwart, R.; Cadena, C. An online multi-robot SLAM system for 3D LiDARs. In Proceedings of the 2017 IEEE/RSJ International Conference on Intelligent Robots and Systems (IROS), Vancouver, BC, Canada, 24–28 September 2017; pp. 1004–1011. [CrossRef]
27. Karrer, M.; Schmuck, P.; Chli, M. CVI-SLAM—Collaborative Visual-Inertial SLAM. *IEEE Robot. Autom. Lett.* **2018**, *3*, 2762–2769. [CrossRef]
28. Xu, H.; Liu, P.; Chen, X.; Shen, S. D^2 SLAM: Decentralized and Distributed Collaborative Visual-inertial SLAM System for Aerial Swarm. *arXiv* **2022**, arXiv:2211.01538.
29. Ramtoula, B.; De Azambuja, R.; Beltrame, G. CAPRICORN: Communication aware place recognition using interpretable constellations of objects in robot networks. In Proceedings of the 2020 IEEE International Conference on Robotics and Automation (ICRA), Paris, France, 31 May–31 August 2020; pp. 8761–8768.
30. Grisetti, G.; Stachniss, C.; Burgard, W. Improved techniques for grid mapping with rao-blackwellized particle filters. *IEEE Trans. Robot.* **2007**, *23*, 34–46. [CrossRef]
31. Hörner, J. *Map-Merging for Multi-Robot System*; Charles University: Prague, Czech Republic, 2016.

sensors

MDPI

Article

Differential Positioning with Bluetooth Low Energy (BLE) Beacons for UAS Indoor Operations: Analysis and Results

Salvatore Ponte [1], Gennaro Ariante [2,*], Alberto Greco [2] and Giuseppe Del Core [2]

1 Department of Engineering, University of Campania "L. Vanvitelli", 81031 Aversa, Italy; salvatore.ponte@unicampania.it

2 Department of Science and Technology, University of Naples "Parthenope", 80133 Naples, Italy; alberto.greco@uniparthenope.it (A.G.); giuseppe.delcore@uniparthenope.it (G.D.C.)

* Correspondence: gennaro.ariante@studenti.uniparthenope.it

Abstract: Localization of unmanned aircraft systems (UASs) in indoor scenarios and GNSS-denied environments is a difficult problem, particularly in dynamic scenarios where traditional on-board equipment (such as LiDAR, radar, sonar, camera) may fail. In the framework of autonomous UAS missions, precise feedback on real-time aircraft position is very important, and several technologies alternative to GNSS-based approaches for UAS positioning in indoor navigation have been recently explored. In this paper, we propose a low-cost IPS for UAVs, based on Bluetooth low energy (BLE) beacons, which exploits the *RSSI* (received signal strength indicator) for distance estimation and positioning. Distance information from measured *RSSI* values can be degraded by multipath, reflection, and fading that cause unpredictable variability of the *RSSI* and may lead to poor-quality measurements. To enhance the accuracy of the position estimation, this work applies a differential distance correction (DDC) technique, similar to differential GNSS (DGNSS) and real-time kinematic (RTK) positioning. The method uses differential information from a reference station positioned at known coordinates to correct the position of the rover station. A mathematical model was established to analyze the relation between the *RSSI* and the distance from Bluetooth devices (Eddystone BLE beacons) placed in the indoor operation field. The master reference station was a Raspberry Pi 4 model B, and the rover (unknown target) was an Arduino Nano 33 BLE microcontroller, which was mounted on-board a UAV. Position estimation was achieved by trilateration, and the extended Kalman filter (EKF) was applied, considering the nonlinear propriety of beacon signals to correct data from noise, drift, and bias errors. Experimental results and system performance analysis show the feasibility of this methodology, as well as the reduction of position uncertainty obtained by the DCC technique.

Keywords: UAS; indoor positioning system; BLE beacon; *RSSI*; trilateration; extended Kalman filter; differential distance correction

1. Introduction

The possibility of gathering real-time information from the environment by means of sensors onboard smart devices and objects has paved the way for the concept of Internet of Things (IoT), with which all smart things, such as devices, sensors, and industrial and utility components, are interconnected through networks, changing our way of living, working, and studying [1]. IoT is becoming important for context awareness [2], network management and security, health monitoring [3], personal delivery [4], location-based services (LBS), cellular network-based indoor positioning [5], and so on.

Unmanned aircraft systems (UASs) are being used in several operational tasks and in civil, military, and scientific contexts, thanks to their flexibility, versatility, low cost, and ease of use, especially in environments that are very dangerous or impossible for human intervention [6]. UASs also show the potential to save lives, increase safety and efficiency, and enable more effective science and engineering research. Knowing the aircraft position in real time during a mission, particularly during beyond visual line of sight (BVLOS) or full autonomy operations, is essential [7–9]. As is well known, in outdoor operations, UAS positioning depends on onboard GNSS receivers that provide location coordinates in a geographical reference system. In many cases, the GNSS positioning accuracy (typically less than 5 m) is not sufficient for precision navigation (e.g., maneuvering in crowded areas, landing in impervious sites, or precision landing), due to attenuation, multipath errors, or signal reflections into urban canyons. On-board sensors and detect-and-avoid systems (DAA) can be used for supporting UAS navigation in harsh environments or sites where GNSS systems have low accuracy or in the absence of GNSS signals [10–17]. Moreover, positioning ability is limited in indoor scenarios or shelters because of blocked GNSS signals [18].

Indoor navigation systems have a wide range of applications: for example, wayfinding for humans in railway stations, bus stations, shopping malls, museums, airports, and libraries. Visually impaired individuals also could significantly benefit from indoor navigation systems. Needless to say, navigation through indoor spaces is more challenging [19]. Currently, UAS usage is not limited to outdoor operations, and unmanned aircraft can be useful in unknown or challenging indoor environments like hospitals, production companies, greenhouses, and manufacturing facilities. Many researchers are focusing their attention on the development of technical solutions for accurate and reliable UAS positioning during indoor operations, exploiting alternative technologies that must be compared to existing methods in terms of accuracy, precision, coverage, computational complexity, cost, and compatibility. Most of the solutions proposed in the literature are based on multi-sensor data fusion between INS/IMU (inertial navigation systems/inertial measurement unit) and other sensors such as camera, visual laser LiDAR, camera radar, LiDAR, or ultrasonic [20–23]. Alternative methodologies are based on image fusion with ultra-wide band (UWB) systems [24,25], on Wi-Fi access points (APs) [26], infrared (IR), radio frequency identification systems (RFID) [27], and computer vision [28]. Table 1 shows a comparison among typical technologies for indoor positioning [29] compared with our method.

Table 1. Overview of indoor positioning technologies.

Technology	Accuracy (m)	Power Consumption	Extra Device	Cost
Wi-Fi	~2–5	High	No	Low
Bluetooth	2–5	Low	No	Low
BLE	1–5	Very low	No	Very low
RFID	~1–3	Low	Yes	Moderate
UWB	~0.1–0.5	Low	Yes	High
Infrared	0.5–3	Low	Yes	Moderate
Acoustic signal	0.3–0.8	Low	No	Moderate
BLE + DDC (our method)	0.4–0.6	Very low	Yes	Very low

In this work, we propose a low-cost UAS indoor positioning system based on Bluetooth low energy (BLE) beacons. A differential distance correction method (DDC) is proposed to improve the positioning accuracy. Sharing the same indoor propagation characteristics as 2.4-GHz Wi-Fi transceivers, BLE is the new specification of Bluetooth technology that guarantees transmission of small amounts of data and ultra-low power consumption [30,31]; consumption is up to 1% of the classic Bluetooth (with typical power consumption of 2.5 mW [32]. BLE beacons are a promising method for indoor positioning, especially in applications of position-based services, thanks to their low deployment cost, low power and dimensions, and suitability for a wide range of mobile devices. Table 2 shows some differences between classic Bluetooth and BLE.

Table 2. Classic Bluetooth vs. BLE technology.

Propriety	Bluetooth	BLE
Frequency	2.4 GHz	2.4 GHz
Data rate	1 to 3 Mbps	1 Mbps
Range	Up to 10 m	Up to 40 m
Power consumption	Low	Very low
Battery life	Multiple weeks	Multiple months
Cost	Low	Very low
Accuracy	2–5 m	1–5 m

The beaconing, or advertising, mode of BLE devices allows the hardware transmitter to broadcast advertising packets, i.e., short unsolicited messages, at flexible update rates [33]. Among the kinds of information permitted in the BLE standard, the received signal strength indicator (*RSSI*), which decreases with increasing distance, can be used to allow a receiving device to detect proximity to a specific BLE beacon based on the received signal strength (RSS). The flexibility of BLE device deployment can also allow good signal geometries for radio positioning; on the contrary, the location of Wi-Fi access points, typically near power sources, is chosen in order to maximize the signal coverage of the indoor area, rather than to optimize the wireless positioning of objects in the field.

BLE-based positioning methods typically consist of two approaches: range-based and fingerprint-based. The range-based method uses a predefined radio frequency (RF) path loss model to estimate the distance between the receiver (user) and the beacons. Assuming a minimum of three *RSSI* measurements, the user's position can be solved by trilateration [34]. The fingerprint-based methods refer to the pattern of RSS measurements at a given location and consist of signal identity information (e.g., Wi-Fi MAC addresses or cellular IDs) and RSS values. Fingerprinting involves an offline and an online phase. During the offline phase, fingerprints at different places are collected to create a reference fingerprint map (RFM). In the online phase, a fingerprint collected at an unknown place is compared to the fingerprints in the RFM to solve for the user position [35].

This work applies to a range-based methodology by exploiting *RSSI* measurements for estimation of the distance between the receiver and a series of transmitters (beacons) using trilateration. To improve the accuracy, an extended Kalman filter (EKF) was employed on the noisy *RSSI* values. This study used a differential distance correction (DDC) methodology to correct the measured distances of a reference station from the beacons, and the position estimation of an object (rover) was calculated by trilateration and the *RSSI*-derived measurements. The differential correction was sent to the rover to attain a refined estimate of its location. During experimental tests, a Raspberry Pi 4 model B board was used as the reference (or master) station, and the Arduino Nano 33 BLE as the rover, whereas the anchor points (transmitters) were BLE Eddystone beacons. Experimental data were collected during several indoor tests conducted by the PFDL (Parthenope Flight Dynamics Labs) team of the University of Naples "Parthenope" (Italy).

The paper is organized as follows. In Section 2, the theoretical framework (*RSSI* distance method, trilateration method, EKF and DDC techniques) is presented. Section 3 describes the hardware used for the prototype of our positioning system. Simulations and results are shown and discussed in Section 4. Section 5 concludes the paper with final considerations and future work directions.

2. Theoretical Framework

The observation geometry of the BLE-based indoor positioning system (IPS) prototype mounted on a UAV is depicted in Figure 1. BLE was introduced in the Bluetooth 4.0 Standard, allowing advanced use for indoor localization technology by introducing a new type of device called "BLE beacons". This new technology reduces costs and power consumption with respect to the "classical" Bluetooth; indeed, unlike the devices using the previous standard, the new ones have the option of transmitting at set intervals, which contributes significantly to the energy efficiency of the system, also improving the hardware and the immunity to interference. BLE has many similarities with Wi-Fi (in the 2.4-GHz band), and it is often used for indoor positioning in the same way as Wi-Fi, i.e., applying *RSSI*-based techniques. The BLE advertisement channels are labelled 37, 38, and 39 and are centered on 2402 MHz, 2426 MHz, and 2480 MHz, with 2 MHz bandwidth. Frequency hopping is used to communicate, each advertisement is repeated on each of the three channels, and the receiving device cycles over the advertising channels listening to the sent packets.

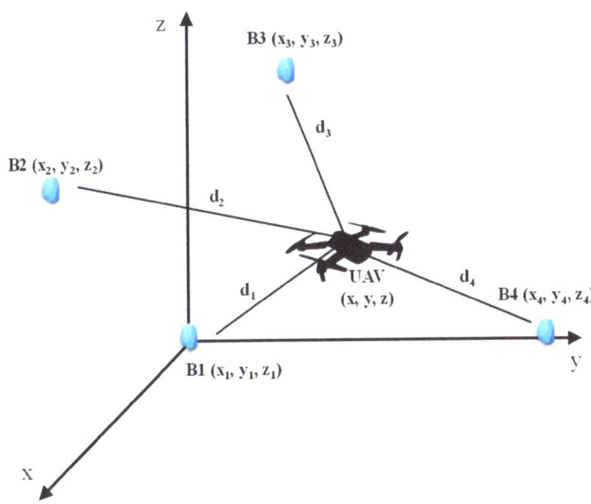

Figure 1. UAS indoor 3-D positioning system with four BLE devices. The ideal aircraft position is provided by the intersection of four spheres with centers on the known positions of the beacons B1, …, B4.

2.1. RSSI Distance Model

The principle of distance measurements by *RSSI* consists of transforming the signal attenuation into distance from the signal source, using the following empirical formula, which relates the BLE *RSSI* with the transmission distance based on the commonly used logarithmic attenuation model [36–39]:

$$L_d = L_l + 10\, n log_{10} d + \nu \tag{1}$$

$$L_l = 10\, log_{10}\left[G_t G_r \left(\frac{c/f}{4\pi}\right)^2\right] \tag{2}$$

where G_t and G_r are the transmitting and receiver antenna gain, respectively; c is the speed of light; f is the carrier frequency; n is the channel attenuation coefficient (typically in a range from 2 to 6); ν is the noise (modeled as zero-mean, Gaussian, with variance σ^2, i.e., $\mathcal{N}(0, \sigma^2)$); d is the distance between receiver and transmitter; and L_d is the channel loss after d meters. It is important to mention that the measurement error in *RSSI* does not regularly produce a Gaussian distribution, but for simplicity, the *RSSI* measurement error is treated as a Gaussian random variable [40,41]. A simplified relationship, based on Equation (1), in which we set the reference distance equal to 1 m, is [42]:

$$P_r\ (d) = A - 10\ n\ log_{10}d \tag{3}$$

where P_r represents the *RSSI* (in dBm); *A* (dBm) is the *RSSI* obtained when the signal transmits from 1 m distance from the receiver; and *n* is the propagation factor (also known as environmental factor or attenuation factor), which, as well as the noise standard deviation σ, depends on environmental conditions. When a BLE beacon is used, it periodically broadcasts an advertisement packet containing a unique ID and a calibrated *RSSI* value, relative to the distance with respect to receiver. This value allows us to determine the distance between a beacon and a device using the model in Equation (3), where *n* can also be interpreted as a calibration parameter for the path loss exponent. Obviously, the approach assumes that all installed beacons possess their predefined location information, including their exact coordinates. The general expression relating the *RSSI* to the distance is:

$$RSSI = \ A - 10\ n\ log_{10}\left(\frac{d}{d_0}\right) \tag{4}$$

where d_0 is the reference distance value.

Time and space variations of the signal environment inevitably degrade the environmental factor of BLE reference nodes. When a BLE receiver (rover) enters the coverage area, there are not only reference nodes but also walls, other wireless devices, obstacles, and so on. The *n* factor describes the influence of walls and other obstacles inside the scenario. The environmental factor n_i between two reference nodes can be estimated by:

$$n_i = \left(\frac{RSSI - A}{10log_{10}d_i}\right) \tag{5}$$

where *RSSI* is the received power of the *i*-th reference node and d_i is the known distance between the two reference nodes. At the reference distance d_0 (1 m), proximity and distance estimation can be calculated from Equation (4) as follows:

$$d = 10^{\frac{A - RSSI}{10n}} \tag{6}$$

The localization algorithm can then be applied to use this distance and estimate the position using trilateration.

2.2. Extended Kalman Filter (EKF)

To obtain accurate position estimates from the ideal values of *RSSI*, filtering techniques are applied in order to mitigate noise effects and *RSSI* signal drift, resulting in better proximity estimation [43–45]. In this study, EKF was adopted for noise reduction and bias corrections considering the nonlinear characteristics of the beacon signal [46]. During the initial simulation tests, it was assumed that the UAV maintains a constant altitude (zero for simplicity). Consequently, dynamic behavior was equivalently characterized on a 2-D surface. EKF was designed to implement a measurement model, where the distance measurement based on the *RSSI* from the beacons exhibits a nonlinear relationship with the state values (UAS position). The measurement process model is

$$x_{k+1} = f(x_k,\ u_k, v_k) \tag{7}$$

$$z_k = g(x_k,\ u_k,\ v_k) \tag{8}$$

where k is the generic time index; x is the state; u is the input; z is the measured output; and υ and v are the process and measurement noises, respectively. We assume normal distribution for the noises υ_k and v_k, with known covariance matrices Q_k and R_k:

$$\upsilon_k \sim \mathcal{N}(0,\ Q_k) \qquad v_k \sim \mathcal{N}(0,\ R_k) \tag{9}$$

The algorithm is divided into two phases, prediction and correction. At each time step, the prediction phase provides the a priori state estimation $\hat{x}_{k+1}^-$ and the state covariance matrix P_{k+1}^-, whereas during the correction phase (update phase), the a posteriori state estimate $\hat{x}_{k+1}^+$ and the state covariance matrix P_{k+1}^+ are carried out.

The a priori estimation of the state $\hat{x}_{k+1}^-$ is performed by Equation (7), considering null process noise $(\upsilon_k = 0)$:

$$\hat{x}_{k+1}^- = f(\hat{x}_k^+,\ u_k, 0) \tag{10}$$

The a priori state covariance matrix P_{k+1}^- is obtained as follows:

$$P_{k+1}^- = A_k\, P_k^+\, A_k^T + W_k\, Q_k\, W_k^T \tag{11}$$

where

$$A_k = \left(\frac{\partial f}{\partial x}\right)_{\hat{x}_k^+,\ u_k, 0} \qquad W_k = \left(\frac{\partial f}{\partial \upsilon}\right)_{\hat{x}_k^+,\ u_k, 0}$$

are the Jacobian matrices of the function f, and the P_k^+ is the a posteriori state covariance matrix at time index k. Once a new measurement has been acquired, the a posteriori state estimation can be computed by updating the a priori state estimation:

$$\hat{x}_k^+ = \hat{x}_k^- + K_k(z_k - h(\hat{x}_k^-)) \tag{12}$$

where the Kalman gain (K_k) is given by

$$K_k = P_k^-\ H_k^T \left(H_k P_k^- H_k^T + V_k R_k V_k^T\right)^{-1} \tag{13}$$

H_k is the observation matrix:

$$H_k = \left(\frac{\partial z}{\partial x}\right)_{\hat{x}_k^-,\ u_k, 0} \qquad V_k = \left(\frac{\partial z}{\partial v}\right)_{\hat{x}_k^-,\ u_k, 0}$$

Finally, the a posteriori state matrix is updated as follows:

$$P_k^+ = (I - K_k H_k) P_k^- \tag{14}$$

where I is the identity matrix.

2.3. Trilateration Method and Least-Squares Solution

Trilateration is a classical signal-based positioning technique that utilizes the estimated distances from known reference points to determine the target location [47]. At least three reference nodes are required to identify a unique solution. Each line of position is a circle with radius equal to the exact distance from the reference point, and the intersection of the circles is the target position, represented in Figure 2a for an ideal scenario. In real scenarios, errors in distance measurements will transform the intersection point into an overlap among the three circumferences, creating an uncertainty area (Figure 2b).

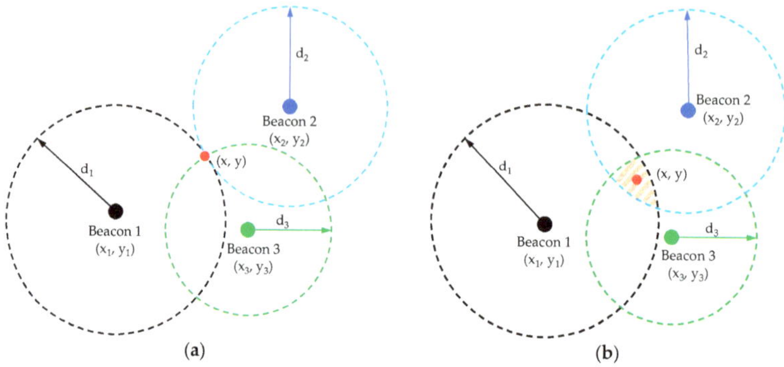

Figure 2. The 2-D trilateration method in an ideal (**a**) and a real (**b**) scenario.

Consequently, there will be various possible solutions in the intersection area. When the number of distance measurements N is ≥ 3, the least squares method is adopted to calculate the optimal position of the target by minimizing the square of the offset in the user's position relative to a linearization point (initial estimate of the user position), that is, the 2×1 vector $\delta \vec{r}$, which is related to the $n \times 1$ vector $\delta \vec{d}$ (distance error, or offset in the distance values) by [48]

$$\delta \vec{d} = d_i - \hat{d}_\imath = \boldsymbol{H}\delta \vec{r}, \; i = 1, \ldots, N \tag{15}$$

The parameters present in Equation (15) are described below:

$$d_i = \sqrt{(x - x_i)^2 + (y - y_i)^2} \text{ (real beacon-target distance)} \tag{16}$$

$$\hat{d}_\imath = \sqrt{(\hat{x} - x_i)^2 + (\hat{y} - y_i)^2} \text{ (estimated distance)} \tag{17}$$

$$\boldsymbol{H} = \begin{bmatrix} \frac{\hat{x}-x_1}{\hat{d}_1} & \frac{\hat{y}-y_1}{\hat{d}_1} \\ \frac{\hat{x}-x_2}{\hat{d}_2} & \frac{\hat{y}-y_2}{\hat{d}_2} \\ \vdots & \vdots \\ \frac{\hat{x}-x_n}{\hat{d}_n} & \frac{\hat{y}-y_n}{\hat{d}_n} \end{bmatrix} \tag{18}$$

$$\delta \vec{r} = \begin{bmatrix} x - \hat{x} \\ y - \hat{y} \end{bmatrix} = \begin{bmatrix} \delta x \\ \delta y \end{bmatrix} \tag{19}$$

where $(x, y)^T$ are the real target coordinates, or user's position, where the superscript T stands for transpose; $(x_i, y_i)^T$ are the known i-th beacon coordinates; $(\hat{x}, \hat{y})^T$ are the estimated target coordinates; and $\boldsymbol{H}$ is the $\in \mathbb{R}^{Nx2}$ design matrix derived by linearizing Equation (18) about an initial position guess. The least-squares solution provides the offset of the user's position from the linearization point expressed as a linear function of $\delta \vec{d}$:

$$\delta \vec{r} = \left(H^T H\right)^{-1} H^T \delta \vec{d} \tag{20}$$

Once $\delta \vec{r}$ is calculated, a new estimate of the user's position is obtained from Equation (19), and a linearization around the new estimate is performed, obtaining the optimal solution by the iteration

$$\begin{bmatrix} \hat{x} \\ \hat{y} \end{bmatrix}_{k+1} = \begin{bmatrix} \hat{x} \\ \hat{y} \end{bmatrix}_k + \begin{bmatrix} \delta x \\ \delta y \end{bmatrix}_k \tag{21}$$

where $k+1$ denotes the updated coordinate vector and k denotes the vector at the previous iteration. The acceptable displacement $(\delta x, \delta y)^T$ is related to the accuracy requirements. The 3-D positioning by trilateration allows one to find the user position using the *RSSI* signal received from at least four non-coplanar beacons at known positions (the fourth equation removes the ambiguity between two possible solutions given by three spheres intersecting in two points):

$$\begin{cases} (x-x_1)^2 + (y-y_1)^2 + (z-z_1)^2 = d_1^2 \\ (x-x_2)^2 + (y-y_2)^2 + (z-z_2)^2 = d_2^2 \\ (x-x_3)^2 + (y-y_3)^2 + (z-z_3)^2 = d_3^2 \\ (x-x_4)^2 + (y-y_4)^2 + (z-z_4)^2 = d_4^2 \end{cases} \tag{22}$$

where (x, y, z) are the unknown 3-D coordinates of the target Bluetooth receiver.

During preliminary simulation tests, the z coordinates were set to zero to evaluate the accuracy and stability of the method. Considering 2-D positioning, it is possible to obtain the (x, y) coordinates of the receiver by

$$x = \frac{(x_1^2 + y_1^2 - d_1^2)(y_3 - y_2) + (x_2^2 + y_2^2 - d_2^2)(y_1 - y_3) + (x_3^2 + y_3^2 - d_3^2)(y_2 - y_1)}{2[y_1(x_3 - x_2) + y_2(x_1 - x_3) + y_3(x_2 - x_1)]} \tag{23}$$

$$y = \frac{(x_1^2 + y_1^2 - d_1^2)(x_3 - x_2) + (x_2^2 + y_2^2 - d_2^2)(x_1 - x_3) + (x_3^2 + y_3^2 - d_3^2)(x_2 - x_1)}{2[y_1(x_3 - x_2) + y_2(x_1 - x_3) + y_3(x_2 - x_1)]} \tag{24}$$

With the known locations of the beacons $B1 = (0,0); B2 = (0, y_2); B3 = (x_3, 0)$, as shown in Figure 3, Equations (23) and (24) are simplified as follows [49]:

$$x = \frac{x_3^2 + \left(d_1^2 - d_2^2\right)}{2x_3} \tag{25}$$

$$y = \frac{y_2^2 + \left(d_1^2 - d_3^2\right)}{2y_2} \tag{26}$$

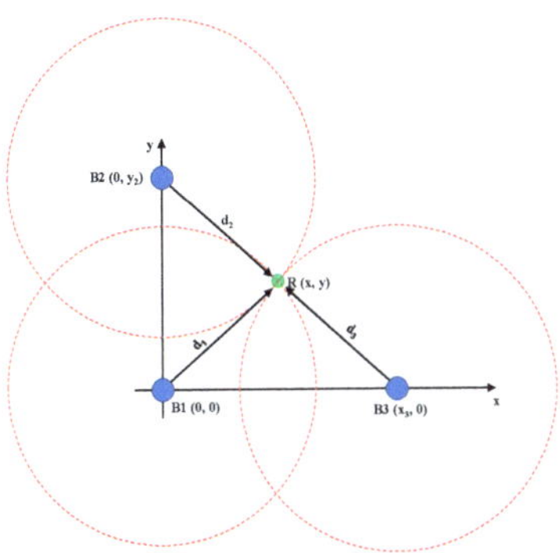

Figure 3. Positioning based on 2-D trilateration with three BLE beacons.

2.4. *Differential Distance Correction Technique*

The distance converted from *RSSI* is an estimated value influenced by environmental issues such as multipath, reflections, and shadowing. The DDC methodology uses multiple reference stations to estimate the distance error [50]. The concept is similar to differential GNSS (DGNSS) and real-time kinematic (RTK) positioning that use the differential information from the reference station (Master), i.e., the error $\delta\vec{r}$, to correct the position of the rover, in the hypothesis that the measurements performed by the master station(s) and the rover are affected by the same environmental errors (i.e., with low temporal and spatial latency). Figure 4 shows the schematic diagram of differential correction [51,52] with four beacons located at known positions. The coordinates of the master station are also known in advance.

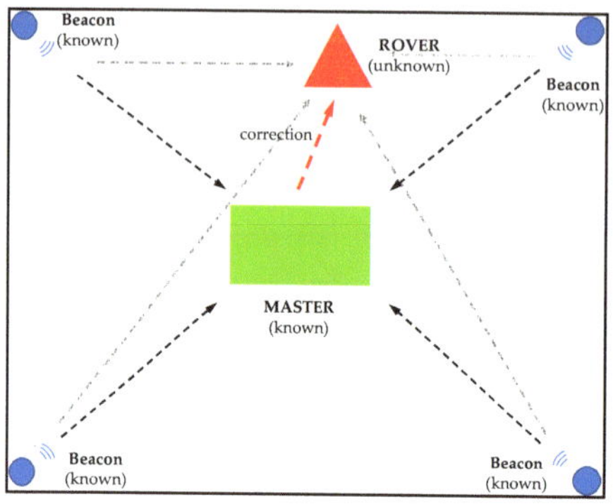

Figure 4. Schematic diagram of DDC methodology with a single master station of known position.

The master station compares the estimated position to its known coordinates to evaluate the errors and sends the corrections to be applied to the *RSSI* values measured by the rover. To estimate the residual at every location in the field, the inverse distance weighted (IDW) interpolation method is adopted. The estimated residual r_{est} at unknown coordinate (x, y) by IDW can be calculated from the equation as follows [50,52]:

$$r_{est} = \sum_{i=1}^{N} w_i \times r_i \tag{27}$$

where r_{est} is the estimated residual from IDW, r_i is the estimated residual of the i-th reference station, and w_i is the weight of the i-th reference station.

3. Hardware Description

The prototype of our experimental BLE-based IPS is composed of

- Raspberry Pi 4 model B is used as the master station.
- Arduino Nano 33 BLE used as the rover.
- BLE Eddystone beacons as transmitting nodes.

3.1. Master Station: Raspberry Pi 4 Model B

The Raspberry Pi 4 Model B used for our prototype system (Figure 5) is a board with features like a camera connector, 802.11ac Wi-Fi, Bluetooth 5, full gigabit Ethernet (throughput not limited), two USB 2.0 ports, two USB 3.0 ports, 1–8 GB of RAM, dual-monitor support via a pair of micro-HDMI (HDMI type D) portsFirst bullet, GPIO pins for interfacing sensors and switches, USB ports to connect to external devices (keyboard, mouse, Wi-Fi adapter, etc.), and an audio jack [53]. The board has no internal mass storage or built-in operating system, requiring an SD card preloaded with a version of the Linux Operating System. Due to its relatively low current consumption, Raspberry Pi can be powered using a 5 V USB Power Bank, and it can be carried around by a subject or placed on the object to be tracked. The main algorithm is hosted on the 1.5 GHz 64-bit quad core ARM Cortex-A72 processor on-board the Pi 4.

Figure 5. Raspberry Pi 4 model B.

The master station is connected to a PC-based ground control station (GCS) by hotspot connection to send and store *RSSI* values of each single transmitter. The main task of the Raspberry Pi microcomputer is to receive and process the messages transmitted by the BLE beacons, in particular, the beacon identifier address and the *RSSI* value. From these values, the algorithm estimates the real-time position and compares it to its real position in order to calculate the error (residual) and send the correction to the rover.

3.2. Rover: Arduino Nano 33 BLE

Arduino Nano 33 BLE is a multiprotocol microcontroller (Figure 6, Table 3), supporting only 3.3 V I/Os. In addition, the 5 V pin is connected, through a jumper, to the USB power input. There are two 15-pin connectors on the board, one on each side, pin to pin compatible with the original Arduino Nano. The main processor is the nRF52840 (Nordic Semiconductors), a 32-bit ARM Cortex-M4 running at up to 64 MHz. The main processor includes other features like Bluetooth pairing via NFC (near field communication) and ultra-low-power-consumption modes. It has an embedded 9-axis inertial sensor that makes this board ideal for wearable devices but also for a large range of scientific experiments in the need of short-distance wireless communication. Most of its pins are connected to the external headers; however, some are reserved for internal communication with the wireless module and the on-board internal I^2C peripherals (IMU and Crypto) [54,55].

Figure 6. Arduino nano 33 BLE.

Table 3. Arduino nano 33 BLE main characteristics.

Type Characteristics	Nano 33 BLE
Microcontroller	nRF52480
Clock speed	64 MHz
Flash	1 MB
RAM	256 KB
Connectivity	BLE
Sensors	9-axis IMU

The Arduino nano 33 BLE is powered by two 2600-mAh Li-ion 18650 3.7-V batteries and has a microSD connected to the board to store *RSSI* values. Figure 7 shows the prototype of the rover station. The algorithm running on the Arduino can receive data from the beacons and save them on the micro-SD.

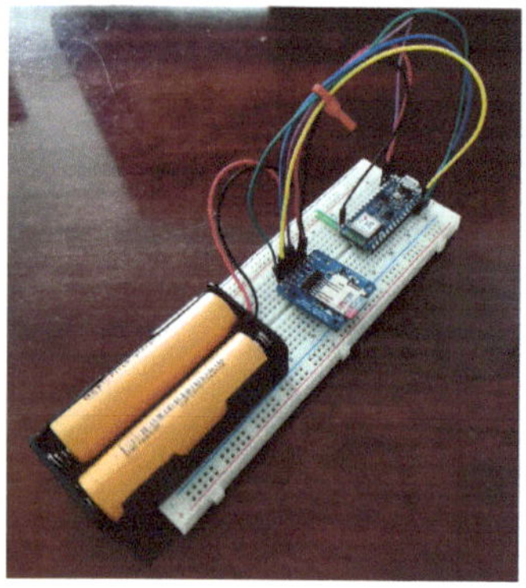

Figure 7. Rover station prototype.

3.3. Transmitters: BLE Eddystone Beacons

A beacon is a hardware transmitter capable of broadcasting periodically a short message to nearby portable electronic devices. The basic information of the advertising packet from the beacon is shown in Table 4.

Table 4. Beacon advertising packet [51].

Beacon prefix	Fixed by beacon protocol and makes information follow the protocol.
UUID identifier	Identifier that should be used to distinguish the different classes of beacons in a wide range.
Major	Identifier for determining a different group of beacons.
Minor	Identifier for determining individual beacons.
RSSI	The radio signal strength indicator (*RSSI*) is a measurement of the power present in a received radio signal.
Measured power	The *RSSI* value, which is measured at 1 m away from a beacon.

We used Eddystone BLE beacons with an embedded nRF51822 chip (Figure 8). Eddystone is a protocol specification that defines a BLE message format for proximity beacon messages. It describes several different frame types that may be used individually or in combinations to create beacons that can be used for a variety of applications [56].

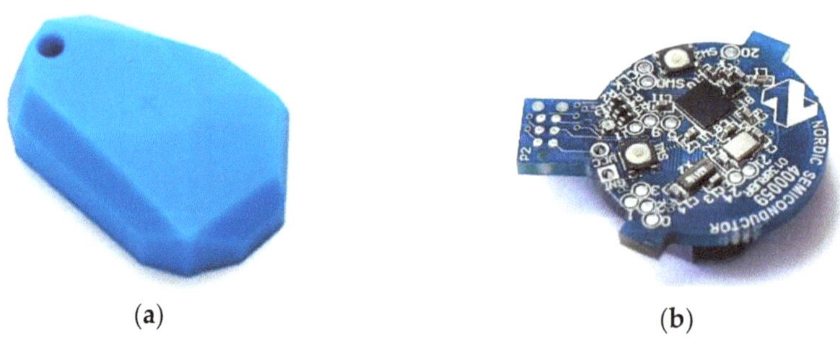

Figure 8. Eddystone beacon: (**a**) silicon cover, (**b**) chip nRF51822.

The nRF51822 is a powerful multi-protocol single chip solution for wireless applications. It incorporates a 32-bit ARM Cortex M0 CPU and 256 kB flash + 16 kB RAM memory, and it supports Bluetooth low energy and 2.4 GHz protocol stacks [57]. The programmable peripheral interconnect (PPI) system provides a 16-channel bus for direct and autonomous system peripheral communication without CPU intervention. This brings predictable latency times for peripheral-to-peripheral interaction and power saving benefits associated with leaving the CPU idle.

4. Simulation and Results

The software for IPS is divided into three phases, shown in Figure 9, each running on its own platform. MATLAB 2024b® runs on the GCS and is used for non-real-time algorithms (post-processing).

Figure 9. Phases of the proposed IPS.

4.1. Calibration Phase

During the first phase, environment calibration is performed, necessary for environmental factor (n) modelling and Kalman filter calibration. Preliminary data collection consisted of indoor *RSSI* measurements at known distances between the beacons and the receiver. The distance range considered was 0.25–5 m, with steps of 0.25 m for the range 0.25–3 m, and with steps of 0.50 m for the range 3–5 m. The acquisition time was about 10 min in static conditions. The measurement procedure was performed in a clean environment (without metallic objects and the presence of other nearby devices). This first phase (environmental calibration, n factor modeling and Kalman filter calibration) is presented in [58], where the data collections and preliminary results are shown in detail. Noting the high presence of instability in the *RSSI* signal, a simple smoothing filter based on a moving average was applied before implementing the EKF algorithm. Figures 10 and 11 show data collections (raw *RSSI* values and smoothing filtered data), during the calibration phase, where the distances between receiver and transmitters (four beacons were used) were 3 and 0.75 m, respectively. Figure 12 shows the comparison between variances of the raw and filtered *RSSI* values, and Figure 13 shows the *RSSI* mean values of the filtered data.

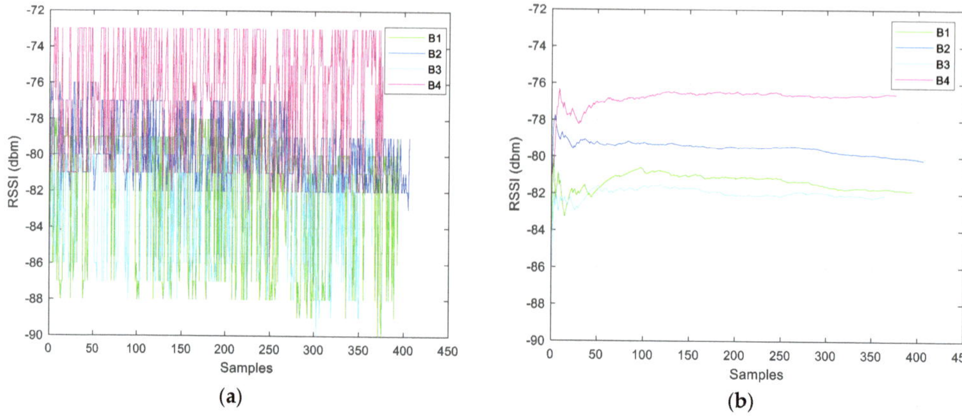

Figure 10. *RSSI* values, raw (**a**) and filtered (**b**), at 3 m distance.

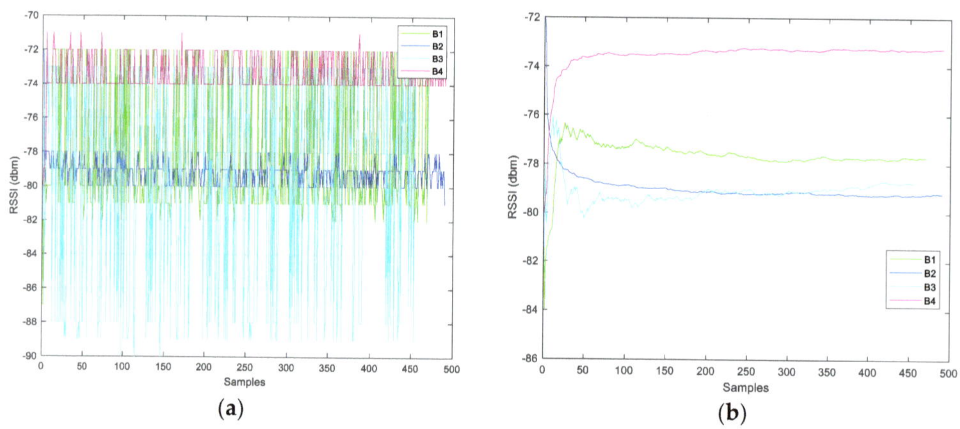

Figure 11. *RSSI* values, raw (**a**) and filtered (**b**), at 0.75 m.

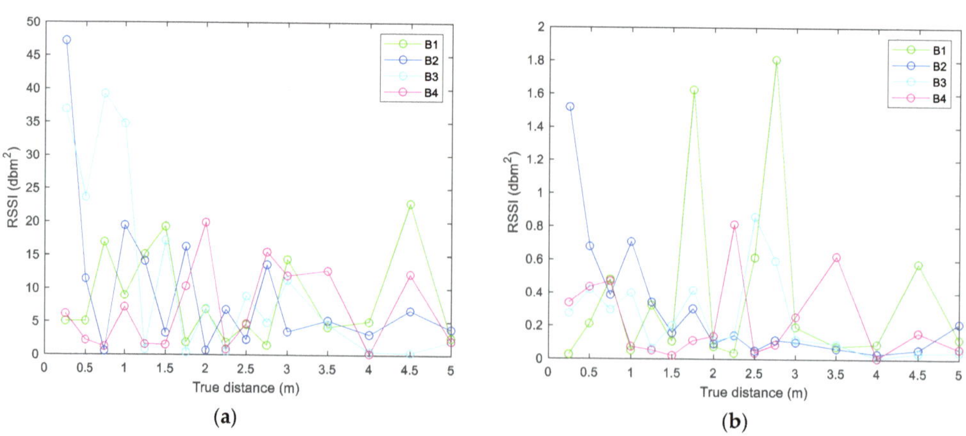

Figure 12. Variance of the *RSSI* raw (**a**) and filtered (**b**) values.

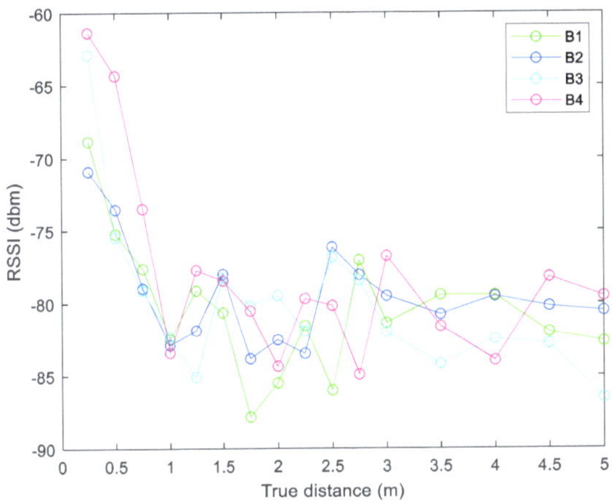

Figure 13. Mean *RSSI* values of the filtered data.

After data collection, the environmental factor n was evaluated for each known distance and for each beacon, following Equation (5). This factor describes the influence of walls and other obstacles inside the scenario (Figure 14).

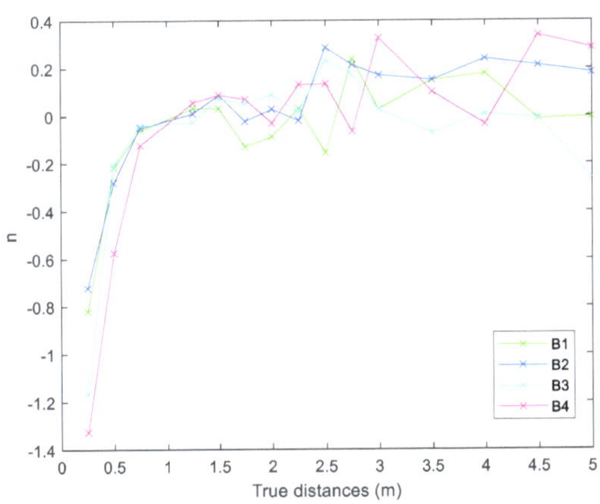

Figure 14. Trend of the measured environmental factor.

4.2. Experimental Tests: 2-D Scenario

After the calibration phase, experimental tests were conducted by positioning the beacons and the master station within a designated area. The rover, simulating a UAV maintaining a constant altitude (in this case equal to zero simulating take-off positioning area, reducing dynamic behavior on a 2-D surface), was placed at unknown distances from the transmitters and the master station within this area. Experimental indoor tests were performed in a free area inside the University of Studies of Naples "Parthenope" (Italy). The field has a length of 3 m and width of 3 m. The arrangement of the experimental field is depicted in Figure 15, where there were four beacons (B1, …, B4) and one reference station in the field. Table 5 shows the coordinates of the work area, delimited by the beacon positions, and of the master station.

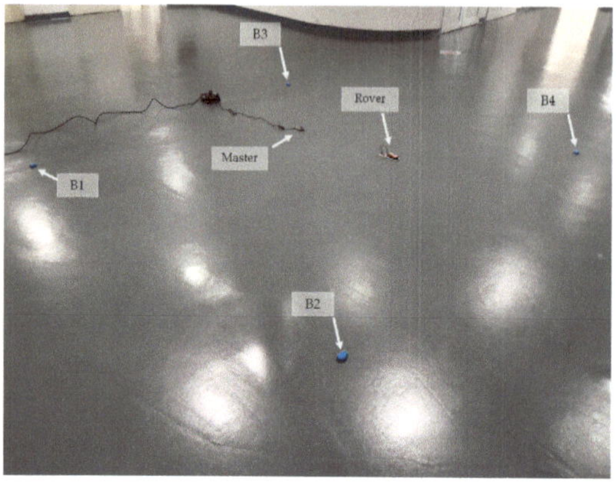

Figure 15. Experimental setup area.

Table 5. Work area: beacon and maser coordinates (origin in B1 position).

Devices	Coordinates (m)
Master	(1, 2)
B1	(0, 0)
B2	(3, 0)
B3	(0, 3)
B4	(3, 3)

During the static tests, the master station was connected to the GCS by hotspot connection to store *RSSI* data from the beacons, whereas the rover was equipped with a micro-SD, allowing *RSSI* data storage. The acquisition time was about 10 min.

In post-processing, the raw master receivers' coordinates were calculated following Equations (25) and (26), where d_1, d_2 and d_3 were the unknown distances from B1, B2, and B3, respectively (B4 was added in case of failure of one of the other beacons). These distances were derived from Equation (6), where n (see Figure 14) was obtained by using Equation (5). Figure 16 shows the raw coordinates of the master station compared to its actual position.

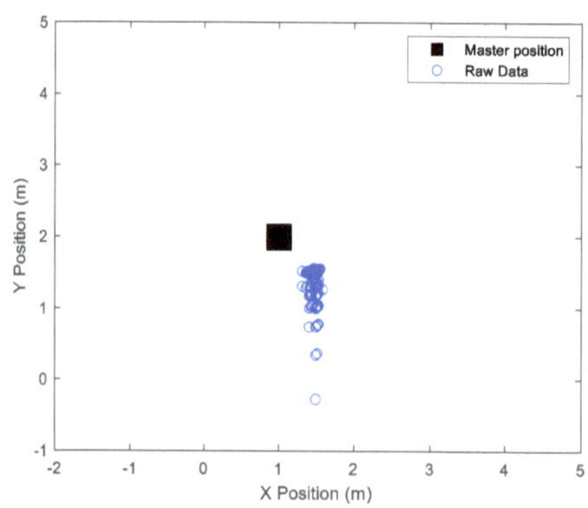

Figure 16. Comparison among real and raw master coordinates.

In the successive step, EKF was applied to mitigate noise, bias, and drift issues. Table 6 shows the parameters used during the EKF initialization phase.

Table 6. EKF initial parameters.

Parameter	Value
Initialization state x	[0, 0]
σ_R^2 (dbm^2)	2
σ_Q^2 (dbm^2)	1
P	I(2)
R	$\sigma_R^2 \cdot$ I(4)
Q	$\sigma_Q^2 \cdot$ I(2)

Figures 17 and 18 show a comparison between raw data, EKF data, and the actual position of the master station and rover station, respectively.

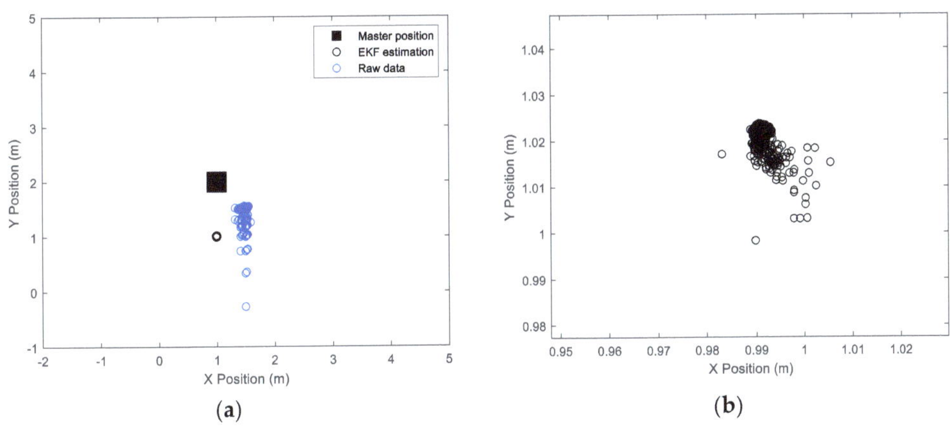

Figure 17. (**a**) Raw and EKF coordinates estimation, compared to the real position of the master station. (**b**) Zoom view of the EKF data estimation.

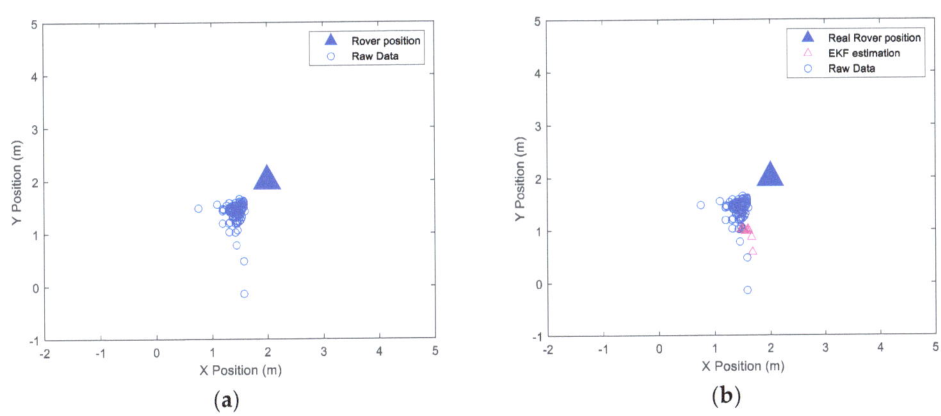

Figure 18. (**a**) Raw and EKF coordinates estimation, compared with the real position of the rover station. (**b**) Zoom view of the EKF data estimation.

Table 7 presents the real positions of the master and rover stations, along with their corresponding raw and EKF coordinate mean values.

Table 7. Comparison among real, raw, and EKF coordinates of the master station and rover.

Master Station Coordinates			Rover Coordinates		
Real (m)	Raw (m)	EKF (m)	Real (m)	Raw (m)	EKF (m)
(1, 2)	(1.48, 1.37)	(0.99, 1.02)	(2, 2)	(1.45, 1.41)	(1.51, 1.01)

Following these steps, the coordinates of the master station were calculated and compared to the real coordinates to evaluate the residual (correction), used to correct the rover position (Figure 19). Then, the rover coordinates were calculated, and the residual was added, correcting its position.

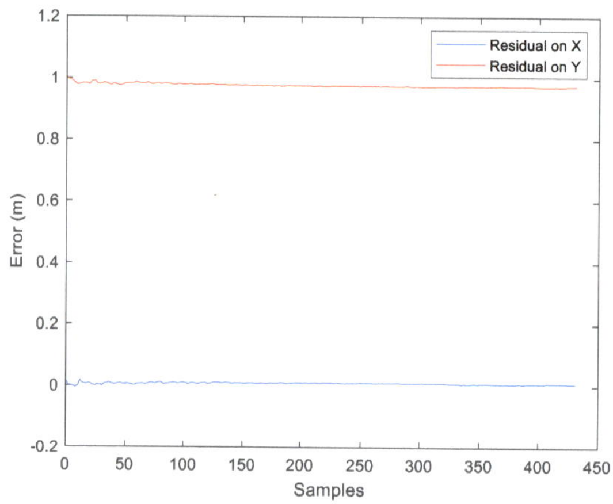

Figure 19. Error calculated on known master coordinates, used to correct the rover position by the DDC method.

Figure 20 shows the configuration area and the comparison between the testing point's estimated position and the real position of the rover.

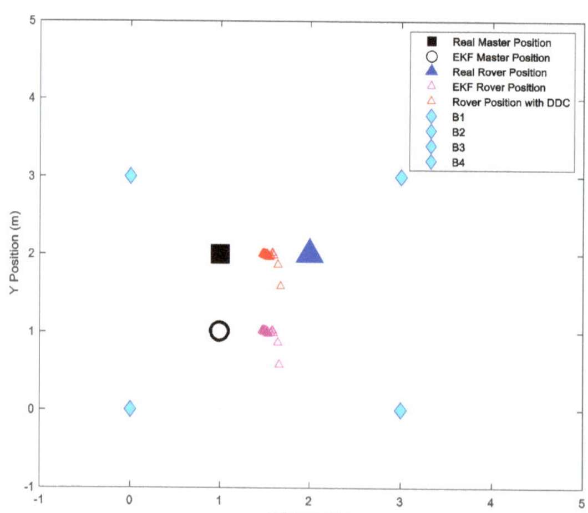

Figure 20. Configuration area representing final positioning results.

Table 8 compares raw and filtered data in terms of the standard deviation of the error.

Table 8. Error standard deviation (raw and EKF data).

Master Station		Rover	
STD Raw Data (m)	STD EKF (m)	STD Raw Data (m)	STD EKF (m)
(0.05, 0.59)	(0.002, 0.003)	(0.10, 0.18)	(0.02, 0.03)

Table 9 shows the results indicating the final position corrected using the DDC method, in terms of mean value.

Table 9. Positioning results with DDC methodology during the first test.

Rover Coordinates	
Real (m)	With DDC (m)
(2, 2)	(1.52, 1.99)

The second test was conducted using the same configuration area, beacons, and master positions as the first test, but with a different rover position. Figure 21 illustrates the configuration area, and Table 10 presents the obtained results.

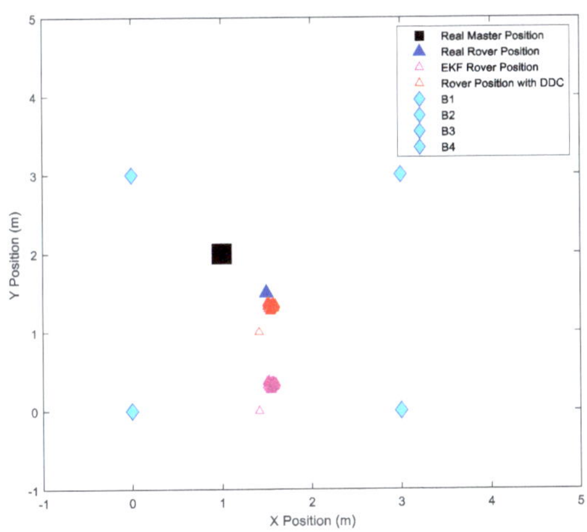

Figure 21. Configuration area representing final positioning results during the second test.

Table 10. Positioning results with the DDC method during the second test.

Rover Coordinates	
Real (m)	With DDC (m)
(1.5, 1.5)	(1.54, 1.28)

4.3. Experimental Test: 3-D Scenario

In this section, a 3-D test is presented, considering the same scenario depicted during the tests performed and described in the previous section. For this test, the rover station was positioned on a vertical structure, at 1.50 m above the ground (as shown in Figure 22), simulating a drone in hovering phase.

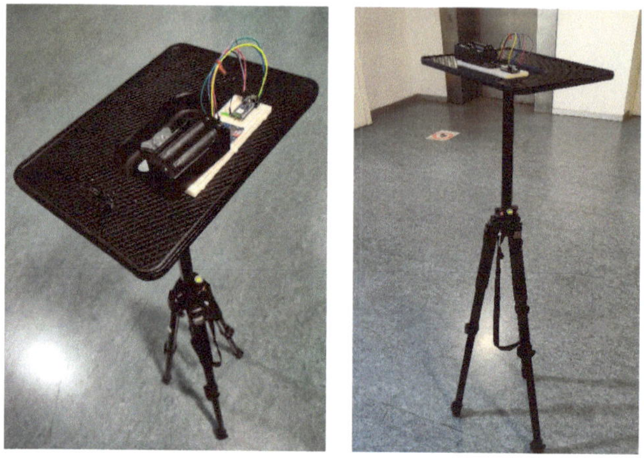

Figure 22. Rover prototype placed on a vertical structure.

The beacons were positioned at the same coordinates as in the 2-D tests. To include the third dimension, they were placed one meter above the ground. The master station's coordinates remained similar, simulating a ground station. Table 11 shows the coordinates of the work area during the 3-D test, whereas Table 12 shows the parameters used during the EKF initialization phase for the 3-D scenario.

Table 11. Work area: beacon and maser coordinates (origin in B1 position) during the 3-D test.

Devices	Coordinates (m)
Master	(1, 2, 0)
B1	(0, 0, 1)
B2	(3, 0, 1)
B3	(0, 3, 1)
B4	(3, 3, 1)

Table 12. EKF initial parameters.

Parameter	Value
Initialing state x	[0, 0, 0]
σ_R^2 (dbm^2)	2
σ_Q^2 (dbm^2)	1
P	I(3)
R	$\sigma_R^2 \cdot$ I(4)
Q	$\sigma_Q^2 \cdot$ I(3)

Figure 23 depicts the filtered coordinates of the master station, which will be compared to its real position to calculate the residual and subsequently correct the rover's position, as illustrated in Figure 24.

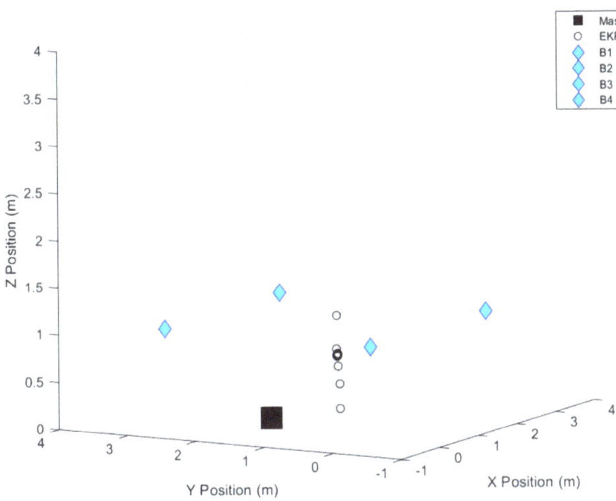

Figure 23. EKF data estimation of the master station in the 3-D scenario.

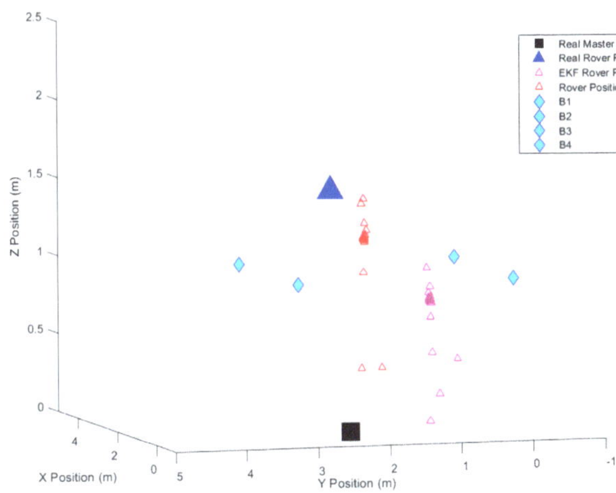

Figure 24. Configuration area representing final positioning results during the 3-D scenario.

Table 13 shows the results indicating the final position corrected using the DDC method in terms of mean value.

Table 13. Comparison among real, EKF, and DDC methods for the rover coordinates.

Rover Coordinates		
Real (m)	**EKF (m)**	**With DDC (m)**
(2, 2, 1.5)	(1.54, 0.74, 0.81)	(1.50, 1.67, 1.22)

5. Conclusions and Further Work

This research proposes a preliminary indoor positioning system prototype utilizing BLE technology for indoor missions using small UAVs. The methodology is based on the *RSSI* measured from BLE beacons and on trilateration. To improve the performance of the positioning method, extended Kalman filtering (EKF) and the differential distance correction (DDC) method were used. The DDC technique, similar to differential GNSS and RTK positioning, uses differential information from the reference station to correct the position of the rover station. The reference station (master) in a known position computes the residual of distance, sending a correction to be applied to the distance observation made by the receiving station (rover). The DDC-corrected distance value is used in the trilateration scheme to obtain a better estimate of the rover's position. In order to reduce noise effects, bias, and *RSSI* signal drift, EKF was applied on raw data. The components of the IPS prototype consist of the transmitters (Eddystone BLE beacons) placed in the indoor operating field, a master station (Raspberry Pi 4 model B), and an autonomous rover station (Arduino Nano 33 BLE), which will be mounted on-board the UAV, added to the standard instrumentation, to enhance the UAS position during indoor operations.

A first analysis was performed to verify the device characteristics and the functionality and accuracy of the methodologies applied, considering only 2-D positioning in static conditions. The most critical aspect was found to be the accurate modeling of the environmental factor, which can cause signal fluctuations due to multipath, increasing the measurement errors. To reduce multipath effects, reflection, and fading, the tests were carried out in a clean environment (without metallic objects and other devices in the test area). The results demonstrate an improved position estimate after the DDC correction. As shown in Table 9, the estimated residual errors were 0.48 m on the x-coordinate and 0.01 m on the y-coordinate. The rover position estimate improved from an error of 27.5% and 29.5% for the x- and y-coordinates, respectively, before DDC correction (Table 7), to an error of 24% on the x-coordinate and 0.5% on the y-coordinate after the application of the EKF and DDC methodology during test 1. Additionally, an error of 2.7% on the x-coordinate and 14.7% on the y-coordinate was obtained during the second test. Finally, a test considering a 3-D scenario was performed. In this test, a hovering phase within a terminal area was simulated by positioning the rover prototype on a vertical structure at 1.50 m above the ground. Following the same process described during the 2-D tests, the application of EKF and DDC methods resulted in an error of 25% on the x-coordinate, 16.5% on the y-coordinate, and 18.6% on the z-coordinate.

Further research will focus on conducting other experiments in an empty room, as well as new tests with obstacles and a greater number of beacons and master stations, performing 3-D positioning in dynamic conditions, to compare the influence of the obstacles in the experimental environment, and mounting the rover device on-board a UAV. Based on these promising results, our future objective is to expand coverage to larger areas by implementing a "mosaic" strategy. This would entail covering extensive areas by creating multiple smaller, controlled zones similar to the test area used in this study, ensuring consistent performance across a broader field. Additionally, future developments will concern analysis of the accuracy of BLE *RSSI* values as a function of the environmental changes (for example, considering more targets, changes in the density of people present in the test area, position of furniture or walls, and changes that will require regular recalibration of the IPS to ensure the accuracy of the methodology), as well as the study of BLE data fusion with other systems, for example, Wi-Fi for increasing the number of reference nodes, or on-board devices as IMUs to reduce positioning errors. In the future, the proposed IPS will be integrated into a safe landing area determination system recently developed by the authors [17].

Author Contributions: Conceptualization, S.P. and G.A.; methodology, S.P. and G.A.; software, G.A. and A.G.; validation, G.A.; formal analysis, G.A.; investigation, G.A. and S.P.; resources, G.D.C.; data curation, G.A.; writing—original draft preparation, G.A. and S.P.; writing—review and editing, S.P. and G.A.; supervision, project administration, and funding acquisition, G.D.C. All authors have read and agreed to the published version of the manuscript.

Funding: This work was supported by the University of Naples "Parthenope" (Italy) Internal Research Project DIC (Drone Innovative Configurations, CUP I62F17000060005-DSTE 338).

Institutional Review Board Statement: Not applicable.

Informed Consent Statement: Not applicable.

Data Availability Statement: Data are contained within the article.

Conflicts of Interest: The authors declare no conflicts of interest.

References

1. Sun, X.; Ansari, N. EdgeIoT: Mobile edge computing for the Internet of Things. *IEEE Commun. Mag.* **2016**, *54*, 22–29. [CrossRef]
2. Patterson, C.A.; Muntz, R.R.; Pancake, C.M. Challenges in location-aware computing. *IEEE Pervasive Comput.* **2003**, *2*, 80–89. [CrossRef]
3. Rodriguez, M.D.; Favela, J.; Martínez, E.A.; Muñoz, M.A. Location-aware access to hospital information and services. *IEEE Trans. Inf. Technol. Biomed.* **2004**, *8*, 448–455. [CrossRef] [PubMed]
4. Lau, B.P.L.; Marakkalage, S.H.; Zhou, Y.; Hassan, N.U.; Yuen, C.; Zhang, M.; Tan, U.X. A survey of data fusion in smart city applications. *Inf. Fusion* **2019**, *52*, 357–374. [CrossRef]
5. Paul, A.K.; Sato, T. Localization in Wireless Sensor Networks: A Survey on Algorithms. Measurement Techniques, Applications and Challenges. *J. Sens. Actuator Netw.* **2017**, *6*, 24. [CrossRef]
6. Austin, R. *Unmanned Aircraft Systems*; Wiley: Hoboken, NJ, USA, 2010.
7. Ariante, G.; Papa, U. *Embedded Platforms for UAS Landing Path and Obstacle Detection II: Navigation Support Systems for Urban Air Mobility Operations*; Springer Nature: Berlin/Heidelberg, Germany, 2024; Volume 530.
8. *NATO STANAG 4670–ATP-3.3.7*; Guidance for the Training of Unmanned Aircraft Systems (UAS) Operators. 3rd ed. NATO Standardization Agency: Brussels, Belgium, 2004. Available online: http://everyspec.com/NATO/NATO-STANAG/SRANAG-4670_ED-3_52054 (accessed on 30 October 2021).
9. Menichino, A.; Di Vito, V.; Ariante, G.; Del Core, G. AAM/goods delivery: Main enablers for BVLOS routine operations within environment at low and medium risk. *Aircr. Eng. Aerosp. Technol.* **2023**, *95*, 1578–1587. [CrossRef]
10. Papa, U.; Ariante, G.; Del Core, G. UAS aided landing and obstacle detection through LIDAR-sonar data. In Proceedings of the 2018 5th IEEE International Workshop on Metrology for AeroSpace (MetroAeroSpace), Rome, Italy, 20–22 June 2018; pp. 478–483.
11. Wang, J.; Liu, Y.; Song, H. Counter-unmanned aircraft system (s)(C-UAS): State of the art, challenges, and future trends. *IEEE Aerosp. Electron. Syst. Mag.* **2021**, *36*, 4–29. [CrossRef]
12. Ariante, G.; Papa, U.; Ponte, S.; Del Core, G. UAS for positioning and field mapping using LIDAR and IMU sensors data: Kalman filtering and integration. In Proceedings of the 2019 IEEE 5th International Workshop on Metrology for AeroSpace (MetroAeroSpace), Turin, Italy, 19–21 June 2019; pp. 522–527.
13. Mostafa, M.; Zahran, S.; Moussa, A.; El-Sheimy, N.; Sesay, A. Radar and visual odometry integrated system aided navigation for UAVS in GNSS denied environment. *Sensors* **2018**, *18*, 2776. [CrossRef]
14. Ariante, G. Embedded System for Precision Positioning, Detection, and Avoidance (PODA) for Small UAS. *IEEE Aerosp. Electron. Syst. Mag.* **2020**, *35*, 38–42. [CrossRef]
15. Ponte, S.; Ariante, G.; Papa, U.; Del Core, G. An embedded platform for positioning and obstacle detection for small unmanned aerial vehicles. *Electronics* **2020**, *9*, 1175. [CrossRef]
16. Ariante, G.; Papa, U.; Ponte, S.; Del Core, G. Real-Time Obstacle Detection and Field Mapping System Using LIDAR-ToF Sensors for Small UAS. In *Sensors and Microsystems, Proceedings of the AISEM Annual Conference on Sensors and Microsystems, Portici, Italy, 28 February 2020*; Springer: Cham, Switzerland, 2020; pp. 9–16.
17. Ariante, G.; Ponte, S.; Papa, U.; Greco, A.; Del Core, G. Ground control system for UAS safe landing area determination (SLAD) in urban air mobility operations. *Sensors* **2022**, *22*, 3226. [CrossRef] [PubMed]
18. Lachapelle, G. GNSS indoor location technologies. *J. Glob. Position. Syst.* **2004**, *3*, 2–11. [CrossRef]
19. Kunhoth, J.; Karkar, A.; Al-Maadeed, S.; Al-Ali, A. Indoor positioning and wayfinding systems: A survey. *Hum.-Centric Comput. Inf. Sci.* **2020**, *10*, 1–41. [CrossRef]
20. Obeidat, H.; Shuaieb, W.; Obeidat, O.; Abd-Alhameed, R. A Review of Indoor Localization Techniques and Wireless Technologies. *Wirel. Pers. Commun.* **2021**, *119*, 289–327. [CrossRef]
21. Moshe, B.B.; Shvalb, N.; Baadani, J.; Nagar, I.; Levy, H. Indoor positioning and navigation for micro UAV drones—Work in progress. In Proceedings of the 2012 IEEE 27th Convention of Electrical and Electronics Engineers in Israel, Eilat, Israel, 14–17 November 2012; pp. 1–5.
22. Wang, C.; Li, K.; Liang, G.; Chen, H.; Huang, S.; Wu, X. A heterogeneous sensing system-based method for unmanned aerial vehicle indoor positioning. *Sensors* **2017**, *17*, 1842. [CrossRef] [PubMed]
23. Duran, D.; Johnson, M.; Stansbury, R.S. Towards Collaborative Obstacle Avoidance using Small UAS in Indoor Environments. In Proceedings of the 2020 IEEE/ION Position, Location and Navigation Symposium (PLANS), Portland, OR, USA, 20–23 April 2020; pp. 1596–1605.
24. Mustafah, Y.M.; Azman, A.W.; Akbar, F. Indoor UAV positioning using stereo vision sensor. *Procedia Eng.* **2012**, *41*, 575–579. [CrossRef]
25. Masiero, A.; Fissore, F.; Antonello, R.; Cenedese, A.; Vettore, A. A comparison of UWB and motion capture UAV indoor positioning. *Int. Arch. Photogramm. Remote Sens. Spat. Inf. Sci.* **2019**, *42*, 1695–1699. [CrossRef]

26. Bagosi, T.; Baruch, Z. Indoor localization by WiFi. In Proceedings of the 2011 IEEE 7th International Conference on Intelligen Computer Communication and Processing, Cluj-Napoca, Romania, 25–27 August 2011; pp. 449–452.
27. Mautz, R. Indoor Positioning Technologies. Habilitation Thesis, ETH Zurich, Department of Civil, Environmental and Geomati Engineering, Institute of Geodesy and Photogrammetry, Zürich, Switzerland, 2012. [CrossRef]
28. Kim, J.; Jun, H. Vision-based location positioning using augmented reality for indoor navigation. *IEEE Trans. Consum. Electron* **2008**, *54*, 954–962. [CrossRef]
29. Zou, H.; Jiang, H.; Luo, Y.; Zhu, J.; Lu, X.; Xie, L. Bluedetect: An ibeacon-enabled scheme for accurate and energy-efficient indoor-outdoor detection and seamless location-based service. *Sensors* **2016**, *16*, 268. [CrossRef]
30. Heydon, R.; Hunn, N. Bluetooth Low Energy. CSR Presentation, Bluetooth SIG. 2012. Available online: https://picture.iczhiku com/resource/eetop/SyITTzRUzEfGEmCn.pdf (accessed on 4 November 2024).
31. Faragher, R.; Harle, R. An analysis of the accuracy of Bluetooth low energy for indoor positioning applications. In Proceedings of the 27th International Technical Meeting of The Satellite Division of the Institute of Navigation (ION GNSS+ 2014), Tampa, FL, USA, 8–12 September 2014; pp. 201–210.
32. Putra, G.D.; Pratama, A.R.; Lazovik, A.; Aiello, M. Comparison of energy consumption in Wi-Fi and Bluetooth communication in a Smart Building. In Proceedings of the 2017 IEEE 7th Annual Computing and Communication Workshop and Conference (CCWC), Las Vegas, NV, USA, 9–11 January 2017; pp. 1–6.
33. Hashmi, A. A novel drone-based search and rescue system using bluetooth low energy technology. *Eng. Technol. Appl. Sci. Res.* **2021**, *11*, 7018–7022. [CrossRef]
34. Zhuang, Y.; Yang, J.; Li, Y.; Qi, L.; El-Sheimy, N. Smartphone-based indoor localization with Bluetooth low energy beacons. *Sensors* **2016**, *16*, 596. [CrossRef] [PubMed]
35. Zuo, Z.; Liu, L.; Zhang, L.; Fang, Y. Indoor positioning based on Bluetooth low-energy beacons adopting graph optimization. *Sensors* **2018**, *18*, 3736. [CrossRef]
36. Zhan, J.; Liu, H. Research on ranging accuracy based on *RSSI* of wireless sensor network. In Proceedings of the 2nd International Conference on Information Science and Engineering, Hangzhou, China, 4–6 December 2010; pp. 2338–2341.
37. Andersen, J.B.; Rappaport, T.S.; Yoshida, S. Propagation measurements and models for wireless communications channels. *IEEE Commun. Mag.* **1995**, *33*, 42–49. [CrossRef]
38. Cantón Paterna, V.; Calveras Auge, A.; Paradells Aspas, J.; Perez Bullones, M.A. A Bluetooth low energy indoor positioning system with channel diversity, weighted trilateration and Kalman filtering. *Sensors* **2017**, *17*, 2927. [CrossRef]
39. Kalbandhe, A.A.; Patil, S.C. Indoor positioning system using bluetooth low energy. In Proceedings of the 2016 International Conference on Computing, Analytics and Security Trends (CAST), Pune, India, 19–21 December 2016; pp. 451–455.
40. Li, G.; Geng, E.; Ye, Z.; Xu, Y.; Lin, J.; Pang, Y. Indoor positioning algorithm based on the improved *RSSI* distance model. *Sensors* **2018**, *18*, 2820. [CrossRef]
41. Nagah Amr, M.; ELAttar, H.M.; Abd El Azeem, M.H.; El Badawy, H. An enhanced indoor positioning technique based on a novel received signal strength indicator distance prediction and correction model. *Sensors* **2021**, *21*, 719. [CrossRef]
42. Zhu, J.; Luo, H.; Chen, Z.; Li, Z. *RSSI* based Bluetooth low energy indoor positioning. In Proceedings of the 2014 International Conference on Indoor Positioning and Indoor Navigation (IPIN), Busan, Republic of Korea, 27–30 October 2014; pp. 526–533.
43. Mackey, A.; Spachos, P.; Song, L.; Plataniotis, K.N. Improving BLE beacon proximity estimation accuracy through Bayesian filtering. *IEEE Internet Things J.* **2020**, *7*, 3160–3169. [CrossRef]
44. Zhou, C.; Yuan, J.; Liu, H.; Qiu, J. Bluetooth indoor positioning based on *RSSI* and Kalman filter. *Wirel. Pers. Commun.* **2017**, *96*, 4115–4130. [CrossRef]
45. Sung, Y. *RSSI*-based distance estimation framework using a Kalman filter for sustainable indoor computing environments. *Sustainability* **2016**, *8*, 1136. [CrossRef]
46. Lee, S.H.; Lim, I.K.; Lee, J.K. Method for improving indoor positioning accuracy using extended Kalman filter. *Mob. Inf. Syst.* **2016**, *2016*, 2369103. [CrossRef]
47. Boukerche, A.; Oliveira, H.A.; Nakamura, E.F.; Loureiro, A.A. Localization systems for wireless sensor networks. *IEEE Wirel. Commun.* **2007**, *14*, 6–12. [CrossRef]
48. Murphy, W.S. Determination of a Position Using Approximate Distances and Trilateration. Master's Thesis, Colorado School of Mines, Golden, CO, USA, 2007.
49. Pascale, F.; Adinolfi, E.A.; Avagliano, M.; Giannella, V.; Salas, A. A low energy IoT application using beacon for indoor localization. *Appl. Sci.* **2021**, *11*, 4902. [CrossRef]
50. Kuo, Y.-T.; Liao, J.-K.; Chiang, K.-W. BLE-based Indoor Positioning using Differential Distance Correction Technique. In Proceedings of the International Symposium on GNSS 2017 (ISGNSS 2017), Hong Kong, China, 10–13 December 2017; Available online: https://www.researchgate.net/publication/322594329_BLE-based_Indoor_Positioning_using_Differential_Distance_Correction_Technique (accessed on 4 November 2024).
51. Kuo, Y.T.; Liao, J.K.; Chiang, K.W. Accuracy Analysis of Differential Distance Correction using Bluetooth Low Energy Technology on Indoor Positioning System. In Proceedings of the GSDI 15 World Conference, Taipei, Taiwan, 28 November–2 December 2016.
52. Tang, Y.T.; Kuo, Y.T.; Liao, J.K.; Chiang, K.W. Improved indoor positioning using BLE differential distance correction and pedestrian dead reckoning. *Int. Arch. Photogramm. Remote Sens. Spat. Inf. Sci.* **2020**, *43*, 193–198. [CrossRef]

53. Raspberry Pi Ltd. Raspberry pi 4 Model b Datasheet. 2022. Available online: https://www.raspberrypi.com/products/raspberry-pi-4-model-b/specifications/ (accessed on 4 November 2024).
54. Al-Mimi, H.; Al-Dahoud, A.; Fezari, M.; Daoud, M.S. A Study on New Arduino NANO Board for WSN and IoT Applications. *Int. J. Adv. Sci. Technol.* **2020**, *29*, 10223–10230.
55. Arduino Srl. Arduino Nano 33 BLE, Datasheet. 2022. Available online: https://docs.arduino.cc/resources/datasheets/ABX00030-datasheet.pdf (accessed on 4 November 2024).
56. Google/Eddystone. Eddystone Protocol Specification. 2018. Available online: https://github.com/google/eddystone/blob/master/protocol-specification.md (accessed on 4 November 2024).
57. Nordic Semiconductor ASA. nRF51822 Multiprotocol Bluetooth® Low Energy/2.4-GHz RF System on Chip—Product Specification v. 3.3. 2014. Available online: https://www.digikey.it/htmldatasheets/production/1732580/0/0/1/nrf51822-dk.html (accessed on 4 November 2024).
58. Ariante, G.; Ponte, S.; Del Core, G. Bluetooth Low Energy based Technology for Small UAS Indoor Positioning. In Proceedings of the 2022 IEEE 9th International Workshop on Metrology for AeroSpace (MetroAeroSpace), Pisa, Italy, 27–29 June 2022; pp. 113–118.

sensors

MDPI

Article

A Novel Multi-Objective Trajectory Planning Method for Robots Based on the Multi-Objective Particle Swarm Optimization Algorithm

Jiahui Wang [1], Yongbo Zhang [1,2,*], Shihao Zhu [1] and Junling Wang [3]

1 School of Aeronautic Science and Engineering, Beihang University, Beijing 100191, China
2 Aircraft and Propulsion Laboratory, Ningbo Institute of Technology, Beihang University, Ningbo 315100, China
3 School of Reliability and Systems Engineering, Beihang University, Beijing 100191, China
* Correspondence: zhangyongbo@buaa.edu.cn

Abstract: The three performance indexes of the space robot, travel time, energy consumption, and smoothness, are the key to its important role in space exploration. Therefore, this paper proposes a multi-objective trajectory planning method for robots. Firstly, the kinematics and dynamics of the Puma560 robot are analyzed to lay the foundation for trajectory planning. Secondly, the joint space trajectory of the robot is constructed with fifth-order B-spline functions, realizing the continuous position, velocity, acceleration, and jerk of each joint. Then, the improved multi-objective particle swarm optimization (MOPSO) algorithm is used to optimize the trajectory, and the distribution uniformity, convergence, and diversity of the obtained Pareto front are good. The improved MOPSO algorithm can realize the optimization between multiple objectives and obtain the trajectory that meets the actual engineering requirements. Finally, this paper implements the visualization of the robot's joints moving according to the optimal trajectory.

Keywords: Puma560 robot; multi-objective trajectory planning; MOPSO; B-spline

1. Introduction

With the rapid development of science and technology in today's world, humankind's space exploration has shown unprecedented momentum, and space technology has gradually risen to the focus of public attention. Among them, space robots have become indispensable to space activities due to their powerful functions and high adaptability to the space environment. A series of challenging tasks, such as assembling satellite parts, capturing space targets, and monitoring alien spacecraft, cannot be realized without the support of space robots. These tasks and harsh working environments also set higher requirements for space robots' travel time, energy consumption, and smoothness.

The primary method to improve each performance index is to design and plan the robot's trajectory rationally. Trajectory planning can be performed both in Cartesian space and in joint space. The latter plans the trajectory of each joint of the robot, which has a small amount of calculation and enables the real-time control of the robot. Trajectory planning can usually be divided into two steps. The first step is to interpolate between given path points using interpolation algorithms to obtain a trajectory-time sequence. The second step is to optimize the trajectory in terms of single or multiple performance indexes within the constraints of the kinematics and dynamics of the robot [1].

The mainstream interpolation algorithms in joint space are polynomial interpolation and spline curve interpolation, and the former is mainly used in early research [2–5]. To obtain the robot's trajectory, Ref. [6] used a cubic polynomial to connect the path points. This method is simple to calculate, but the acceleration curve obtained is not continuous, the smoothness could be better, and it tends to cause rigid impacts. Ref. [7] used quintic polynomial interpolation to ensure the continuity of acceleration, but it increased the amount of calculation, and there was still a problem of easy distortion. Compared to the fifth-degree polynomial, the seventh-degree polynomial adds constraints to the jerk at the start and termination points, realizing the continuity of the jerk. However, the eight boundary conditions increase the difficulty of the solution, and the high-order polynomial interpolation may cause the Runge phenomenon [8]. With the deepening of research, some scholars have applied spline curves to the trajectory planning of robots [9–14]. Ref. [15] used the fifth-order B-spline curves to interpolate the joint space trajectory, which realized the continuity of the jerk and set the velocity and acceleration at the start and stop time to be 0. When interpolating with seventh-order B-spline curves, it is possible to specify the acceleration at the start and stop time, but the calculation process is complicated [1].

Trajectory optimization mostly takes a single performance index as the optimization objective. Robot efficiency, the shortest time required by the robot to perform a task, was the earliest goal of trajectory planning [16–19]. Ref. [20] constructed the trajectory with a quintic polynomial and reduced the robot's travel time by 74.35% through the improved MOPSO algorithm under the constraints of each joint's angles, velocities, and accelerations. The time-optimal trajectory often leads to a large impact on the robot, affecting its motion accuracy and shortening the service life of the robot structure. Many scholars have solved this problem by optimizing the jerk. Ref. [7] effectively increased the smoothness of the robot by optimizing the maximum value of joint jerks. Ref. [21] combined the PSO algorithm with K-means clustering to achieve a fast solution for joint trajectories with minimal shocks. Energy consumption optimization is also an important issue for robots working in unique environments such as oceans, deserts, and space [22–25]. Ref. [26] obtained a parameterized dynamic robot model through identification experiments and used a sequential quadratic programming solver to minimize a mechanical energy-based cost function under consideration of physical constraints. These three performance indexes all play an important role in the motion of the space robot, and the single optimization objective ignores the intricate balance between them. Therefore, there are studies on the comprehensive optimization of multiple objectives [27–29]. Ref. [30] realized the comprehensive optimization of time, energy, and smoothness by a differential evolution algorithm. Ref. [31] used the NSGA-II algorithm to optimize the same three objectives and obtained Pareto optimal solution sets, thus obtaining the high-order continuous optimal trajectories.

There are few studies on multi-objective optimization problems, so this paper proposes a trajectory planning method that can make the robot's travel time, energy consumption, and smoothness achieve the integrated optimal state when performing the task. In this paper, continuous and smooth joint space trajectories are constructed using fifth-order B-spline functions, which also realize the specification of the velocity and acceleration of the robot at the start/stop moment. The trajectories are then optimized using the improved MOPSO algorithm to obtain the Pareto optimal solution sets, from which suitable solutions are selected according to practical needs.

The rest of the paper is organized as follows. Section 2 analyzes the kinematics and dynamics of the Puma 560 robot manufactured by Unimation, USA. Section 3 uses fifth-order B-spline curves to construct the joint space trajectories of the robot, builds a mathematical model for the multi-objective optimization problem based on Section 2, and solves the model with the improved MOPSO algorithm. Section 4 performs simulation experiments on multi-objective trajectory planning. Section 5 summarizes this article.

2. Kinematics and Dynamics Analysis

2.1. Kinematics Analysis

This section gives the kinematics model of the Puma560 robot and analyzes its forward and inverse kinematics. The MDH (Modified Denavit–Hartenberg) coordinate system [32] shown in Figure 1 is established on the Puma560 robot with six rotary joints. The link parameters from the MATLAB R2020b Robot Toolbox are shown in Table 1.

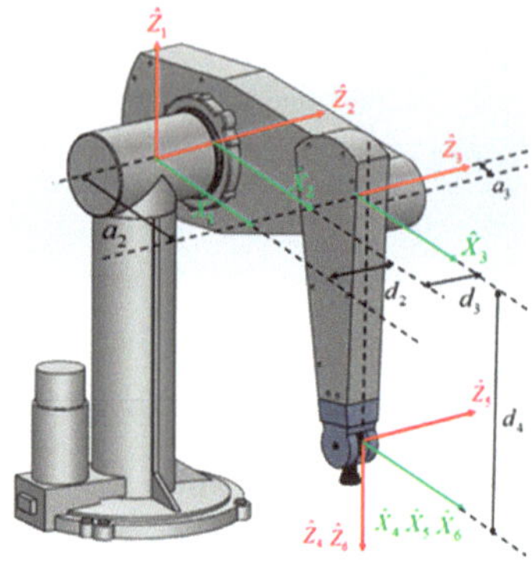

Figure 1. The MDH coordinate system of the Puma560 robot.

Table 1. Link parameters of the Puma560 robot.

Link i	$\alpha_{i\text{-}1}$ (rad)	$a_{i\text{-}1}$ (m)	d_i (m)	θ_i (rad)
1	0	0	0	θ_1
2	−1.5708	0	0.2435	θ_2
3	0	0.4318	−0.0934	θ_3
4	1.5708	−0.0203	0.4331	θ_4
5	−1.5708	0	0	θ_5
6	1.5708	0	0	θ_6

The simulation model of the Puma560 robot is established by using the Robot Toolbox in MATLAB, as shown in Figure 2.

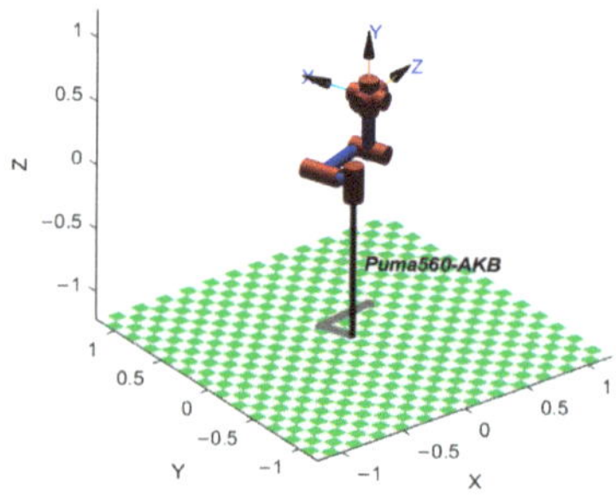

Figure 2. Robot model in MATLAB.

2.1.1. Forward Kinematics Analysis

Forward kinematics analysis refers to obtaining the end-effector pose relative to the base according to the angle of each robot joint. The transformation matrix between neighboring links, that is, the coordinate system $\{i\}$ relative to the coordinate system $\{i-1\}$, can be represented by

$$ {}^{i-1}_{i}T = \begin{bmatrix} \cos\theta_i & -\sin\theta_i & 0 & a_{i-1} \\ \sin\theta_i\cos\alpha_{i-1} & \cos\theta_i\cos\alpha_{i-1} & -\sin\alpha_{i-1} & -\sin\alpha_{i-1}d_i \\ \sin\theta_i\sin\alpha_{i-1} & \cos\theta_i\sin\alpha_{i-1} & \cos\alpha_{i-1} & \cos\alpha_{i-1}d_i \\ 0 & 0 & 0 & 1 \end{bmatrix} \quad (1) $$

Substituting the parameters in Table 1 into (1), the six homogeneous transformation matrixes of the Puma560 robot can be obtained by

$$
\begin{aligned}
{}^0_1T &= \begin{bmatrix} \cos\theta_1 & -\sin\theta_1 & 0 & 0 \\ \sin\theta_1 & \cos\theta_1 & 0 & 0 \\ 0 & 0 & 1 & 0 \\ 0 & 0 & 0 & 1 \end{bmatrix}
{}^1_2T = \begin{bmatrix} \cos\theta_2 & -\sin\theta_2 & 0 & 0 \\ 0 & 0 & 1 & d_2 \\ -\sin\theta_2 & -\cos\theta_2 & 0 & 0 \\ 0 & 0 & 0 & 1 \end{bmatrix}
{}^2_3T = \begin{bmatrix} \cos\theta_3 & -\sin\theta_3 & 0 & a_2 \\ \sin\theta_3 & \cos\theta_3 & 0 & 0 \\ 0 & 0 & 1 & d_3 \\ 0 & 0 & 0 & 1 \end{bmatrix} \\
{}^3_4T &= \begin{bmatrix} \cos\theta_4 & -\sin\theta_4 & 0 & a_3 \\ 0 & 0 & -1 & -d_4 \\ \sin\theta_4 & \cos\theta_4 & 0 & 0 \\ 0 & 0 & 0 & 1 \end{bmatrix}
{}^4_5T = \begin{bmatrix} \cos\theta_5 & -\sin\theta_5 & 0 & 0 \\ 0 & 0 & 1 & 0 \\ -\sin\theta_5 & -\cos\theta_5 & 0 & 0 \\ 0 & 0 & 0 & 1 \end{bmatrix}
{}^5_6T = \begin{bmatrix} \cos\theta_6 & -\sin\theta_6 & 0 & 0 \\ 0 & 0 & -1 & 0 \\ \sin\theta_6 & \cos\theta_6 & 0 & 0 \\ 0 & 0 & 0 & 1 \end{bmatrix}.
\end{aligned}
\tag{2}
$$

The transformation matrixes in (2) can be multiplied together to find the transformation matrix of the end-effector coordinate system {6} relative to the base coordinate system {0}

$$
{}^0_6T = {}^0_1T\,{}^1_2T\,{}^2_3T\,{}^3_4T\,{}^4_5T\,{}^5_6T = \begin{bmatrix} r_{11} & r_{12} & r_{13} & p_x \\ r_{21} & r_{22} & r_{23} & p_y \\ r_{31} & r_{32} & r_{33} & p_z \\ 0 & 0 & 0 & 1 \end{bmatrix} \tag{3}
$$

where p_x, p_y, p_z represent the position of the end-effector, and r_{11}, r_{12}, r_{13}, r_{21}, r_{22}, r_{23}, r_{31}, r_{32}, r_{33} represent the orientation of the end-effector. These 12 elements are calculated by the joint angles.

2.1.2. Inverse Kinematics Analysis

Inverse kinematics analysis refers to the inverse solution of the angle of each joint by using the end-effector pose relative to the base. Analyzing the structural characteristics of the Puma560 robot, it is easy to know that its last three axes intersect at one point, and the six joints of the robot are all rotary joints. So, the Pieper method [33] can solve the inverse kinematics of the Puma560 robot. When $\sin(\theta_5) \neq 0$, the joint angles can be obtained by

$$
\theta_1 = \mathrm{A}\tan 2\left(\frac{g_1 y - g_2 x}{g_1^2 + g_2^2}, \frac{g_1 x + g_2 y}{g_1^2 + g_2^2}\right) \tag{4}
$$

$$
\theta_2 = \mathrm{A}\tan 2\left(\frac{-z}{\rho_2}, \pm\sqrt{1 - \frac{z^2}{\rho_2^2}}\right) - \mathrm{A}\tan 2(f_2, f_1) \tag{5}
$$

$$
\theta_3 = \mathrm{A}\tan 2\left(\frac{C - r}{\rho_3}, \pm\sqrt{1 - \frac{(C - r)^2}{\rho_3^2}}\right) - \mathrm{A}\tan 2(a_3, d_4) \tag{6}
$$

$$
\theta_4 = \mathrm{A}\tan 2\left(\frac{x_{33}}{\sin(\theta_5)}, \frac{x_{13}}{\sin(\theta_5)}\right) \tag{7}
$$

$$
\theta_5 = \mathrm{A}\tan 2\left(\pm\sqrt{x_{21}^2 + x_{22}^2}, -x_{23}\right) \tag{8}
$$

$$
\theta_6 = \mathrm{A}\tan 2\left(\frac{-x_{22}}{\sin(\theta_5)}, \frac{x_{21}}{\sin(\theta_5)}\right) \tag{9}
$$

where $\mathrm{A}\tan 2$ is predefined in many programming language libraries; its function is to judge the quadrant of the angle according to the positive and negative of x and y while calculating $\tan^{-1}\left(\frac{y}{x}\right)$. Unknown in (4)–(9), such as g_1, g_2, are the intermediate quantities in the derivation process of the Pieper method, which are defined as

$$
f_1 = a_2 + a_3 c_3 + d_4 s_3,\ f_2 = a_3 s_3 - d_4 c_3,\ f_3 = d_3 \tag{10}
$$

$$
g_1 = f_1 c_2 - f_2 s_2,\ g_2 = f_3 + d_2,\ g_3 = -f_1 s_2 - f_2 c_2 \tag{11}
$$

$$r = f_1^2 + f_2^2 + f_3^2 + d_2^2 + 2f_3d_2,\ \ C = r - \left(a_2^2 + a_3^2 + d_4^2 + d_2^2 + d_3^2 + 2d_2d_3\right) \tag{12}$$

$$x = (f_1c_2 - f_2s_2)c_1 - (f_3 + d_2)s_1,\ \ y = (f_1c_2 - f_2s_2)s_1 + (f_3 + d_2)c_1,\ \ z = -f_1s_2 - f_2c_2 \tag{13}$$

$$\rho_2 = \sqrt{f_1^2 + f_2^2},\ \ \rho_3 = 2a_2\sqrt{a_3^2 + d_4^2} \tag{14}$$

$$\phi_2 = \mathrm{A}\tan 2(f_2, f_1),\ \ \phi_3 = \mathrm{A}\tan 2(a_3, d_4) \tag{15}$$

$$x_{13} = r_{13}c_{23}c_1 - r_{33}s_{23} + r_{23}c_{23}s_1,\ \ x_{21} = -r_{31}c_{23} - r_{11}c_1s_{23} - r_{21}s_1s_{23} \tag{16}$$

$$x_{22} = -r_{32}c_{23} - r_{12}c_1s_{23} - r_{22}s_1s_{23},\ \ x_{23} = -r_{33}c_{23} - r_{13}c_1s_{23} - r_{23}s_1s_{23},\ \ x_{33} = r_{23}c_1 - r_{13}s_1 \tag{17}$$

where $c_i = \cos\theta_i$, $s_i = \sin\theta_i$, $s_{ij} = \sin(\theta_i + \theta_j)$, $c_{ij} = \cos(\theta_i + \theta_j)$, $i, j = [1, 2, 3, 4, 5, 6]$, and $i \neq j$.

If $\sin(\theta_5) \neq 0$ and $\theta_5 = 0$, then the solutions of $\theta_1, \theta_2, \theta_3$ do not change, and $\theta_4, \theta_5, \theta_6$ become

$$\theta_4 = 0 \tag{18}$$

$$\theta_5 = 0 \tag{19}$$

$$\theta_6 = \mathrm{A}\tan 2(-x_{12}, x_{11}) \tag{20}$$

where $x_{11} = r_{11}c_{23}c_1 - r_{31}s_{23} + r_{21}c_{23}s_1$, $x_{12} = r_{12}c_{23}c_1 - r_{32}s_{23} + r_{22}c_{23}s_1$.

If $\sin(\theta_5) \neq 0$ and $\theta_5 = \pi$, then the solutions of $\theta_1, \theta_2, \theta_3$ do not change, and $\theta_4, \theta_5, \theta_6$ become

$$\theta_4 = 0 \tag{21}$$

$$\theta_5 = \pi \tag{22}$$

$$\theta_6 = \mathrm{A}\tan 2(x_{12}, -x_{11}) \tag{23}$$

2.2. Dynamics Analysis

The iterative Newton–Euler dynamics algorithm [34] is computationally efficient, suitable for real-time control, and commonly used for modeling robot dynamics. The algorithm is composed of two parts. First, link velocities and accelerations are iteratively calculated from the base. Second, starting from the end-effector, the force and torque of each link are calculated in reverse. The specific iterative calculation process is as follows.

Outward iterations: $i : 0 \rightarrow n_l - 1$

$${}^{i+1}w_{i+1} = {}^{i+1}_{i}R\,{}^{i}w_i + \dot{\theta}_{i+1}\,{}^{i+1}\hat{Z}_{i+1} \tag{24}$$

$${}^{i+1}\dot{w}_{i+1} = {}^{i+1}_{i}R\,{}^{i}\dot{w}_i + {}^{i+1}_{i}R\,{}^{i}w_i \times \dot{\theta}_{i+1}\,{}^{i+1}\hat{Z}_{i+1} + \ddot{\theta}_{i+1}\,{}^{i+1}\hat{Z}_{i+1} \tag{25}$$

$${}^{i+1}\dot{v}_{i+1} = {}^{i+1}_{i}R\left[{}^{i}\dot{w}_i \times {}^{i}P_{i+1} + {}^{i}w_i \times \left({}^{i}w_i \times {}^{i}P_{i+1}\right) + {}^{i}\dot{v}_i\right] \tag{26}$$

$${}^{i}\dot{v}_{C_i} = {}^{i}\dot{w}_i \times {}^{i}P_{C_i} + {}^{i}w_i \times \left({}^{i}w_i \times {}^{i}P_{C_i}\right) + {}^{i}\dot{v}_i \tag{27}$$

$$F_i = m\dot{v}_{C_i} \tag{28}$$

$$N_i = {}^{C_i}I\dot{w}_i + w_i \times {}^{C_i}Iw_i \tag{29}$$

where n_l is the number of links, ${}^{i+1}\hat{Z}_{i+1}$ is the unit vector of the coordinate system $\{i + 1\}$ on the Z axis, F_i and N_i are, respectively, the inertia force and torque acting on the mass center of the link i.

Inward iterations: $i : n_l \rightarrow 1$

$${}^{i}f_i = {}^{i}_{i+1}R\,{}^{i+1}f_{i+1} + {}^{i}F_i \tag{30}$$

$${}^{i}n_i = {}^{i}N_i + {}^{i}_{i+1}R\,{}^{i+1}n_{i+1} + {}^{i}P_{C_i} \times {}^{i}F_i + {}^{i}P_{i+1} \times {}^{i}_{i+1}R\,{}^{i+1}f_{i+1} \tag{31}$$

$$\tau_i = {}^{i}n_i^{T}\,{}^{i}\hat{Z}_i \tag{32}$$

where ${}^{i}f_i$, ${}^{i}n_i$ are, respectively, the force and torque acting on the link i, and τ_i is the driving force of the joint motor.

3. Multi-Objective Trajectory Planning

3.1. Construction of Joint Space Trajectory

A kth-degree B-spline curve [35] is defined by

$$p(u) = \sum_{i=0}^{n} d_i N_{i,k}(u) \tag{33}$$

where d_i is the control point, n+1 is its number, $p(u)$ is the path point at node u, $N_{i,k}(u)$ is the kth-degree B-spline basis function, and its specific definition is

$$\begin{cases} N_{i,0}(u) = \begin{cases} 1, & u_i \le u < u_{i+1} \\ 0, & \text{others} \end{cases} \\ N_{i,k}(u) = \frac{u-u_i}{u_{i+k}-u_i} N_{i,k-1}(u) + \frac{u_{i+k+1}-u}{u_{i+k+1}-u_{i+1}} N_{i+1,k-1}(u) \\ \frac{0}{0} = 0. \end{cases} \tag{34}$$

The interval of $N_{i,k}(u)$, $u \in [u_i, u_{i+k+1}]$, contains $k+1$ node intervals. It can be seen from (34) that for any node $u \in [u_i, u_{i+k+1}]$ on the parameter axis, there are only up to $k+1$ nonzero basis functions $N_{r,k}(u)(r = i-k, i-k+1, \ldots, i)$. This is the local support property of the B-spline curve. Therefore, the B-spline curve can also be expressed as

$$p(u) = \sum_{r=i-k}^{i} d_r N_{r,k}(u) \tag{35}$$

In this paper, the joint space trajectory of the robot is obtained by fifth-order B-spline curve interpolation, so $k = 5$. Assuming that the position-time series of a certain joint is $P = (p_j, t_j), j = 0, 1, \ldots, m$, then the node vector is $U = [u_0, u_1, \ldots, u_{m+2k}]$, and $n = m + k - 1$.

In order to make the B-spline curve pass through the first and end position points of the joint, the node repetition degree of these two positions needs to be defined as

$$u_0 = u_1 = \ldots = u_5 = 0 \tag{36}$$

$$u_{m+5} = u_{n+6} = \ldots = u_{m+10} = 1 \tag{37}$$

Moreover, the accumulative chord length parameterization method normalizes the remaining $m-1$ inner nodes

$$u_i = u_{i-1} + \frac{|\Delta t_{i-6}|}{\sum\limits_{r=0}^{m-1} |\Delta t_r|}, i = 6, 7, \ldots, m+4 \tag{38}$$

The $n+1$ equations are needed to solve $n+1$ control points, where $m+1$ equations can be given by

$$p(u_{i+5}) = \sum_{r=i}^{i+5} d_r N_{r,5}(u_{i+5}) = p_i, \;\; u_{i+5} \in [u_5, u_{m+5}], i = 0, 1, \ldots, m \tag{39}$$

Additional conditions determine the other $k-1$ equations. For the fifth-order B-spline curve, specifying the velocity and acceleration of the joint at the start and end points can add four additional equations

$$\begin{cases} p'(u)|_{u=u_5} = v_s, & p'(u)|_{u=u_{m+5}} = v_e \\ p''(u)|_{u=u_5} = a_s, & p''(u)|_{u=u_{m+5}} = a_e \end{cases} \tag{40}$$

where $p'(u)$, $p''(u)$ are, respectively, the first and second derivatives of the B-spline curve, representing the velocity and acceleration of the joint. The deBoor–Cox recurrence formula can calculate the lth derivative of the B-spline curve

$$\begin{cases} p^l(u) = \sum\limits_{r=i-k+l}^{i} d_r^l N_{r,k-l}(u), \ u_i \le u < u_{i+1} \\ d_r^l = \begin{cases} d_j, \ l = 0 \\ (k+1-l)\left(d_r^{l-1} - d_{r-1}^{l-1}\right)/(u_{r+k+1-l} - u_r), \ l = 1,2,\ldots,r. \end{cases} \end{cases} \tag{41}$$

Expression (39) is combined with (40) to obtain

$$A_n d = p \tag{42}$$

where $d = [d_0, d_1, \ldots, d_{n-1}, d_n]^T$, $p = [p_0, p_1, \ldots, p_m, v_s, v_e, a_s, a_e]^T$, and the coefficient matrix is

$$A_n = \begin{bmatrix} 1 & & & & & & & & \\ & N_{1,5}(u_6) & N_{2,5}(u_6) & \cdots & N_{5,5}(u_6) & & & & \\ & & N_{2,5}(u_7) & N_{3,5}(u_7) & \cdots & N_{6,5}(u_7) & & & \\ & & \ddots & & & \ddots & & & \\ & & & N_{m-2,5}(u_{m+3}) & N_{m-1,5}(u_{m+3}) & \cdots & N_{m+2,5}(u_{m+3}) & & \\ & & & & N_{m-1,5}(u_{m+4}) & N_{m,5}(u_{m+4}) & \cdots & N_{m+3,5}(u_{m+4}) & \\ & & & & & & & & 1 \\ c_{s1} & c_{s2} & & & & & & & \\ & & & & & & & c_{e1} & c_{e2} \\ a_{s1} & a_{s2} & & & & & & & \\ & & & & & & a_{e1} & a_{e2} & a_{e3} \end{bmatrix}$$

Some parameters in the coefficient matrix are defined by

$$\begin{gathered} c_{s1} = -5/(u_6 - u_1) \\ c_{s2} = 5/(u_6 - u_1) \\ c_{e1} = -5/(u_{m+9} - u_{m+4}) \\ c_{e2} = 5/(u_{m+9} - u_{m+4}) \\ a_{s1} = 20/[(u_6 - u_2)(u_6 - u_1)] \\ a_{s2} = -20\{1/[(u_6 - u_2)(u_6 - u_1)] + 1/[(u_6 - u_2)(u_7 - u_2)]\} \\ a_{s3} = 20/[(u_6 - u_2)(u_7 - u_2)] \\ a_{e1} = 20/[(u_{m+8} - u_{m+4})(u_{m+8} - u_{m+3})] \\ a_{e2} = -20\{1/[(u_{m+8} - u_{m+4})(u_{m+8} - u_{m+3})] + 1/[(u_{m+8} - u_{m+4})(u_{m+9} - u_{m+4})]\} \\ a_{e3} = 20/[(u_{m+8} - u_{m+4})(u_{m+9} - u_{m+4})]. \end{gathered} \tag{43}$$

From (42), all the control points can be obtained by

$$d = A_n^{-1} p \tag{44}$$

Bringing the control points back to (35), the angular displacement curve of each robot joint can be obtained. Then, the angular velocity, angular acceleration, and angular jerk curves of each joint are obtained by (41).

3.2. Establishing the Multi-Objective Optimization Model

This model consists of objective functions and constraint conditions. Firstly, the specific expressions of travel time, energy consumption, and smoothness are shown in

(45)–(47). It is assumed that the trajectory of each joint of the Puma560 is divided into m segments, which means that each joint is assigned $m + 1$ angle values in turn.

The total travel time of the robot is the sum of the travel time of each trajectory, so its expression is

$$S_1 = T = \sum_{j=1}^{m} \Delta t_j = \sum_{j=1}^{m} (t_j - t_{j-1}) \tag{45}$$

where t_{j-1}, t_j, Δt_j are, respectively, the starting time, ending time, and travel time of the jth trajectory, and T is the total travel time.

The average accelerations of joints represent the energy consumption

$$S_2 = \sum_{i=1}^{6} \sqrt{\frac{1}{T} \int_0^T \ddot{\theta}_i^2 dt} \tag{46}$$

where $\ddot{\theta}_i$ is the acceleration of the ith joint.

The average jerks of joints measure the smoothness

$$S_3 = \sum_{i=1}^{6} \sqrt{\frac{1}{T} \int_0^T \dddot{\theta}_i^2 dt} \tag{47}$$

where $\dddot{\theta}_i$ is the jerk of the ith joint.

Constraints of kinematics and dynamics of the robot need to be considered when the robot performs tasks. Kinematic constraints mainly include the limitation of joint angle, velocity, and acceleration. Dynamic constraints mostly refer to the restriction of the joint torque. Therefore, the multi-objective optimization model can be established as

$$\begin{aligned}
&\text{Minimize} \\
&\qquad S = [S_1, S_2, S_3]^T \\
&\text{Subject to} \\
&g_{i,1} = \theta_i^{\min} - \min(\theta_i(t)) \leq 0 \\
&g_{i,2} = \max(\theta_i(t)) - \theta_i^{\max} \leq 0 \\
&g_{i,3} = \dot{\theta}_i^{\min} - \min\left(\dot{\theta}_i(t)\right) \leq 0 \\
&g_{i,4} = \max\left(\dot{\theta}_i(t)\right) - \dot{\theta}_i^{\max} \leq 0 \\
&g_{i,5} = \ddot{\theta}_i^{\min} - \min\left(\ddot{\theta}_i(t)\right) \leq 0 \\
&g_{i,6} = \max\left(\ddot{\theta}_i(t)\right) - \ddot{\theta}_i^{\max} \leq 0 \\
&g_{i,7} = \tau_i^{\min} - \min(\tau_i(t)) \leq 0 \\
&g_{i,8} = \max(\tau_i(t)) - \tau_i^{\max} \leq 0.
\end{aligned} \tag{48}$$

where $\theta_i(t)$, $\dot{\theta}_i(t)$, $\ddot{\theta}_i(t)$, $\tau_i(t)$ are, respectively, the angle, velocity, acceleration, and driving torque of the ith joint at t time, and $\theta_i^{\max}$, $\theta_i^{\min}$, $\dot{\theta}_i^{\max}$, $\dot{\theta}_i^{\min}$, $\ddot{\theta}_i^{\max}$, $\ddot{\theta}_i^{\min}$, $\tau_i^{\max}$, $\tau_i^{\min}$ are, respectively, the upper and lower limits of the angle, velocity, acceleration, and driving torque of the ith joint.

3.3. Solving the Multi-Objective Optimization Model

The three optimization objectives conflict with each other, and there is a complex balance between them, so they cannot achieve the best solution simultaneously. When the improved MOPSO algorithm is used to solve the previous model, the result is no longer a single optimal solution, but an optimal solution set called the Pareto optimal solution set [36]. There is no good or bad solution in the Pareto optimal solution set, and the appropriate solution can be selected according to the actual engineering needs.

The MOPSO algorithm [37] originated from the study of bird foraging behavior and has the advantages of fast convergence speed, strong global search capability, and a wide range of application. In this algorithm, each particle's position represents a solution to the problem. Under the influence of the individual optimal position and group optimal position, particles update their velocity and position

$$v_{id}^{k+1} = wv_{id}^{k} + c_1 r_1 \left(p_{id,pbest}^{k} - x_{id}^{k} \right) + c_2 r_2 \left(p_{d,gbest}^{k} - x_{id}^{k} \right) \tag{49}$$

$$x_{id}^{k+1} = x_{id}^{k} + v_{id}^{k+1} \tag{50}$$

where w is the inertia weight, c_1 and c_2 are the individual and group learning factors, r_1 and r_2 are random numbers in the range of [0,1], k is the current iteration number, d is the vector's dimension number, x_{id}^{k} and v_{id}^{k} are the position and velocity of particle i, and $p_{id,pbest}^{k}$ and $p_{d,gbest}^{k}$ are the optimal position of individual and group.

Furthermore, the algorithm determines the dominance relationship between particles by comparing the objective function values of particles. To better manage and preserve the non-dominated solution, it is saved to the external archive.

The traditional MOPSO algorithm determines the global leader and deletes the redundant particles in the external archive by random selection. The convergence, distribution uniformity, and accuracy of the Pareto front are not good. When facing complex problems, the algorithm easily falls into the local optimum.

In order to solve these problems, this paper uses the adaptive grid technology and roulette strategy to change the selection of the global leader and redundant particles in the external archive, applies the adaptive mutation technique to the position of particles, and makes the inertia weight, individual, and group learning factors change nonlinearly with the iterations number. The workflow flow chart of the improved MOPSO algorithm is show in Figure 3.

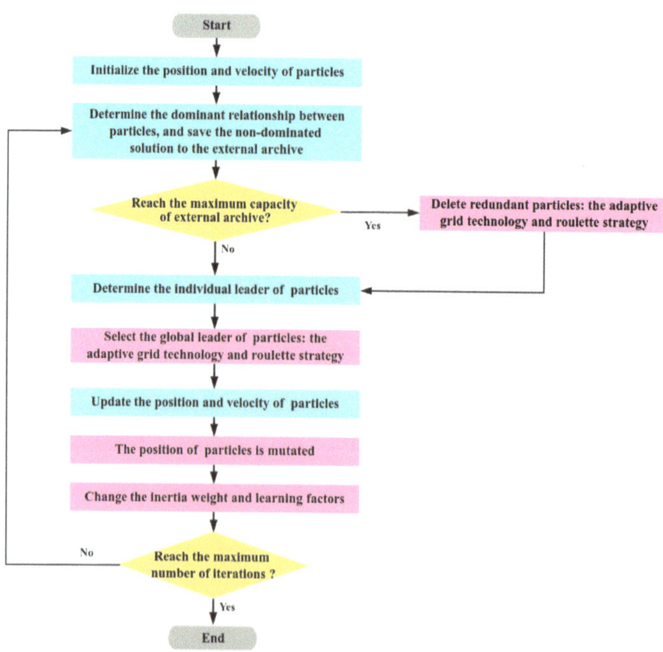

Figure 3. The improved MOPSO algorithm.

The combination of the adaptive grid technology and roulette strategy enables the algorithm to select particles in a suitable sub-grid. For each iteration, the maximum value f_{imax} and minimum value f_{imin} of the ith objective function on all particles are found as the upper and lower bounds of the initial grid. In order to cover the boundary particles, the range of the grid is expanded according to

$$LB_i = f_{imin} - \alpha(f_{imax} - f_{imin}), UB_i = f_{imax} + \alpha(f_{imax} - f_{imin}) \tag{51}$$

where LB_i, UB_i are the new upper and lower bounds of the grid, and α is the expansion ratio.

The number of grids for each dimension is set to n_d, and the grid is equally divided into n_d^e sub-grids, where e is the number of objective function. Figure 4 shows the general process of adaptive grid technology.

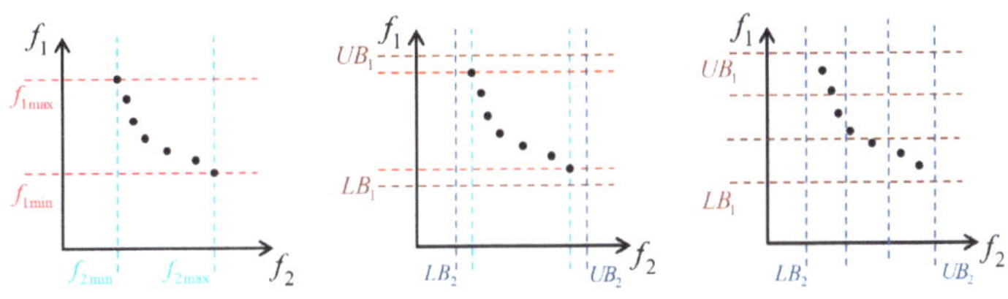

Figure 4. The adaptive grid technology in two-dimensional case. (**a**) The minimum and maximum values of each objective function; (**b**) the enlarged grid range; (**c**) the uniform distribution of grids when $n_d = 3$.

Suppose that there are n_s sub-grids containing particles, and the number of particles in the ith sub-grid is N_i. When selecting the global leader of particles, the probability that the ith sub-grid is selected is

$$P_{i,1} = \frac{e^{-\beta N_i}}{e^{-\beta N_1} + e^{-\beta N_2} + \cdots + e^{-\beta N_b}} \tag{52}$$

where $i = 1, 2, \ldots, n_s$, and β is a non-negative leader selection pressure parameter. It can be seen from (52) that the global leader of particles is more likely to come from sub-grids with fewer particles, which encourages the algorithm to explore areas that have been searched less before and increase the diversity of solutions.

When the number of particles in the external archive is greater than the set number, it is necessary to delete the redundant particles. At this time, the probability that the ith sub-grid is selected is

$$P_{i,2} = \frac{e^{\sigma N_i}}{e^{\sigma N_1} + e^{\sigma N_2} + \cdots + e^{\sigma N_b}} \tag{53}$$

where σ is a non-negative delete selection pressure parameter. It can be seen from (63) that the deleted redundant particles are more likely to come from the sub-grids with more particles, promoting the uniform distribution of the solutions.

Expressions (52)–(53) are the individual selection probabilities in the roulette strategy, and the cumulative probabilities of each grid are defined as

$$Q_{i,1} = \sum_{j=1}^{i} P_{j,1}, \; Q_{i,2} = \sum_{j=1}^{i} P_{j,2}. \tag{54}$$

The individual selection strategy is to generate a random number $r \in [0, 1]$, compare it with $Q_{i,1}$ or $Q_{i,2}$, find the first cumulative probability exceeding r, and select its sub-grid.

When the particle's velocity and position are updated, the particle's position is adaptively mutated. The mutation rate in the kth iteration is

$$pm_k = \left(1 - \frac{k-1}{M-1}\right)^{\frac{1}{h}} \tag{55}$$

where k is the current iteration number, M is the maximum iteration number, and h is a given constant.

The mutation step of the particle i is defined as

$$\Delta x_i^k = pm_k \cdot (x_{\max} - x_{\min}) \tag{56}$$

where $x_{\max}, x_{\min}$ are the limit values of particle position. When pm_k decreases nonlinearly with the increase in the iterations number, Δx_i^k also decreases.

Now, it is randomly specified that the dth dimensional component x_{id}^k of the position vector is mutated. The upper and lower bounds of the range of mutation are

$$lb = x_{id}^k - \Delta x_i^k, ub = x_{id}^k + \Delta x_i^k \tag{57}$$

A random number is generated in the continuous distribution of lb and ub as the value of the position vector after mutation in this dimension. This makes the algorithm jump out of the local optimum in the early stage and converge to the global optimal solution better in the later stage.

The relationship between the inertia weight and the number of iterations is

$$w = w_{\max} - (w_{\max} - w_{\min})\left(\frac{k}{M}\right)^2 \tag{58}$$

where $w_{\max}, w_{\min}$ are the upper and lower limits of w. As the number of iterations k increases, the inertia weight decreases nonlinearly.

The individual and group learning factors are defined as

$$c_1 = c_{1s} + (c_{1e} - c_{1s}) \sin\left(\frac{\pi k}{2M}\right), c_2 = c_{2s} + (c_{2e} - c_{2s}) \sin\left(\frac{\pi k}{2M}\right) \tag{59}$$

where c_{1s} and c_{2s} are, respectively, the initial values of c_1 and c_2, and c_{1e} and c_{2e} are, respectively, the final values of c_1 and c_2. As the number of iterations increases, c_1 decreases from large to small, and c_2 is the opposite.

These give the particle a strong global search capability at the beginning of the algorithm iterations to avoid falling into local optimum and a strong local search capability at the later stages to improve convergence accuracy.

4. Simulation

In this section, the multi-objective trajectory planning simulation of the Puma560 robot is carried out in MATLAB. The constraints of each joint are shown in Table 2.

Table 2. Kinematic and dynamic constraints.

Constraints	Joint 1	Joint 2	Joint 3	Joint 4	Joint 5	Joint 6
Angle/(rad)	3.100	3.100	3.100	3.100	3.100	3.100
Velocity/$(\mathrm{rad \cdot s^{-1}})$	0.876	0.876	1.598	0.876	0.926	0.926
Acceleration/$(\mathrm{rad \cdot s^{-2}})$	0.725	0.725	2.378	0.725	1.450	1.450
Torque/$(\mathrm{N \cdot m})$	44.940	44.940	8.866	44.940	0.050	0.050

The target captured by the robot is a small ball moving at a constant speed of 0.5 m/s along the Z-axis, and its trajectory is known. Six key positions of each joint in the process of catching the ball are given, as shown in Table 3.

Table 3. Position sequence of each joint.

Node	Joint 1/(rad)	Joint 2/(rad)	Joint 3/(rad)	Joint 4/(rad)	Joint 5/(rad)	Joint 6/(rad)
1	0.5821	−0.3805	−0.8168	0.6283	−0.9390	0.2531
2	0.4829	−0.3735	−0.7981	0.6299	−0.9245	0.2621
3	0.0383	−0.1212	0.0608	0.4289	−0.4005	0.6502
4	−0.5872	0.1770	1.0702	0.1995	0.2189	1.1091
5	−1.0317	0.3890	1.7877	0.0364	0.6592	1.4352
6	−1.1310	0.4363	1.9478	0	0.7547	1.5080

The MOPSO algorithm takes the time interval of each trajectory Δt_j as the decision variable, which is in the range of [0.75, 7]. The population size and the maximum iteration number of the traditional and improved MOPSO algorithms are both 200, and the size of the Pareto optimal solution set is 100. In addition, this paper sets n_d in the improved MOPSO algorithm to 5, β to 2, and σ to 2.

The Pareto front obtained by the traditional MOPSO algorithm is shown in Figure 5a. It falls into the local optimum, and the distribution and convergence of the Pareto front are also poor. The Pareto front obtained by the improved MOPSO algorithm is shown in Figure 5b. It jumps out of the local optimum, and the convergence and distribution are significantly improved, which proves the effectiveness of the improved MOPSO algorithm.

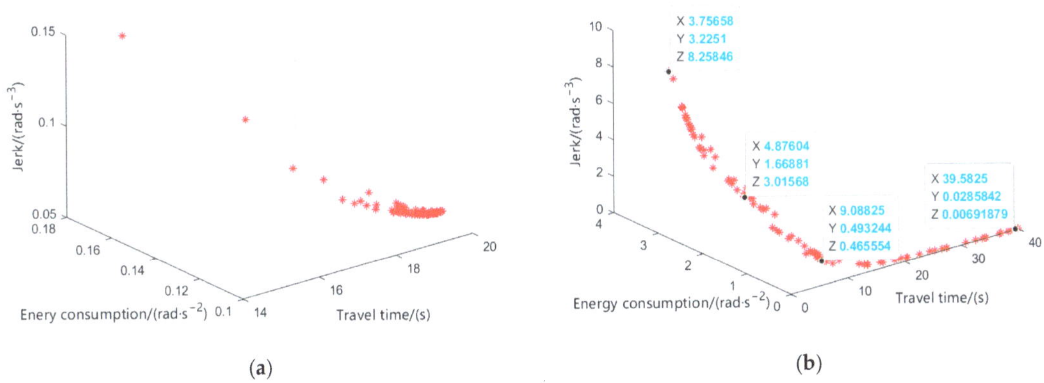

Figure 5. The Pareto front. (**a**) The traditional MOPSO algorithm and (**b**) the improved MOPSO algorithm.

Four points are taken on the Pareto front, which, from top to bottom, are A, B, C, and D. The closer to A, the shorter the travel time, the more the energy consumption, and the greater the jerk; the closer to D, the less the energy consumption, the smaller the jerk, and the longer the travel time. It follows that smoothness is positively correlated with energy consumption, while they are negatively correlated with travel time. The values of the three objective functions for A, B, C, and D are shown in Table 4.

Table 4. The partial optimum solution.

Solution	Travel Time/(s)	Energy Consummation/(rad·s^{-2})	Jerk/(rad·s^{-3})
A	3.7566	3.2251	8.2585
B	4.8760	1.6688	3.0157
C	9.0883	0.4932	0.4656
D	39.5825	0.0286	0.0069

Taking B and C points as examples, the travel time of point B is 4.8760 s, 46.35% less than that of point C. The energy consumption of point C is 0.4932 rad $\cdot$ s^{-2}, 70.45% lower than that of point B. The jerk of point C is 0.4656 rad $\cdot$ s^{-3}, 84.56% lower than that of point B.

C is selected as the actual solution of the project, and its time series is [0, 1.1990, 3.6445, 5.3612, 7.2749, 9.0883]. As shown in Figure 6, the curves of joint angles, velocities, accelerations, and jerks varying with time can be obtained by interpolating fifth-order B-spline curves.

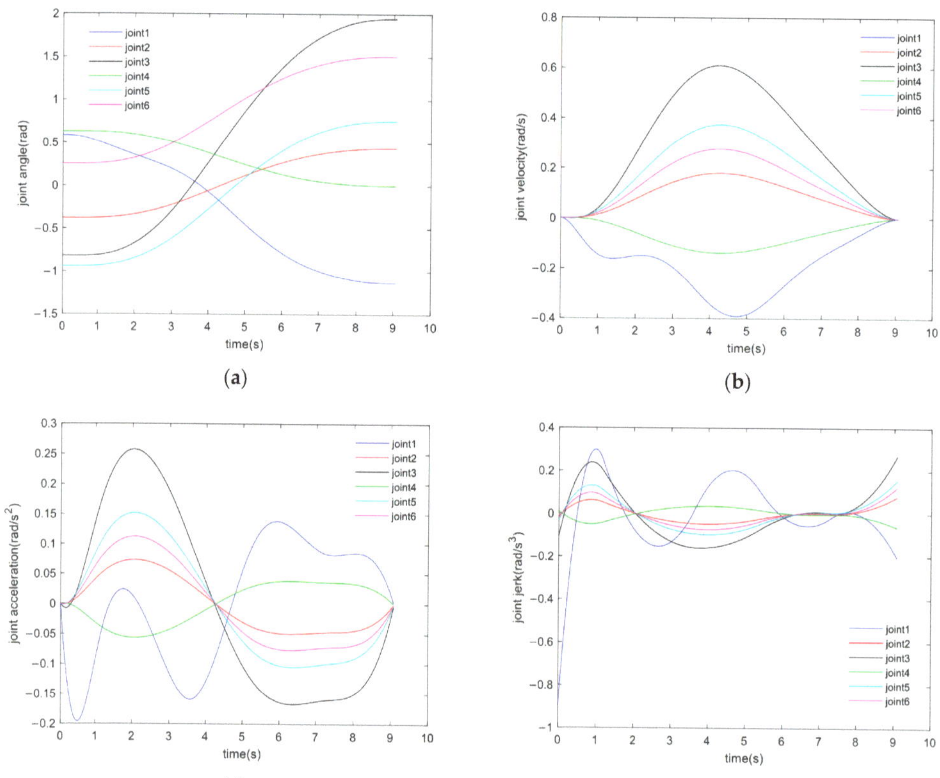

Figure 6. (a–d) are the angle, velocity, acceleration, and jerk curves of the joints under solution C.

A dominant solution E is randomly selected outside the Pareto optimal solution set as the time series [0, 1.3, 2.4, 5.3, 8.4, 10.4] before the trajectory optimization. Under this time series, the three performance indexes of the robot are shown in Table 5.

Table 5. Comparison before and after optimization.

Solution	Travel Time/(s)	Energy Consumption/(rad·s^{-2})	Jerk/(rad·s^{-3})
C	9.0883	0.4932	0.4656
E	10.4000	1.1457	1.8733

According to the data in Table 5, the travel time of point C is 12.61% less than that of E, the energy consumption is decreased by 56.95%, and the jerk is decreased by 75.15%. The three objective function values of point C are better than the results before optimization, improving the robot's comprehensive performance.

Taking robot joint 2 as an example, Figure 7 shows that the trajectories after optimization are smoother and more continuous than before optimization, especially the curves of joint velocity, acceleration, and jerk.

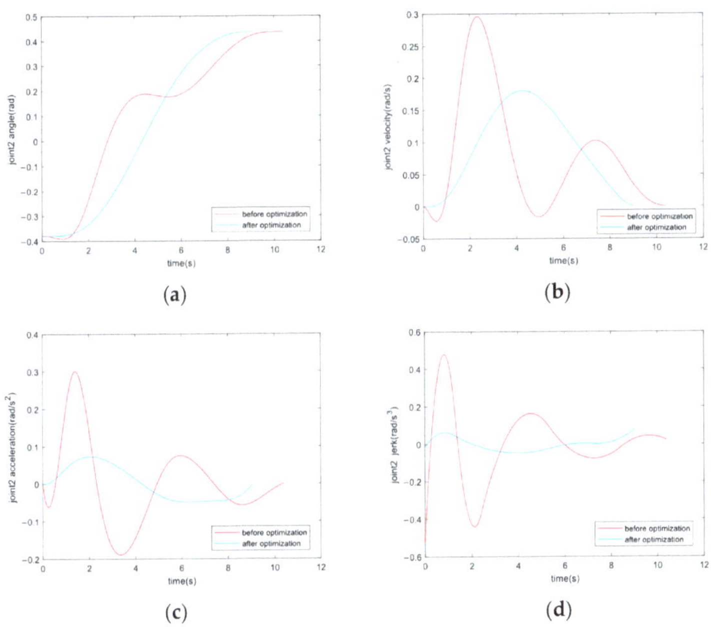

Figure 7. (**a**–**d**) show the curves of angle, velocity, acceleration, and jerk of robot joint 2 before and after optimization.

An optimal solution is randomly selected in the Pareto optimal solution set, called D, and its time series is [0, 1.7255, 4.5821, 6.1810, 7.7798, 10.0000]. The motion of the Puma560 robot under this solution can be visualized by the Robot Toolbox of MATLAB, as shown in Figure 8.

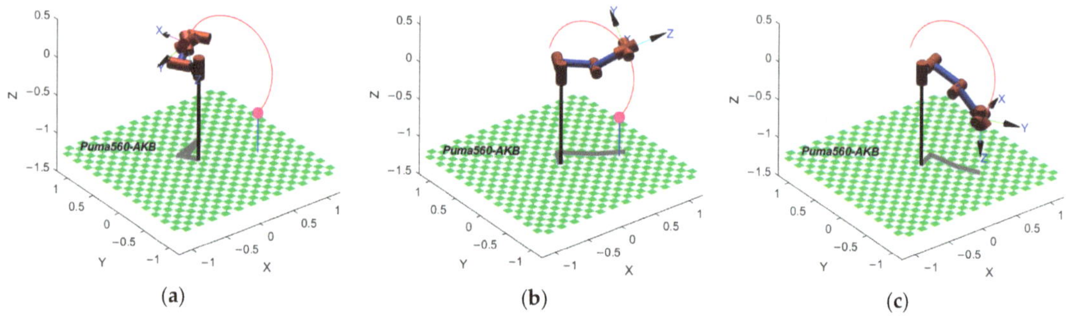

Figure 8. (**a**) The pose of the robot at the start moment; (**b**) the pose of the robot at the middle moment; and (**c**) the pose of the robot at the end moment.

5. Conclusions

This paper investigates a trajectory planning method for a robot which enables it to reach a comprehensive optimal state of travel time, energy consumption, and smoothness when executing a task. In order to fully understand the kinematics and dynamics characteristics of the robot and lay a solid theoretical foundation for follow-up research, this paper first deduces the position and orientation of the end-effector relative to the base and uses the Pieper method to calculate the closed solutions of the inverse kinematics. Finally, the dynamic model of the robot is established by the iterative Newton–Euler dynamics algorithm.

The joint space trajectory of the Puma560 robot is constructed using fifth-order B-spline curves, which has the advantages of continuous jerk and zero velocity and acceleration at the start/stop time. Then, the improved MOPSO algorithm is used to optimize the trajectory of the robot with the time interval between the path points as the decision variable. The convergence and distribution of the Pareto front are good, and the different solutions in the Pareto optimal solution set correspond to different engineering needs. In addition, by comparing the robot's travel time, energy consumption, and smoothness before and after optimization, it can be seen that its three performances have improved. This paper also visualizes the robot movement according to the planned trajectory in the Robot Toolbox of MATLAB.

Author Contributions: Conceptualization, J.W. (Jiahui Wang) and Y.Z.; methodology, J.W. (Jiahui Wang); software, J.W. (Jiahui Wang); validation, J.W. (Jiahui Wang), Y.Z. and S.Z.; formal analysis, J.W. (Jiahui Wang); investigation, J.W. (Jiahui Wang); resources, J.W. (Jiahui Wang) and S.Z.; data curation, J.W. (Jiahui Wang); writing—original draft preparation, J.W. (Jiahui Wang); writing—review and editing, J.W. (Jiahui Wang), Y.Z. and J.W. (Junling Wang); visualization, J.W. (Jiahui Wang); supervision, Y.Z.; project administration, Y.Z.; funding acquisition, Y.Z. All authors have read and agreed to the published version of the manuscript.

Funding: This research was funded by Science, Technology and Innovation Commission of Shenzhen Municipality, grant number No. JCYJ20200109141201714 ("Research on Flight Test Methods for Aerodynamics and Flight Control of Shipborne UAV").

Institutional Review Board Statement: Not applicable.

Informed Consent Statement: Not applicable.

Data Availability Statement: Data are contained within the article.

Conflicts of Interest: The authors declare no conflicts of interest.

References

1. Lan, J.; Xie, Y.; Liu, G.; Cao, M. A Multi-Objective Trajectory Planning Method for Collaborative Robot. *Electronics* **2020**, *9*, 859. [CrossRef]
2. Bailon, W.P.; Cardiel, E.B.; Campos, I.J.; Paz, A.R. Mechanical energy optimization in trajectory planning for six DOF robot manipulators based on eighth-degree polynomial functions and a genetic algorithm. In Proceedings of the 7th International Conference on Electrical Engineering Computing Science and Automatic Control, Tuxtla Gutierrez, Mexico, 8–10 September 2010; pp. 446–451.
3. Liu, J.; Wang, H.; Li, X.; Chen, K.; Li, C. Robotic arm trajectory optimization based on multiverse algorithm. *Math. Biosci. Eng.* **2023**, *20*, 2776–2792. [CrossRef] [PubMed]
4. Machmudah, A.; Parman, S.; Zainuddin, A.; Chacko, S. Polynomial joint angle arm robot motion planning in complex geometrical obstacles. *Appl. Soft Comput.* **2013**, *13*, 1099–1109. [CrossRef]
5. Porawagama, C.D.; Munasinghe, S.R. Reduced jerk joint space trajectory planning method using 5-3-5 spline for robot manipulators. In Proceedings of the 7th International Conference on Information and Automation for Sustainability, Colombo, Sri Lanka, 22–24 December 2014; pp. 1–6.
6. Kim, K.W.; Kim, H.S.; Choi, Y.K.; Park, J.H. Optimization of cubic polynomial joint trajectories and sliding mode controllers for robots using evolution strategy. In Proceedings of the IECON'97 23rd International Conference on Industrial Electronics, Control, and Instrumentation, New Orleans, LA, USA, 14 November 1997; pp. 1444–1447.
7. Lu, S.; Ding, B.; Li, Y. Minimum-jerk trajectory planning pertaining to a translational 3-degree-of-freedom parallel manipulator through piecewise quintic polynomials interpolation. *Adv. Mech. Eng.* **2020**, *12*, 1687814020913667. [CrossRef]
8. Boryga, M.; Grabo, A. Planning of manipulator motion trajectory with higher-degree polynomials use. *Mech. Mach. Theory* **2009**, *44*, 1400–1419. [CrossRef]
9. Chen, D.; Li, S.; Wang, J.; Feng, Y.; Liu, Y. A multi-objective trajectory planning method based on the improved immune clonal selection algorithm. *Robot. Comput. Integr. Manuf.* **2019**, *59*, 431–442. [CrossRef]
10. Gasparetto, A.; Zanotto, V. Optimal trajectory planning for industrial robots. *Adv. Eng. Softw.* **2010**, *41*, 548–556. [CrossRef]
11. Wang, Z.; Li, Y.; Sun, P.; Luo, Y.; Chen, B.; Zhu, W. A multi-objective approach for the trajectory planning of a 7-DOF serial-parallel hybrid humanoid arm. *Mech. Mach. Theory* **2021**, *165*, 104423. [CrossRef]
12. Gao, Y.; Xie, W.; Li, Q.; Li, X.; Hu, M.; Zhao, L. Time-Jerk Optimal Trajectory Planning of Industrial Robot based on Hybrid Particle Swarm Optimization Algorithm. In Proceedings of the 2021 China Automation Congress (CAC), Beijing, China, 22–24 October 2021; pp. 6327–6331.
13. Hansen, C.; Öltjen, J.; Meike, D.; Ortmaier, T. Enhanced approach for energy-efficient trajectory generation of industrial robots. In Proceedings of the IEEE International Conference on Automation Science and Engineering (CASE), Seoul, Republic of Korea, 20–24 August 2012; pp. 1–7.
14. Shi, B.; Zeng, H. Time-Optimal Trajectory Planning for Industrial Robot based on Improved Hybrid-PSO. In Proceedings of the 40th Chinese Control Conference (CCC), Shanghai, China, 26–28 July 2021; pp. 3888–3893.
15. Gasparetto, A.; Zanotto, V. A new method for smooth trajectory planning of robot manipulators. *Mech. Mach. Theory* **2007**, *42*, 455–471. [CrossRef]
16. Yao, J.; Sun, C.; Zhang, L.; Xiao, C.; Yang, M.; Zhang, S. Time optimal trajectory planning based on simulated annealing algorithm for a train uncoupling robot. In Proceedings of the 29th Chinese Control and Decision Conference (CCDC), Chongqing, China, 28–30 May 2017; pp. 5781–5785.
17. Bianco, C.G.L.; Piazzi, A. A genetic/interval approach to optimal trajectory planning of industrial robots under torque constraints. In Proceedings of the 1999 European Control Conference (ECC), Karlsruhe, Germany, 31 August–3 September 1999; pp. 942–947.
18. Mora, P.R. On the Time-optimal Trajectory Planning along Predetermined Geometric Paths and Optimal Control Synthesis for Trajectory Tracking of Robot Manipulators. Ph.D. Thesis, University of California, Berkeley, CA, USA, 2013.
19. Abu-Dakka, F.J.; Assad, I.F.; Alkhdour, R.M. Statistical evaluation of an evolutionary algorithm for minimum time trajectory planning problem for industrial robots. *Int. J. Adv. Manuf. Technol.* **2017**, *89*, 389–406. [CrossRef]
20. Zhang, W.; Fu, S. Time-optimal Trajectory Planning of Dulcimer Music Robot Based on PSO Algorithm. In Proceedings of the 2020 Chinese Control and Decision Conference (CCDC), Hefei, China, 22–24 August 2020; pp. 4769–4774.
21. Lin, H.I. A fast and unified method to find a minimum-jerk robot joint trajectory using particle swarm optimization. *J. Intell. Robot. Syst. Theory Appl.* **2014**, *75*, 379–392. [CrossRef]
22. Zhou, Y.; Han, G.; Wei, Z.; Huang, Z.; Chen, X.; Wu, J. Optimal trajectory planning of robot energy consumption based on improved sparrow search algorithm. *Meas. Control* **2024**, *57*, 1014–1021. [CrossRef]
23. Yokose, Y. Energy-saving trajectory planning for robots using the genetic algorithm with assistant chromosomes. *Artif. Life Robot.* **2020**, *25*, 89–93. [CrossRef]
24. Luo, L.-P.; Yuan, C.; Yan, R.-J.; Yuan, Q.; Wu, J.; Shin, K.-S.; Han, C.-S. Trajectory planning for energy minimization of industry robotic manipulators using the Lagrange interpolation method. *Int. J. Precis. Eng. Manuf.* **2015**, *16*, 911–917. [CrossRef]
25. Mohammed, A.; Schmidt, B.; Wang, L.; Gao, L. Minimizing Energy Consumption for Robot Arm Movement. *Procedia CIRP* **2014**, *25*, 400–405. [CrossRef]
26. Paes, K.; Dewulf, W.; Elst, K.V. Energy efficient trajectories for an industrial ABB robot. *Procedia CIRP* **2014**, *15*, 105–110. [CrossRef]

27. Ye, J.; Hao, L.; Cheng, H. Multi-objective optimal trajectory planning for robot manipulator attention to end-effector path limitation. *Robotica* **2024**, *42*, 1761–1780. [CrossRef]
28. Chen, W.; Wang, H.; Liu, Z.; Jiang, K. Time-energy-jerk optimal trajectory planning for high-speed parallel manipulator based on quantum-behaved particle swarm optimization algorithm and quintic B-spline. *Eng. Appl. Artif. Intell.* **2023**, *126*, 10722 [CrossRef]
29. Cao, X.; Yan, H.; Huang, Z.; Ai, S.; Xu, Y.; Fu, R.; Zou, X. A Multi-Objective Particle Swarm Optimization for Trajectory Planning of Fruit Picking Manipulator. *Agronomy* **2021**, *11*, 2286. [CrossRef]
30. Saravanan, R.; Ramabalan, S. Evolutionary Minimum Cost Trajectory Planning for Industrial Robots. *J. Intell. Robot. Syst.* **2008**, *52*, 45–77. [CrossRef]
31. Shi, X.; Fang, H.; Guo, L. Multi-objective optimal trajectory planning of manipulators based on quintic NURBS. In Proceedings of the 2016 IEEE International Conference on Mechatronics and Automation, Harbin, China, 7–10 August 2016.
32. Craig, J.J. *Introduction to Robotics: Mechanics and Control*, 3rd ed.; Pearson Education, Inc.: London, UK, 2005; pp. 73–76.
33. Pieper, D.L. The Kinematics of Manipulators Under Computer Control. Ph.D. Thesis, Stanford University, Stanford, CA, USA, 1968.
34. Luh, J.Y.S.; Walker, M.W.; Paul, R.P.C. On-Line Computational Scheme for Mechanical Manipulators. *ASME J. Dyn. Sys. Meas. Control* **1980**, *102*, 69–76. [CrossRef]
35. Piegl, L.; Tiller, W. *The Nurbs Book*, 2nd ed.; Springer: Berlin/Heidelberg, Germany, 1997; pp. 81–100.
36. Sathiya, V.; Chinnadurai, M. Evolutionary Algorithms-Based Multi-Objective Optimal Mobile Robot Trajectory Planning. *Robotica* **2019**, *37*, 1363–1382. [CrossRef]
37. Coello, C.A.C.; Lechuga, M.S. MOPSO: A proposal for multiple objective particle swarm optimization. In Proceedings of the 2002 Congress on Evolutionary Computation, Honolulu, HI, USA, 12–17 May 2002.

sensors

MDPI

Article

Enhanced Camera Relocalization Through Optimized Accelerated Coordinate Encoding Network and Pose Solver

Xinbo Chai [1], Zhen Yang [2], Xinrong Tan [1], Mengyang Zhu [1], Changbin Zhong [1] and Jianping Shi [1,*]

1 School of Electrical and Automation Engineering, Nanjing Normal University, Nanjing 210046, China; 231812078@njnu.edu.cn (X.C.); 241812081@njnu.edu.cn (X.T.); 231812035@njnu.edu.cn (M.Z.); 231812043@njnu.edu.cn (C.Z.)

2 School of Instrument and Engineering, Southeast University, Nanjing 210096, China; 223367@seu.edu.cn

* Correspondence: jpshi@njnu.edu.cn

Abstract: This paper presents an improved approach for scene-aware camera relocalization using RGB images and poses. Building upon the ACE network, we proposed a refined head structure that integrates skip and dense connections alongside channel attention mechanisms. Additionally, we introduced modifications to the loss function and pose solver, leveraging SQPnP and iterative optimization. These enhancements led to significant improvements in the localization accuracy and speed, as evidenced by our experiments on the 7scenes, 12scenes, and wayspots datasets. Here, we show that the average localization errors were reduced by up to 30% and the computational times were cut by approximately 10% compared to the original ACE network, demonstrating the practicality and robustness of our approach.

Keywords: ACE network; scene-aware; camera relocation; pose calculation

1. Introduction

Visual relocalization is a task within the field of computer vision; it is aimed at determining the camera's position and orientation (i.e., its location and direction) by analyzing visual information within a scene. Currently, the use of camera relocalization algorithms to predict six-degrees-of-freedom camera poses from input images plays an important role in various application areas, such as autonomous driving, robotics, and augmented reality (AR)/mixed reality (MR). Camera relocalization methods can be mainly divided into two broad categories: those based on regression and those based on structure. Early relocalization techniques primarily relied on regression-based methods, which could directly deduce camera poses from images; however, due to their susceptibility to image retrieval influences, they offered a lower accuracy, leading to a trend towards structure-based camera relocalization methods [1]. Structure-based camera relocalization methods can primarily be divided into two stages: establishing a coordinate mapping relationship between 2D pixel coordinates and 3D spatial coordinates through matching or regression, and then estimating the camera pose using the perspective-n-point (PnP) [2] series of pose-solving methods in conjunction with the random sample consensus (RANSAC) [3] algorithm.

Structure-based camera relocalization methods consist of two types: sparse feature matching and scene coordinate regression. Sparse feature matching relies on using local descriptors to establish correspondences between 2D map inputs and a given explicit 3D model, mainly comprising two steps: feature detection with feature detectors and feature decoding with descriptors. The learning approaches can be classified into three types based

https://doi.org/10.3390/s25061920

on the sequence of these steps: detect-then-describe, detect-and-describe-simultaneously, and describe-then-detect [4]. Scene coordinate regression eliminates the need for explicit 3D map construction and descriptor extraction, enabling an implicit transformation to be learned from 2D pixels to 3D point coordinates, which is relatively more efficient. Although scene coordinate regression can achieve an accuracy and relocalization times comparable to sparse feature matching, this method requires retraining with data when applied to new scenes, making it imperative to accelerate the retraining process as much as possible [5]. To address this, Brachmann et al. proposed the accelerated coordinate encoding (ACE) network architecture [6], which divides the network into a scene-agnostic backbone and a scene-specific head. When introducing new images, only the head needs to be retrained, effectively achieving the goal of accelerated coordinate encoding.

On one hand, to maintain the advantages of accelerated coordinate encoding, this paper retained the use of the pretrained, scene-transferable backbone from the original model and the concept of decorrelating image feature gradients. On the other hand, to achieve a more efficient and accurate camera coordinate computation, we improved the network structure of the model's head, the computational scheme of the loss component, and the calculation method and early exit conditions for the pose solver.

Firstly, we designed a new network structure for the head section that predicts pixel 3D coordinates with a higher precision and robustness. Subsequently, we modified the loss computation scheme to reduce the computational complexity. Next, we introduced a new pose-solving scheme that, in conjunction with the original approach, enables the pose solver to achieve an equivalent accuracy at a faster rate. Lastly, our experiments indicated that, compared to the previous ACE network, our improved network can achieve camera relocalization with a greater precision and at a faster pace in the majority of scenarios.

2. Related Work

2.1. Feature Extraction

Feature extraction is a key step in the fields of computer vision and image processing, serving to extract useful feature information from raw data to support subsequent tasks such as classification, detection, and recognition. Initially, feature extraction relied on manually designed methods such as edge detection and corner detection, which possess invariance and distinctiveness, requiring extensive expert knowledge and a targeted design [7]. Subsequently, with the development of deep learning, data-driven feature learning has become mainstream. Early classic models such as LeNet [8], AlexNet [9], and VGG [10] that utilize convolutional neural networks learn feature representations from raw data, overcoming the limitations of manual feature extraction and making feature detection learning and usage more efficient and automated. In recent years, with the increase in computing resources and the prevalence of large-scale datasets, transfer learning using pre-trained models has become increasingly popular. For instance, models such as ResNet [11] and EfficientNet [12] include a wealth of pre-trained weights available online, and utilizing these pre-trained weights can significantly enhance the performance for specific tasks while reducing data requirements and computational costs. The backbone used in our model, derived from the ACE model, consisted of eleven convolutional layers and two skip connections. The convolutional layer weights were obtained from a week-long pre-training on the first 100 scenes of the ScanNet dataset [13], enabling the rapid and effective extraction of crucial feature information from images. Moreover, for scene localization, the information from the image feature extraction part is transferable across scenes, and not affected by scene variations in terms of the localization accuracy. Therefore, this structure can effectively reduce the training time required for new scenes, better meeting the practical demands of application scenarios.

2.2. Coordinate Regression

Coordinate regression refers to the generation of 3D coordinates from input images, with conventional methods including random forest regression algorithms and deep learning neural networks. The random forest regression algorithm is a tree-based regression method that constructs multiple uncorrelated decision trees by randomly sampling data and features, obtaining prediction results in parallel. Each decision tree yields a predictive outcome based on the sampled data and features; by aggregating and averaging the results of all trees, the overall regression prediction of the forest is obtained. Deep learning neural networks establish an end-to-end coordinate mapping relationship between 2D pixel coordinates and 3D spatial coordinates and can be divided into two types as mentioned above: sparse feature matching and scene coordinate regression methods [14]. Scene coordinate regression methods, such as SANet [15]; DSAC [16] and its derivatives, DSAC++ [17] and DSAC* [18]; KFNet [19]; etc., are deep learning neural network approaches that implicitly establish coordinate mapping relationships. Compared to sparse feature matching methods that explicitly create 3D models, these approaches not only serve the purpose of privacy protection, but they also have lower storage requirements [20]. The main drawback of scene coordinate regression methods is the considerable time required for mapping in new scenes [21]. The ACE network architecture significantly reduces the training time for new scenes, effectively addressing this issue. Our proposed improvements enabled the model to achieve faster localization speeds during testing while maintaining a training duration similar to that of the ACE, making it better suited to the demands of practical scene localization.

2.3. Pose Estimation

Pose estimation relies on the fundamental principles of analytical geometry, deducing a camera's external parameters through known 3D points and their corresponding points in the camera image [22]. The PnP solution method can estimate the camera pose from the 3D spatial coordinates of at least three known points on an object and their corresponding 2D-pixel coordinates. With the enhancement of computational capabilities, optimization algorithms have been increasingly incorporated into camera pose-estimation processes, such as estimating a camera pose by minimizing the reprojection error, thereby further improving the accuracy and robustness of camera pose estimation. Common PnP algorithms include ITERATIVE [23], [24], which refines the pose using the nonlinear Levenberg–Marquardt minimization scheme; squared quadratically constrained quadratic program for perspective-n-point (SQPnP) [25], which can quickly and globally optimize for the perspective-n-point problem; and accurate and practical three-point perspective (AP3P) [26], which is tailored for the three-point relocalization issue. Building upon this, the RANSAC algorithm was introduced, repeatedly sampling and fitting models to obtain the optimal model parameters, thus mitigating the adverse effects of noise and outliers, estimating the best inlier set, and endowing camera pose estimation with a greater precision and robustness. The pose solver we used originated from DSAC*, where the pose solver in DSAC* first randomly selects 64 sets of pixel points to relocalize the camera using the perspective-3-point (P3P) [27] method, and then scores the relocalization results, applying RANSAC's pose optimization method to the highest-scoring result to involve as many valid predicted points as possible. We effectively improved this pose solver by altering the PnP pose-estimation approach and the iterative optimization process, ultimately achieving a faster execution of pose estimation.

3. Methods

3.1. Overview

In Figure 1, we present the general processing flow of the model on images. Initially, the backbone part extracts and compresses feature information from the image, and then randomly selects some features to feed into the head, predicting the three-dimensional coordinates of pixels in the image. Finally, these coordinates are used to relocalize the camera that captured the image. If needed, the obtained three-dimensional coordinates can be used for scene reconstruction, which, although more time-consuming, allows for a more intuitive acquisition of scene information and the display of the camera pose.

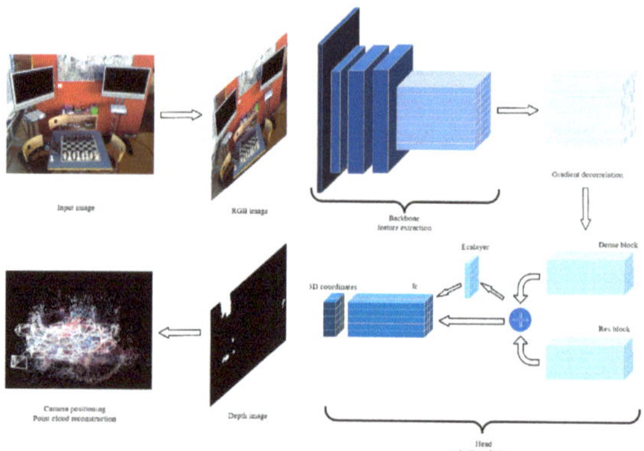

Figure 1. The model's image processing workflow.

Our improved model adopted the backbone part of the ACE network, which was trained for one week on the first 100 scenes of the ScanNet dataset, yielding a transferable feature extraction module. Given its superior performance and training costs, we chose not to attempt further modifications here. The enhancements made to the ACE network are described in the following subsections. Firstly, in Section 3.2, we detail the improvements to the head section of the ACE network, including the stacking of network layers, the types of modules referenced, the activation functions used, and the introduction of channel attention mechanisms. Then, in Section 3.3, we introduce improvements to the ACE loss function. On one hand, by modifying the computation sequence and logic, we ultimately reduced the computational complexity; on the other hand, by combining the Manhattan and Euclidean distances, we adapted the method to different datasets. Lastly, in Section 3.4, we present improvements to the pose solver. We not only changed the computation method for relocalizing the camera using predicted three-dimensional points, but we also added conditions for early termination during camera relocalization optimization, leading to both enhanced relocalization results and significantly shortened localization times.

3.2. Head Network Architecture

In the original ACE, the primary structure comprises eight 1×1 convolutional layers with an input and output channel count of 512 each, where skip connections are introduced after the third and sixth layers. Skip connections facilitate faster information propagation and gradient capture within the network, helping to address the problems of vanishing and exploding gradients. However, the sole use of serial 1×1 convolutions with skip connections still has its limitations, leading us to introduce dense connections in our network [28]. Dense connections not only reinforce gradient flow to mitigate the issue of gradient vanishing, but they also reduce the number of parameters, improving the parameter efficiency.

Initially, we replaced the three layers spanned by skip connections with three layers of dense connections, yet the test results indicated no improvement in the network performance, as the dense connections discarded some features preserved by skip connections. Therefore, considering the need to combine the strengths of both dense and skip connections, we proposed a parallel architecture incorporating three layers of both, allowing for the full exploitation of the advantages of each connection type to enhance the overall feature representation capabilities and optimization efficiency, enabling the model to perform well under various conditions.

Subsequently, after the dense and skip connections, we introduced a channel attention mechanism [29] by drawing on the ECA layer architecture using adaptive average pooling to weight the output channels, thereby further enhancing the feature representation ability and enabling the neural network to more effectively express key features in the data while reducing the impact of other noise.

Next, we switched all the activation functions in the network structure from ReLU to parametric ReLU (PReLU) [30]. The ReLU activation function used in the original model had its limitations, causing neuron death during training, whereas replacing it with PReLU can alleviate this issue, introducing a slight gradient in the negative region to maintain gradient flow.

Finally, the head network returns a 4D tensor $(\dot{x}, \dot{y}, \dot{z}, \hat{w})$, where $\dot{x}, \dot{y}, \dot{z}$ represent the homogeneous coordinates of the predicted three-dimensional scene, and the fourth element $\hat{w}$ is a non-normalized homogeneous parameter. To obtain the normalized homogeneous parameter w, preprocessing was applied to $\hat{w}$ using a biased and clipped Softplus operator, as defined by the following equation.

$$w = \min\left(\frac{1}{S_{min}}, \beta^{-1} \cdot \log(1 + \exp(\beta \cdot \hat{w})) + \frac{1}{S_{max}}\right), \beta = \frac{\log(2)}{1 - S_{max}^{-1}} \tag{1}$$

Here, S_{min} and S_{max} are used to clip the value of the scaling factor determined by w, and w ensures that, when $\hat{w}$ is 0, the output homogeneous parameter w is 1. The normalized homogeneous parameter w, computed as described, enables the dehomogenization of the network's output coordinates according to the formula $y = \frac{\dot{y}}{w}$, resulting in the predicted actual coordinates of the 3D scene. Figure 2 illustrates the head network structure of the meta-model and the improved results of the head network described in this section. It is evident that our designed network, while maintaining a similar parameter count, enhances the connectivity between its upper and lower layers, thereby more effectively extracting and transmitting image features.

3.3. Improvements to the Loss Function

In the original ACE framework, the loss function first converts predicted 3D coordinates into 2D pixel coordinates and computes the reprojection error, using e_π to quantify the robust reprojection error for valid coordinate predictions. All points are classified into two categories based on hyperparameters, including the maximum depth threshold, minimum depth threshold, and maximum reprojection error threshold, resulting in two distinct loss computations. Points within the threshold apply a scaled tanh function to tighten the reprojection error distribution.

$$\hat{e_\pi}\left(x_i, y_i, h_i^*\right) = \tau(t)\tanh\left(\frac{e_\pi\left(x_i, y_i, h_i^*\right)}{\tau(t)}\right) \tag{2}$$

The scaling factor for tanh is dynamically adjusted with a training-dependent threshold.

$$\tau(t) = k(t)\tau_{\max} + \tau_{\min}, \text{ with } k(t) = \sqrt{1 - t^2} \tag{3}$$

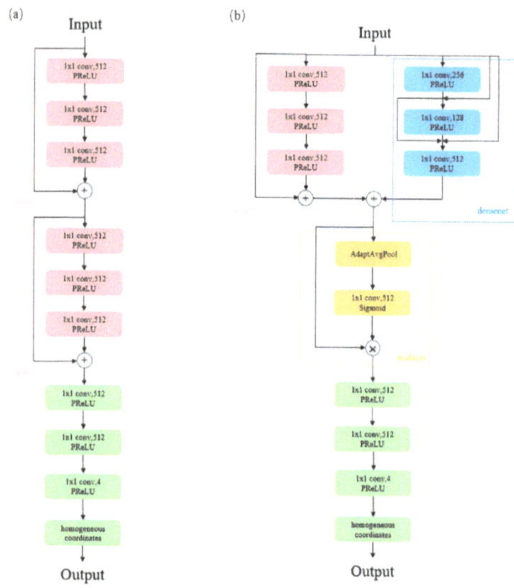

Figure 2. The head section of the original model (**a**). The modified head section (**b**).

In this approach, where $t \in (0, 1)$ represents the relative training progress, the threshold is dynamically adjusted to gradually decrease as training progresses, enabling more effective model convergence and generalization. For points outside the threshold, the Manhattan distance between the predicted and actual three-dimensional coordinates is used as the loss. The two types of losses are summed and divided by the batch size to obtain the final loss value participating in the iteration.

Regarding this loss function, we revised the computation order to filter out some pixel points using the maximum and minimum depth thresholds first, thereby reducing the computational load of subsequent two-dimensional remapping errors. On the other hand, considering the occurrence of overfitting, we introduced an additional penalty term that incorporates the Euclidean distance for pixels outside the threshold in the loss function. The Manhattan distance measures the distance between two points in space along the coordinate axes, while the Euclidean distance calculates the straight-line distance between two points. The former tends to capture local variations, whereas the latter captures global variations. These two metrics are complementary, and combining them in the loss function enables the capture of multi-scale characteristics of the data. As a result, this improvement enhances the compatibility of the loss function across different datasets, leading to a better generalization performance.

3.4. Enhancements to the Pose Solver

In the original ACE, a pose solver from DSAC* is employed, which randomly selects 64 sets of four pixels with reprojection errors not exceeding the threshold for camera relocalization. The scoring of each relocalization result is performed according to Equation (4).

$$\left(1 - \frac{1}{1 + e^{\frac{-5*(err - \tau_{in})}{\tau_{in}}}}\right) * \frac{\alpha_{in}}{width * height} \tag{4}$$

In this approach, err denotes the current reprojection error value, while τ_{in} serves as the threshold for distinguishing between valid inliers and outliers. α_{in} quantifies the weighting factor of the inliers' contributions to scoring, where the width and height represent the image height and width, respectively. The relocalization result with the highest score is selected for iterative pose refinement. Points are iteratively incorporated with reprojection errors within the threshold to refine the camera pose estimation. These refined pose estimates subsequently enhance the 3D point reprojection accuracy. Through successive refinements, the algorithm maximizes the number of inlier points until saturation is achieved.

Initially, we considered replacing the current pose solver with one from SRC [31]. However, experiments demonstrated that this solver's performance was significantly inferior to that of the DSAC*-based solver. A code analysis revealed that, on one hand, this pose solver lacked a threshold constraint for the upper limit of depth error, leading to substantial shifts due to noisy data with large errors. On the other hand, its optimization logic involved calling all points with reprojection errors within the threshold for camera relocalization, iterating until the displacement between the current and previous reprojection results fell below the threshold. In contrast, the DSAC* pose solver ensured that the number of points used for camera relocalization exceeded that of the last iteration, allowing for more stable optimization.

Consequently, we decided to directly optimize the existing DSAC*-based pose solver. For four-point relocalization, we substituted the traditional P3P pose-solving method with the AP3P method, which introduced additional geometric constraints for more accurate, faster, and robust pose estimation. Subsequently, in terms of multi-point relocalization iterative optimization, we initially referred to SQPnP, using the SQPnP pose-estimation method as an optimization scheme for camera relocalization with multiple points within the reprojection error range, replacing the preceding nonlinear Levenberg–Marquardt minimization approach. Compared to the previous method, SQPnP offers a higher computational efficiency and is suitable for sparse feature point sets, significantly improving the model efficiency. Yet, our experiments indicated that directly utilizing SQPnP still had limitations, causing significant declines in localization precision and insufficient robustness in certain scenarios. Therefore, we further refined our approach by integrating the SQPnP pose-estimation method with the nonlinear Levenberg–Marquardt minimization scheme. Specifically, we employed the SQPnP algorithm when the iteration of the camera pose was rapid to achieve a quicker correction, while the nonlinear Levenberg–Marquardt scheme was used for slower iterations to correct the camera pose more accurately and robustly. Lastly, regarding the relocalization iteration conditions, we believe that an optimization criterion solely focusing on involving more points in reconstruction is not prudent. The inclusion of new points may cause significant shifts in the majority of the already-established points, implying erroneous shifts in camera pose estimation as well. Therefore, we recorded the average reprojection error of the points within the threshold for the current reprojection results. If the next iteration led to an average reprojection error exceeding a certain multiple of the previous error, it suggested that subsequent iterations would degrade the accuracy of the camera pose estimation, necessitating the early termination of the iteration process.

4. Experiments

In this section, we provide a detailed description of the network training process. First, in Section 4.1, we introduce the training details and network configuration. In Section 4.2, we present the performance of our improved model compared to the baseline ACE model on two indoor datasets, 7scenes [32] and 12scenes [33], as well as the outdoor dataset wayspots. The results demonstrate the enhancements of our model, with example images from the three datasets shown in Figure 3. Finally, in Section 4.3, we conducted isolated ablation studies on the two improvements, proving their necessity.

Figure 3. Example plots of the 7scenes (**a**), 12scenes (**b**), and wayspots (**c**) datasets.

4.1. Experimental Setup

In the implementation of this work, for the feature extraction component, we retained the ACE backbone network and its pre-trained weights, with a batch size of 5120. For the training phase, we utilized the AdamW optimizer with learning rates ranging between (2.5×10^{-4}) and (7.5×10^{-3}). In the pose-estimation component, we employed an improved DSAC* pose solver with 64 RANSAC hypotheses. Throughout the implementation process, data storage was maintained in a half-precision format as much as possible to reduce memory usage.

4.2. Quantitative Evaluation

Firstly, we compared the test results of various models previously proposed with those of our model on the 7scenes dataset. The 7scenes dataset comprises seven types of indoor localization scenes, providing 7000 mapped images for each scene with areas ranging from 1 m^3 to 18 m^3, making it a commonly used dataset for camera relocalization. Table 1 presents the discrepancies between the predicted and actual camera poses in the 7scenes dataset using different models. It is evident that, compared to previously proposed models, our model achieved a higher precision in camera relocalization tasks.

Table 1. Comparison of relocalization results for the 7scenes dataset between the improved model and previous camera relocalization methods.

	Chess	**Fire**	**Heads**	**Office**	**Pumpkin**	**Redkitchen**	**Stairs**
PoseNet [34]	4.5°/0.13 m	11.3°/0.27 m	13.0°/0.17 m	5.6°/0.19 m	4.8°/0.26 m	5.4°/0.23 m	12.4°/0.35 m
SANet	0.9°/0.03 m	1.1°/0.03 m	1.5°/0.02 m	1.0°/0.03 m	1.3°/0.05 m	1.4°/0.04 m	4.6°/0.16 m
KFNet	0.7°/0.02 m	0.9°/0.02 m	0.8°/0.01 m	0.7°/0.03 m	1.0°/0.04 m	1.2°/0.04 m	1.0°/0.03 m
DSAC++	0.5°/0.02 m	0.9°/0.02 m	0.8°/0.01 m	0.7°/0.03 m	1.1°/0.04 m	1.1°/0.04 m	2.6°/0.09 m
DSAC*	0.7°/0.02 m	1.0°/0.03 m	1.3°/0.02 m	1.0°/0.03 m	1.3°/0.05 m	1.5°/0.05 m	49.4°/1.9 m
Ours	0.7°/0.02 m	0.8°/0.02 m	0.6°/0.01 m	0.8°/0.03 m	1.1°/0.04 m	1.3°/0.04 m	1.1°/0.04 m

4.2.1. Indoor Relocalization

Subsequently, we conducted a more detailed assessment of the improved model on the 7scenes dataset. Table 2 presents the assessment results comparing the improved and original models. The test results include the percentage of scenes with coordinate prediction errors within 10 cm/5 deg, 5 cm/5 deg, 2 cm/2 deg, and 1 cm/1 deg; the median of the camera coordinate prediction error in terms of the angle and distance; and the average time spent on the prediction. The results show that, in most scenarios, our model achieved a higher localization accuracy and a faster localization speed compared to the original model. Additionally, the median errors in the angle and distance predictions were mostly reduced. However, it is important to note that both the improved model and the original model performed poorly in certain scenarios, such as the "pumpkin" scene. This was likely due to interference from factors such as unnatural lighting in the dataset itself. Therefore, the optimization effectiveness should be comprehensively evaluated across multiple scenarios.

Table 2. Test results for the 7scenes dataset using the improved model compared to the original model.

Error Range	Chess	Fire	Heads	Office	Pumpkin	Redkitchen	Stairs
10 cm/5 deg (%)	100/100	99.7/99.8	100/100	98.0/97.8	88.8/87.4	91.9/91.3	93.8/93.1
5 cm/5 deg (%)	96.7/96.6	92.5/93.3	99.7/99.4	86.6/85.0	59.2/58.2	61.2/59.3	68.6/70.2
2 cm/2 deg (%)	55.4/52.3	56.0/54.6	92.4/90.3	30.2/28.8	15.0/15.2	14.1/13.3	9.8/8.3
1 cm/1 deg (%)	18.1/16.9	16.1/13.8	54.0/49.6	6.5/6.0	3.5/3.1	2.6/1.8	1.3/1.2
Rotation (deg)	0.7/0.7	0.8/0.9	0.6/0.7	0.8/0.8	1.1/1.1	1.3/1.4	1.1/1.1
Translation (cm)	1.8/1.9	1.8/1.9	0.9/1.0	2.7/2.8	4.3/4.4	4.2/4.3	3.9/3.8
Avg. time (ms)	30.5/34.6	34.4/37.8	32.6/35.9	33.6/38.6	36.8/40.8	38.5/41.9	41.3/42.6

Subsequently, we replicated the experiments on the 12scenes dataset using the same metrics, with the assessment results of both models presented in Tables 3 and 4. The results demonstrate that, compared to the original model, our model achieved a higher precision and a faster speed in relocalization across most indoor scenes, and it generalized well to different scenarios across various datasets. Furthermore, we observed that the same model achieved a higher relocalization accuracy on the 12scenes dataset compared to the 7scenes dataset, which may be attributed to the characteristics of the dataset itself. Under the premise of easily achieving high-precision localization, our improved model showed a relatively significant enhancement in relocalization speed. This reflects the effectiveness of our approach, which utilized SQPnP and the early termination of iterative pose solvers to reduce the localization time, particularly in scenarios with a lower complexity.

Table 3. Test results for the first six scenes in the 12scenes dataset, comparing the improved model with the original model (ours/ACE).

Error Range	apt1_kitchen	apt1_living	apt2_bed	apt2_kitchen	apt2_living	apt2_luke
10 cm/5 deg (%)	100/100	100/100	100.0/100	100/100	100/100	100/100
5 cm/5 deg (%)	100/100	100/100	100.0/100	100/100	100/100	99.4/99.0
2 cm/2 deg (%)	99.2/98.3	84.8/83.0	91.2/89.2	97.1/96.2	92.8/92.3	78/76.3
1 cm/1 deg (%)	68.3/70.3	47.7/46.7	57.8/52.5	72.4/66.2	62.2/63.6	34.5/31.9
Rotation (deg)	0.4/0.4	0.4/0.4	0.4/0.4	0.4/0.4	0.3/0.3	0.6/0.6
Translation (cm)	0.7/0.7	1.0/1.1	0.9/1.0	0.8/0.8	0.8/0.8	1.3/1.3
Avg. time (ms)	22.4/36.1	21.4/34.4	22.5/38.2	19.9/32.7	21.4/32.6	24.5/38.2

Table 4. Test results for the last six scenes in the 12scene dataset, comparing the improved model with the original model (ours/ACE).

Error Range	Office1-gates362	Office1-gates381	Office1-Lounge	Office1-Manolis	Office2-5a	Office2-5b
10 cm/5 deg (%)	100/100	100/100	100/100	100/100	100/100	100/100
5 cm/5 deg (%)	100/100	99.9/99.1	100/100	100/100	98.2/97.0	99.5/100
2 cm/2 deg (%)	92.0/90.7	86.0/83.4	85.9/83.8	92.6/90.8	81.7/79.5	76.3/74.8
1 cm/1 deg (%)	54.7/54.9	38.7/37.2	35.2/33.9	56.0/56.7	35.6/38.6	36.0/35.1
Rotation (deg)	0.4/0.4	0.5/0.6	0.4/0.4	0.4/0.4	0.5/0.5	0.4/0.4
Translation (cm)	0.9/0.9	1.2/1.2	1.3/1.3	0.9/0.9	1.2/1.2	1.3/1.3
Avg. time (ms)	22.6/37.4	27.8/40.3	21.8/34.4	23.1/34.3	25.3/37.7	23.7/35.2

4.2.2. Outdoor Relocalization

We also conducted equivalent tests on the wayspots outdoor dataset, which is derived from the MapFree dataset [35] and comprises ten consecutive outdoor scenes. Following the testing methodology mentioned in the ACE paper, we utilized scenes 200–209 from the MapFree dataset, with each scene containing 580 training images and 580 testing images. The experimental results are displayed in Tables 5 and 6. Although both our model and the original model struggled to achieve results comparable to those for the indoor datasets in outdoor settings, relatively speaking, our model's camera localization performance proved superior to that of the original model in most parts of the dataset.

Table 5. Test results for the first five scenes in the wayspots dataset, comparing the improved model with the original model (ours/ACE).

Error Range	Bears	Cubes	Inscription	Lawn	Map
10 cm/5 deg (%)	80.0/79.1	87.0/89.7	51.2/42.5	35.1/35.4	56.1/56.1
5 cm/5 deg (%)	67.1/66.2	30.1/41.9	20.7/17.3	23.5/23.7	37.2/34.0
2 cm/2 deg (%)	9.0/6.2	3.8/5.9	1.8/1.8	2.1/0.3	3.4/3.0
1 cm/1 deg (%)	0.5/0.7	0.0/1.0	0.0/0.0	0.0/0.0	0.0/0.0
Rotation (deg)	1.1/1.2	0.8/0.8	1.5/1.6	37.0/30.9	1.2/1.2
Translation (cm)	3.6/3.8	6.5/5.7	9.7/12.0	124.3/128.7	7.0/7.6
Avg. time (ms)	85.5/84.9	40.8/46.4	78.3/81.2	108.3/109.3	46.1/47.3

Table 6. Test results for the last five scenes in the wayspots dataset, comparing the improved model with the original model (ours/ACE).

Error Range	Squarebench	Statue	Tendrils	Therock	Wintersign
10 cm/5 deg (%)	66.6/64.0	0.0/0.0	34.6/33.0	98.4/100	0.9/0.7
5 cm/5 deg (%)	43.0/42.6	0.0/0.0	5.3/5.7	63.7/75.6	0.0/0.0
2 cm/2 deg (%)	6.1/6.4	0.0/0.0	0.0/0.0	19.6/30.1	0.0/0.0
1 cm/1 deg (%)	0.3/0.0	0.0/0.0	0.0/0.0	6.7/7.4	0.0/0.0
Rotation (deg)	0.7/0.7	13.1/15.6	45.2/49.9	0.8/0.8	0.9/1.1
Translation (cm)	5.9/6.0	476.9/576.2	165.8/183.5	3.9/3.1	388.2/484.7
Avg. time (ms)	41.0/42.6	120.0/126.9	104.9/104.0	21.5/29.7	126.7/123.7

4.2.3. Scene Reconstruction Results

We reconstructed the chess scene from the 7scenes dataset using our improved model. Given that scene reconstruction requires depth information predicted from all the pixels, we provisionally employed the previous loss function for this task. As the primary objective of the model is to enable faster and more accurate camera relocalization, which does not necessitate scene reconstruction, we retained the improvements made to the loss component within the model. In Figure 4, we present the process and results of scene reconstruction using the improved model on the chess scene from the 7scenes dataset.

4.3. Ablation Studies

Subsequently, we conducted isolated ablation studies on the two innovative aspects of our model using the 7scenes dataset, to separately validate their effectiveness and analyze the underlying reasons.

4.3.1. Head and Loss Function Improvements

In this section, we discuss improvements made to the model's head and loss function while retaining the original pose solver. Table 7 shows the experimental results of the model on the 7scenes dataset. It is evident that the model achieved an improved accuracy in camera localization. Unlike the original model, which utilized three serial skip connections in its head, our improved version included parallel skip and dense connections. Skip connections help mitigate performance degradation and address potential issues with vanishing or exploding gradients, while dense connections exploit the network's feature information to enhance gradient flow. The parallel structure allowed both to exert their respective advantages, improving the overall representational power and optimization efficiency, thereby enhancing the model's performance. Furthermore, we incorporated the calculation of Euclidean distance into the loss function for pixels outside a certain threshold, in addition to the existing Manhattan distance. The amalgamation of these two distance measures allowed the loss function to achieve a more comprehensive assessment of distances, adapting to various data distributions.

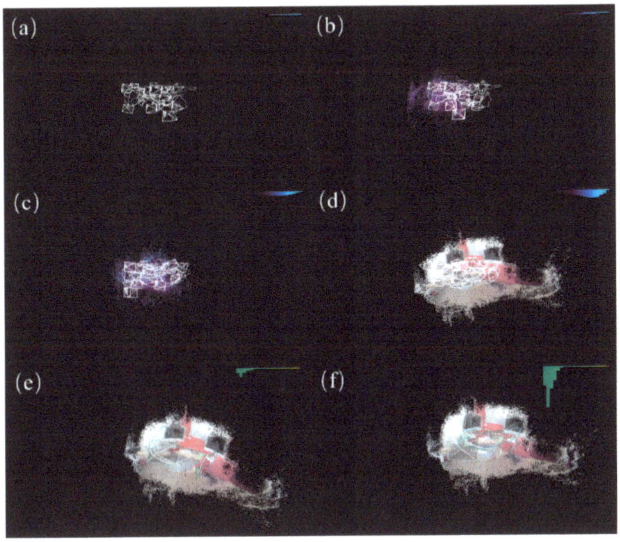

Figure 4. The scene reconstruction process and results for the chess scene of the 7scenes dataset using the improved model. (**a**) shows the scene reconstruction during the initial phase of model training, (**b**,**c**) depict the scene reconstruction during the model-training process, (**d**) illustrates the scene reconstruction after the completion of model training, and (**e**,**f**) display the camera localization and scene reconstruction during model testing, with the bar graph in the top-right corner representing the percentage of pixel offsets within various segmented thresholds.

Table 7. Performance results for the 7scenes dataset using the model with improvements to only the head section and loss function compared to the original model (our_head/ACE).

Error Range	Chess	Fire	Heads	Office	Pumpkin	Redkitchen	Stairs
10 cm/5 deg (%)	100/100	99.8/99.8	100/100	98.0/97.8	88.5/87.4	91.7/91.3	93.4/93.1
5 cm/5 deg (%)	96.8/96.6	93.2/93.3	99.9/99.4	86.2/85.0	59.2/58.2	59.7/59.3	71.8/70.2
2 cm/2 deg (%)	54.3/52.3	56.1/54.6	91.4/90.3	29.1/28.8	16.3/15.2	13.6/13.3	10.1/8.3
1 cm/1 deg (%)	15.7/16.9	16.0/13.8	51.1/49.6	7.1/6.0	3.4/3.1	2.2/1.8	1.3/1.2
Rotation (deg)	0.7/0.7	0.8/0.9	0.6/0.7	0.8/0.8	1.1/1.1	1.3/1.4	1.1/1.1
Translation (cm)	1.9/1.9	1.8/1.9	1.0/1.0	2.7/2.8	4.2/4.4	4.3/4.3	3.7/3.8
Avg. time (ms)	35.6/34.6	38.2/37.8	34.1/35.9	37.4/38.6	40.6/40.8	42.1/41.9	41.8/42.6

However, as one can see from Table 7, the improved head required additional time for some scenes only. One reason for this is the increased number of parameters; we calculated the parameters for the head component and found that, while the original model had approximately 2.1 million parameters in its head, the improved model's parameter count exceeded 2.2 million. On the other hand, compared to residual connections, dense connections require more computational resources, and to enhance the performance further, we integrated a channel attention mechanism into the head, which also added to the time required. Consequently, to compensate for this increased time expenditure and achieve faster computational results, we improved the loss function by restructuring the program to delay pixel offset computations, thereby reducing the overall computational load. Additionally, we enhanced the pose solver, resulting in an accelerated resolution process.

4.3.2. Pose Solver Enhancements

In this section, we report on the enhancements made to the pose solver part of our model, while retaining the original head. Table 8 shows the experimental results for the 7scenes dataset. It is observable that there was an increase in the computational speed during the camera localization by the model. Compared to the DSAC*-based pose solver used in the original model, our improved solver integrated two pose-estimation methods: SQPnP and ITERATIVE. SQPnP can solve for camera poses more quickly and directly via sparse linear systems, but is slightly less accurate than ITERATIVE, which solves for camera poses with a greater precision through iterative optimization, albeit at a cost to real-time performance. The pose-estimation method utilized by this model sequentially invoked both techniques, comparing the pre- and post-optimization offsets with a threshold of 0.01. SQPnP was selected for substantial corrections to the camera pose, while ITERATIVE was employed when incremental adjustments neared the accurate value. Additionally, we imposed further constraints on the early termination of the pose-estimation process. The previous approach terminated early only if the pixel displacement remained below a threshold. We introduced a criterion where, if the mean pixel displacement prior to iteration exceeded 1.2 times that following iteration, the process terminated early as well, effectively addressing issues where pursuing the accuracy of discrete points results in overall larger deviations, thus achieving more precise and faster localization in certain scenarios.

Table 8. Performance results for the 7scenes dataset using the model with improvements to only the pose solver compared to the original model (our_PnP/ACE).

Error Range	Chess	Fire	Heads	Office	Pumpkin	Redkitchen	Stairs
10 cm/5 deg (%)	100/100	99.8/99.8	100/100	97.8/97.8	88.2/87.4	91.9/91.3	93.7/93.1
5 cm/5 deg (%)	96.9/96.6	93.2/93.3	99.6/99.4	85.5/85.0	57.6/58.2	59.3/59.3	67.6/70.2
2 cm/2 deg (%)	52.2/52.3	53.2/54.6	90.6/90.3	29.5/28.8	14.6/15.2	13.1/13.3	9.5/8.3
1 cm/1 deg (%)	17.4/16.9	16.4/13.8	49.6/49.6	6.1/6.0	3.7/3.1	2.0/1.8	1.3/1.2
Rotation (deg)	0.7/0.7	0.8/0.9	0.6/0.7	0.8/0.8	1.1/1.1	1.3/1.4	1.2/1.1
Translation (cm)	1.9/1.9	1.9/1.9	1.0/1.0	2.8/2.8	4.4/4.4	4.4/4.3	4.1/3.8
Avg. time (ms)	30.3/34.6	32.6/37.8	32.3/35.9	33.3/38.6	35.7/40.8	37.0/41.9	37.8/42.6

As indicated by Table 8, the pose solver that integrated ITERATIVE and SQPnP effectively enhanced the speed of pose estimation. It compensated for the computational slowdown caused by the improved head component without noticeably affecting the accuracy of pose estimation.

5. Conclusions

In this paper, we proposed a novel network model that rapidly relocalizes cameras in three dimensions using RGB images. This model improves upon the ACE model with refinements to the head section, the loss computation, and the pose-estimation components, achieving higher camera localization precision and a faster localization speed compared to the original ACE model. Firstly, for the head section, we restructured it by integrating dense connections, skip connections, and channel attention mechanisms, replacing the previous two-layer skip connections. Subsequently, we rewrote the loss calculation code, adjusting the order of coordinate offset computations that map to the pixel coordinate system to reduce the computational load. Simultaneously, we incorporated both Euclidean and Manhattan distances as losses for significant coordinate offsets, enhancing the model's generalizability. Lastly, we modified the pose solver by adopting the AP3P method instead of the conventional P3P within RANSAC pose-estimation hypotheses. In the subsequent iterative optimization process, we integrated SQPnP and ITERATIVE methods, replacing the previous ITERATIVE pose solution, and added early-exit conditions during pose iteration optimization to mitigate the risk of converging to local optima. Ultimately, this allowed us to achieve faster pose estimation while maintaining the accuracy, better suiting the needs of real-world applications. The model was validated on the 7scenes, 12scenes, and wayspots datasets, demonstrating that the improvements in our model exhibited a certain level of generalizability across different scenarios. Furthermore, the ablation experiments elucidated the theoretical basis for the improvements, further confirming the effectiveness of our enhancements. Our work still has limitations. On one hand, the model was only trained and tested on datasets and has not yet been evaluated in real-world application scenarios, which will be the focus of our future work. On the other hand, similar to most previous models, our model still struggles to achieve a high localization accuracy under significant outdoor lighting variations, and it cannot guarantee robust generalizability across diverse environments. In the future, we will further refine the model to enhance its performance under varying lighting conditions and improve its environmental adaptability.

Author Contributions: Conceptualization, X.C. and Z.Y.; methodology, X.C. and Z.Y.; software, X.C. and X.T.; validation, X.T.; data curation, M.Z.; formal analysis, M.Z.; literature review, C.Z.; visualization, C.Z.; resources, J.S.; supervision, J.S.; writing—original draft preparation, X.C.; writing—review and editing, X.C., Z.Y. and J.S. All authors have read and agreed to the published version of the manuscript.

Funding: This research was funded by the Frontier Technologies R&D Program of Jiangsu, grant number BF2024077.

Institutional Review Board Statement: Not applicable.

Informed Consent Statement: Not applicable.

Data Availability Statement: The data are contained within the article.

Conflicts of Interest: The authors declare no conflicts of interest.

References

1. Tang, S.; Tang, C.; Huang, R.; Zhu, S.; Tan, P. Learning Camera Localization via Dense Scene Matching. *arXiv* **2021**, arXiv:2103.16792.
2. Hesch, J.A.; Roumeliotis, S.I. A Direct Least-Squares (DLS) method for PnP. In Proceedings of the IEEE International Conference on Computer Vision (ICCV 2011), Barcelona, Spain, 6–13 November 2021; pp. 383–390.
3. Fischler, M.A.; Bolles, R.C. Random Sample Consensus: A Paradigm for Model Fitting with Applications to Image Analysis and Automated Cartography. In *Readings in Computer Vision*; Fischler, M.A., Firschein, O., Eds.; Morgan Kaufmann: San Francisco CA, USA, 1987; pp. 726–740; ISBN 978-0-08-051581-6.
4. Chen, C.; Wang, B.; Lu, C.X.; Trigoni, N.; Markham, A. Deep learning for Visual Localization and Mapping: A Survey. *IEEE Trans. Neural Netw. Learn. Syst.* **2024**, *35*, 17000–17020. [CrossRef]
5. Liu, D.; Chen, L. SECPNet—Secondary encoding network for estimating camera parameters. *Vis. Comput.* **2021**, *38*, 1689–1702. [CrossRef]
6. Brachmann, E.; Cavallari, T.; Prisacariu, V.A. Accelerated Coordinate Encoding: Learning to Relocalize in Minutes Using RGB and Poses. In Proceedings of the 2023 IEEE/CVF Conference on Computer Vision and Pattern Recognition (CVPR), Vancouver, BC, Canada, 17–24 June 2023; pp. 5044–5053.
7. Wang, J.; Qi, Y. Visual camera relocalization using both hand-crafted and learned features. *Pattern Recognit.* **2023**, *145*, 109914. [CrossRef]
8. Lecun, Y.; Bottou, L.; Bengio, Y.; Haffner, P. Gradient-based learning applied to document recognition. *Proc. IEEE* **1998**, *86*, 2278–2324. [CrossRef]
9. Krizhevsky, A.; Sutskever, I.; Hinton, G.E. Imagenet classification with deep convolutional neural networks. *Commun. the ACM* **2017**, *60*, 84–90. [CrossRef]

10. Simonyan, K.; Zisserman, A. Very Deep Convolutional Networks for Large-Scale Image Recognition. *arXiv* **2014**, arXiv:1409.1556
11. He, K.; Zhang, X.; Ren, S.; Sun, J. Deep Residual Learning for Image Recognition. In Proceedings of the IEEE Conference on Computer Vision and Pattern Recognition, Las Vegas, NV, USA, 27–30 June 2016; pp. 770–778.
12. Tan, M.; Le, Q. EfficientNet: Rethinking Model Scaling for Convolutional Neural Networks. In Proceedings of the 36th International Conference on Machine Learning, Long Beach, CA, USA, 9–15 June 2019; pp. 6105–6114.
13. Dai, A.; Chang, A.X.; Savva, M.; Halber, M.; Funkhouser, T.; Nießner, M. Scannet: Richly-Annotated 3D Reconstructions of Indoor Scenes. In Proceedings of the IEEE Conference on Computer Vision and Pattern Recognition, Honolulu, HI, USA, 21–26 July 2017; pp. 2432–2443.
14. Li, Y.; Luo, F.; Xiao, C. Monocular human depth estimation with 3D motion flow and surface normals. *Vis. Comput.* **2023**, *39*, 3701–3713. [CrossRef]
15. Yang, L.; Bai, Z.; Tang, C.; Li, H.; Furukawa, Y.; Tan, P. SANet: Scene Agnostic Network for Camera Localization. In Proceedings of the IEEE/CVF International Conference on Computer Vision, Seoul, South Korea, 27 October–02 November 2019; pp. 42–51.
16. Brachmann, E.; Krull, A.; Nowozin, S.; Shotton, J.; Michel, F.; Gumhold, S. DSAC-Differentiable RANSAC for Camera Localization. In Proceedings of the IEEE Conference on Computer Vision and Pattern Recognition, Honolulu, HI, USA, 21–26 July 2017; pp. 6684–6692.
17. Brachmann, E.; Rother, C. Expert Sample Consensus Applied to Camera Re-Localization. In Proceedings of the IEEE/CVF International Conference on Computer Vision, Seoul, South Korea, 27 October–2 November 2019; pp. 7525–7534.
18. Brachmann, E.; Rother, C. Visual Camera Re-Localization from RGB and RGB-D Images Using DSAC. *IEEE Trans. Pattern Anal. Mach. Intell.* **2021**, *44*, 5847–5865. [CrossRef] [PubMed]
19. Zhou, L.; Luo, Z.; Shen, T.; Zhang, J.; Zhen, M.; Yao, Y. KFNet: Learning Temporal Camera Relocalization Using Kalman Filtering. In Proceedings of the IEEE/CVF Conference on Computer Vision and Pattern Recognition, Seattle, WA, USA, 13–19 June 2020; pp. 4919–4928.
20. Zhao, T.; Pan, S.; Gao, W.; Sheng, C.; Sun, Y.; Wei, J. Attention Unet++ for lightweight depth estimation from sparse depth samples and a single RGB image. *Vis. Comput.* **2021**, *38*, 1619–1630. [CrossRef]
21. Chen, S.; Cavallari, T.; Prisacariu, V.A.; Brachmann, E. Map-Relative Pose Regression for Visual Re-Localization. *arXiv* **2024**, arXiv:2404.09884.
22. Chatterjee, J.; Vega, M.T. 3D-Scene-Former: 3D scene generation from a single RGB image using Transformers. *Vis. Comput.* **2024**, *41*, 2875–2889. [CrossRef]
23. Eade, E. Gauss-Newton/Levenberg-Marquardt Optimization. Technical Report. 2013. Available online: https://ethaneade.com/optimization.pdf (accessed on 28 February 2025).
24. Ma, Y.; Soatto, S.; Košecká, J.; Sastry, S.S. *An Invitation to 3-D vision: From Images to Geometric Models*; Springer: New York, NY, USA, 2004.
25. Telea, A. An Image Inpainting Technique Based on the Fast Marching Method. *J. Graph. Tools* **2004**, *9*, 23–34. [CrossRef]
26. Ke, T.; Roumeliotis, S.I. An Efficient Algebraic Solution to the Perspective-Three-Point Problem. In Proceedings of the IEEE Conference on Computer Vision and Pattern Recognition, Honolulu, HI, USA, 21–26 July 2017; pp. 7225–7233.
27. Gao, X.-S.; Hou, X.-R.; Tang, J.; Cheng, H.-F. Complete solution classification for the perspective-three-point problem. *IEEE Trans. Pattern Anal. Mach. Intell.* **2003**, *25*, 930–943. [CrossRef]
28. Huang, G.; Liu, Z.; Van Der Maaten, L.; Weinberger, K.Q. Densely Connected Convolutional Networks. In Proceedings of the IEEE Conference on Computer Vision and Pattern Recognition, Honolulu, HI, USA, 21–26 July 2017; pp. 4700–4708.
29. Wang, Q.; Wu, B.; Zhu, P.; Li, P.; Zuo, W.; Hu, Q. ECA-Net: Efficient Channel Attention for Deep Convolutional Neural Networks. In Proceedings of the IEEE/CVF Conference on Computer Vision and Pattern Recognition, Seattle, WA, USA, 13–19 June 2020; pp. 11534–11542.
30. He, K.; Zhang, X.; Ren, S.; Sun, J. Delving deep into rectifiers: Surpassing human-level performance on imagenet classification. In Proceedings of the International Conference on Computer Vision, Las Condes, Chile, 11–18 December 2015; pp. 1026–1034.
31. Dong, S.; Wang, S.; Zhuang, Y.; Kannala, J.; Pollefeys, M.; Chen, B. Visual Localization via Few-Shot Scene Region Classification. In Proceedings of the International Conference on 3D Vision, Prague, Czech Republic, 12–15 September 2022; pp. 393–402.
32. Shotton, J.; Glocker, B.; Zach, C.; Izadi, S.; Criminisi, A.; Fitzgibbon, A. Scene Coordinate Regression Forests for Camera Relocalization in RGB-D Images. In Proceedings of the IEEE Conference on Computer Vision and Pattern Recognition, Portland, OR, USA, 23–28 June 2013; pp. 2930–2937.
33. Valentin, J.; Niebner, M.; Shotton, J.; Fitzgibbon, A.; Izadi, S.; Torr, P. Exploiting uncertainty in regression forests for accurate camera relocalization. In Proceedings of the IEEE Conference on Computer Vision and Pattern Recognition (CVPR), Boston, MA, USA, 7–12 June 2015; pp. 4400–4408.

4. Kendall, A.; Grimes, M.; Cipolla, R. Posenet: A convolutional network for real-time 6-dof camera relocalization. In Proceedings of the IEEE International Conference on Computer Vision, Santiago, Chile, 11–18 December 2015; pp. 2938–2946.
5. Arnold, E.; Wynn, J.; Vicente, S.; Garcia-Hernando, G.; Monszpart, Á.; Prisacariu, V.; Turmukhambetov, D.; Brachmann, E. Map-Free Visual Relocalization: Metric Pose Relative to a Single Image. In Proceedings of the 17th European Conference on Computer Vision, Tel Aviv, Israel, 23–27 October 2022; pp. 690–708.

sensors

MDPI

Article

Generalized q-Method Relative Pose Estimation for UAVs with Onboard Sensor Measurements

Kyl Stanfield [1], Ahmad Bani Younes [2,*] and Mohammad Hayajneh [3]

[1] Guidance, Navigation, & Controls (GNC) Engineer, 5500 Campanile Drive, San Diego, CA 92182, USA; kylstanfield@sbcglobal.net
[2] Department of Aerospace Engineering, San Diego State University, San Diego, CA 92181, USA
[3] Faculty of Engineering, Mechatronics Engineering, The Hashemite University, Zarqa 13133, Jordan; mhayajneh@hu.edu.jo
* Correspondence: abaniyounes@sdsu.edu

Abstract: The q-method for pose estimation utilizes on-board measurement vectors of reference objects to calculate air vehicle position and orientation with respect to an Inertial frame. This new method solves for the quaternion eigenvalue solution of the optimal pose to minimize the error in the derived system of equations. The generalized q-method extends Davenport's q-method for satellite attitude estimation by incorporating inertial position into the relative model and eliminating assumptions throughout the derivation that require spacecraft applications. Thus, the pose estimation model is developed and implemented for UAV applications using an onboard camera to obtain measurements in a controlled environment. Combined with numerical methods, algorithm outputs for position and orientation are validated against truth data to prove accurate estimation despite sensor error.

Keywords: q-method; drone; UAV; pose; estimation; attitude; quaternion; relative; navigation

1. Introduction

In aerospace engineering and robotics, the estimation of attitude and position is an essential factor for precise autonomous operations [1–5]. Attitude estimation enables one to determine the vehicle's orientation in a three-dimensional space, while position estimation determines the vehicle's translational coordinates in a three-dimensional space. Modern methods integrate optimal filtering techniques by fusing measurements from local or global sensors with knowledge of the vehicle's motion dynamics. Traditional methods often involve sensor fusion techniques combining visual, inertial, and GPS data. Visual odometry and simultaneous localization and mapping (SLAM) have been widely adopted for pose estimation using cameras and other imaging sensors, while inertial measurement units (IMUs) are often used to provide complementary motion information [6]. Techniques such as the Kalman filter, Extended Kalman Filter (EKF), and Unscented Kalman Filter (UKF) have been extensively applied. For instance, Zhang et al. [7] demonstrated the effectiveness of EKF in fusing data from inertial measurement units (IMUs) and global navigation satellite systems (GNSSs) to improve the robustness of UAV positioning in urban environments. Similarly, Liu et al. [8] explored the use of particle filters for attitude estimation in multi-sensor setups, highlighting their resilience to sensor noise and dynamic changes. One significant challenge in these methods is to deal with sensor noise and environmental conditions [9] that affect sensor performance, such as poor lighting or featureless scenes. Several researchers have focused on enhancing these techniques through

https://doi.org/10.3390/s25061939

the integration of LiDAR and deep learning-based approaches [10] to improve robustness and real-time performance, particularly in GPS-denied environments. Another challenge is associated with handling the coupled dynamics with the estimated position and attitude simultaneously [11]. Therefore, this research develops a generalized formulation that combines the estimation of position and attitude based on the well-known q-method framework [12].

Using the quaternion (q-method) for attitude estimation was a method developed by Paul Davenport [12,13] as a compact solution for optimal estimation of the rotation of a body relative to an Inertial frame. Quaternions are widely used because of their efficiency and computational advantages [14,15] over other three-parameter attitude representations, such as Euler angles [16,17]. They are ideal for avoiding singularities, such as gimbal lock, which can occur when using Euler angles during large rotations [18]. Quaternions are also computationally lighter and more stable in dynamic conditions, which makes them popular in real-time applications [19,20]. The integration of a quaternion-based algorithm with the Kalman filter has been widely applied for real-time attitude estimation in aerospace applications [21,22]. Another progression has led to the development of more advanced algorithms, such as Unscented Kalman Filters (UKFs) and particle filters, which demonstrate enhanced performance in handling non-linear systems [23]. Additionally, recent advancements have incorporated deep learning approaches to enhance the pose estimation process. For example, methods that fuse quaternions with convolutional neural networks (CNNs) have demonstrated high levels of accuracy in both orientation and position estimation [24]. These developments have researched uncoupled attitude estimation without integrating the translational dynamics. The need for solutions that integrate the coupled dynamics is essential for many motion-based proximity applications.

This paper generalizes Davenport's q-method to solve for the relative pose—the coupled state of attitude and translation—for estimation of UAVs with onboard measurements. This paper first provides a comprehensive overview of mathematical preliminaries and Davenport's q-method for attitude estimation, including a background on the Wahba problem [25] and laying a solid foundation for the subsequent advancements. The development of the quaternion-based approach is explained, highlighting its advantages over traditional methods in terms of simplicity and reduced computational overhead. Following this, the q-method pose estimation model is formulated as a novel, more generalized approach that optimally estimates a body's position and orientation. This approach requires two sets of measurements for reference points, enabling the calculation of the optimal relative pose from an Inertial frame. This framework eliminates various assumptions inherent to the original technique for spacecraft application and provides a precise method for aircraft and spacecraft pose estimation when given known inertial reference data and body-generated measurements. The intricacies and methodologies here are fully documented. Subsequently, a simulation is constructed to corroborate the practical application of the newly devised pose estimation method in a controlled environment. Numerical methods are employed to solve the newly formed six degrees of freedom (6DoF) estimation equations to minimize the system error. This multifaceted analysis ensures empirical evidence of this effective technique in real-world settings.

By unifying the models for attitude and position estimation, the generalized q-method offers a comprehensive solution for the relative pose estimation problem, bridging the gap between quaternion-based rotation estimation and full pose determination. Through rigorous derivation of the governing equations and models, we provide a cohesive framework that extends the reach of the q-method, setting the stage for its practical application in aerospace systems and beyond. The main contribution is summarized as the development of a generalized q-method for pose estimation. This method is based on formulating the coupled dynamics of 6DoF motion in a singularity-free manner. The generalized q-method was tested using relative pose environments in two different attitude configurations with realistically acquired data and an experimental setup in the SPACE lab. The performance of the method was validated against ground truth data obtained from a Vicon system, which provides sub-millimeter accuracy.

2. Mathematical Preliminaries

2.1. Coordinate Frame and System Definition

Illustrated in Figure 1, there are three frames: the Inertial reference (I) frame, the Body reference (B) frame, and the relative reference (D) frame. The Inertial frame remains fixed in space, while the Body frame is free to move throughout the system, relative to the I frame. The B frame is oriented by some rotation C (Direction Cosine Matrix) relative to the I frame. The respective position vectors from the Inertial frame (system origin) are given by $\bar{r}_{B/I}$ and $\bar{r}_{D/I}$. $\bar{r}_{D/B}$ goes to D from B, and can be given with respect to the body-centered frame $\bar{r}^B_{D/B}$, or the inertial-centered frame $\bar{r}^I_{D/B} \because C\bar{r}^I_{D/B} = \bar{r}^B_{D/B}$. This relation will be upheld across estimation methods; the upfront definitions are given because individual notation varies, but the fundamentals bind together the varying models with a unifying system. Ultimately, the final goal is to obtain a solution that yields the rotation C or q and the position of the body with respect to the Inertial frame $\bar{r}^I_{B/I}$. This general system relation will be interrelated across all forms of this problem.

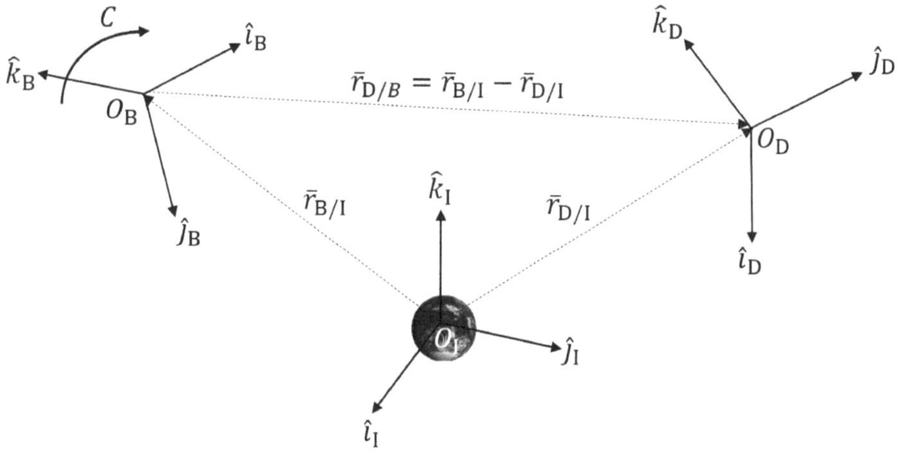

Figure 1. Relative coordinate frame system definition.

2.2. Wahba's Least Squares Optimization Problem

In 1965, mathematician Grace Wahba formulated an intuitive way to describe a rotation between two coordinate systems [25]. This method built the problem as a least squares optimization problem to minimize the error between the known relative reference position vector in the Inertial frame $\bar{r}^I_{D/B}$ and the measurements in the B frame $\bar{r}^B_{D/B}$. This optimization problem can then lead to the optimal estimate of the rotation matrix C that satisfies the relation between the two frames $C\bar{r}^I_{D/B} = \bar{r}^B_{D/B}$. For ease, we will designate $\hat{\mathbf{r}}$ as the unit vector for $\bar{r}^I_{D/B}$ and $\hat{\mathbf{b}}$ as the unit vector for $\bar{r}^B_{D/B}$ used in this model. Also, this model was derived for spacecraft attitude estimation purposes and finds use in [26] with the simplification $\bar{r} = \bar{r}^I_{D/I} \approx \bar{r}^I_{D/B}$ (Referencing Figure 1). This assumption can be made when $\bar{r}^I_{D/I} >> \bar{r}^I_{B/I}$ where the relative distance of a celestial body for D reference measurements (e.g., star catalog) is far greater than the distance of a satellite body to the earth.

Now, the relation for each reference vector $C\hat{\mathbf{r}}_i = \hat{\mathbf{b}}_i$ is applicable in an ideal world where body-frame measurements are without any error. In reality, all N measurements $\hat{\mathbf{b}}_i$ will posses some form of innate sensor error, noise, or uncertainty. The error term for each measurement can be condensed into δ_i and gives the relation $\hat{\mathbf{b}}_i = C\hat{\mathbf{r}}_i + \delta_i$.

This can then be formulated as a least squares problem to find the solution for C to minimize the error δ_i in the system shown in Equation (1). Furthermore, by expanding the least squares it is converted into a minimization problem for $L(C)$ with Equation (2), and solving for the optimal direction cosine matrix C.

$$\delta_i^2 = ||\hat{\mathbf{b}}_i - C\hat{\mathbf{r}}_i||^2 = (\hat{\mathbf{b}}_i - C\hat{\mathbf{r}}_i)^{\mathrm{T}}(\hat{\mathbf{b}}_i - C\hat{\mathbf{r}}_i) \tag{1}$$

$$L(C) = \frac{1}{2}\sum_{i=1}^{n}\alpha_i\delta_i^2 = \frac{1}{2}\sum_{i=1}^{n}\alpha_i\left(\hat{\mathbf{b}}_i^{\mathrm{T}}\hat{\mathbf{b}}_i + \hat{\mathbf{r}}_i^{\mathrm{T}}C^{\mathrm{T}}C\hat{\mathbf{r}}_i - \hat{\mathbf{r}}_i^{\mathrm{T}}C^{\mathrm{T}}\hat{\mathbf{b}}_i - \hat{\mathbf{b}}_i^{\mathrm{T}}C\hat{\mathbf{r}}_i\right) \tag{2}$$

Given that $L(C)$ = min, it can also be modeled as a maximization problem where $G(C)$ = max. The term α_i is a set of non-negative weights for each observation; α_i may remain unweighted with a scalar value of one if desired or uncertain. Through some algebraic simplification, the following equation is obtained:

$$G(C) = \sum_{i=1}^{n}\alpha_i\hat{\mathbf{b}}_i^{\mathrm{T}}C\hat{\mathbf{r}}_i = 1 - L(C) = max \tag{3}$$

This leads to the one final equation in simplified form, maximizing the trace of the direction cosine matrix C, Equation (4), and the summation matrix B, Equation (5).

$$G(C) = \mathrm{tr}[CB^{\mathrm{T}}] = max \tag{4}$$

$$B = \sum_{i=1}^{n}\alpha_i\hat{\mathbf{b}}_i\hat{\mathbf{r}}_i^{\mathrm{T}} \tag{5}$$

This forms the basis of Wahba's problem for optimal attitude estimation. The objective is to solve for a direction cosine matrix C that maximizes the function $G(C)$. However, it is not always the easiest to solve for C directly—although there have since been solutions developed to do so [27]. The formalized solution to obtain the solution using the quaternion is given in the following sections.

2.3. Functional Concepts for Quaternions

The quaternion [28] q is an application of complex numbers on a Clifford algebra in $\mathbb{R}^4$ defined as the following:

$$q = q_0 + q_1 i + q_2 j + q_3 k \tag{6}$$

The group of quaternions as defined by Hamilton in 1843 [29] utilizes the imaginary units that follow the definition $i^2 = j^2 = k^2 = ijk = -1$ and $\{q_0, q_1, q_2, q_3 \in \mathbb{R}\}$. It is also common to represent the quaternion as two components, the vector component (i, j, and k) and the scalar component (denoted by q_0). The purpose of the scalar component is to provide an additional, redundant parameter that keeps the quaternion fully defined in the event that a singularity may occur. This keeps the quaternion singularity free. Another way of thinking of it is thus:

$$q = (q_0, \bar{q}) \tag{7}$$

Note that $q_0 = \cos\left(\frac{\phi}{2}\right) \in \mathbb{R}$ and $\bar{q} = [q_1, q_2, q_3]^T = \bar{e}\sin\left(\frac{\phi}{2}\right), \in \mathbb{R}^3$. Here, $\bar{e}$ is the principal axis unit vector $[i, j, k]$ and ϕ is the principal angle for attitude and rotation representation purposes. Additionally, the quaternion may sometimes be defined with the scalar component last as q_4. However, for the purpose of this paper, the quaternion will always use the scalar component as the first element q_0. In practice, the scalar component tells the angle of rotation, and the normalized vector component provides the direction of the rotation axis.

A quaternion used for attitude representation is a unit quaternion (also called rotation quaternion) with norm $\|q\| = 1$ and satisfies the condition $q^T q = 1$—similarly to how a direction cosine matrix (DCM) possesses its orthogonal property such that $CC^T = I_{3\times3}$. The norm constraint is the additional parameter that fully constrains the quaternion in the event of an angular singularity. A quaternion describing the orientation of the X frame from the Y frame ($q_{X/Y}$) satisfies the condition $(q_{Y/X})^*(q_{Y/X}) = (q_{Y/X})(q_{Y/X})^* = \mathbf{1}_q$, where $\mathbf{1}_q \triangleq (1, \bar{0}_{3\times1})$. See Table 1 for operation definitions.

Due to the quaternion being defined in such a way that it is constructed with a scalar component and a vector component, ordinary linear algebra operations may not apply to the quaternion as they would a typical vector. As such, Table 1 [30] summarizes the operations that are implemented when working with quaternion algebra. For example, the distinction must be made that $(q_{Y/X})^*(q_{Y/X})$ is the quaternion multiplied with its conjugate using the quaternion operations, while $q^T q$ is to be taken as the traditional 4×1 column vector ordinarily multiplied by its transpose.

Table 1. Quaternion operations.

Operation	Definition
Addition	$a + b = (a_0 + b_0, \bar{a} + \bar{b})$
Scalar multiplication	$\lambda a = (\lambda a_0, \lambda \bar{a})$
Multiplication	$ab = (a_0 b_0 - \bar{a} \cdot \bar{b}, a_0 \bar{b} + b_0 \bar{a} + \bar{a} \times \bar{b})$
Conjugate	$a^* = (a_0, -\bar{a})$
Dot product	$a \cdot b = (a_0 b_0 + \bar{a} \cdot \bar{b}, 0_{3\times1})$
Cross product	$a \times b = (0, a_0 \bar{b} + b_0 \bar{a} + \bar{a} \times \bar{b})$
Norm	$\|\| a \|\| = \sqrt{a \cdot a}$

Three-dimensional vectors may also be interpreted as special cases of quaternions. This allows for combined use of quaternions along with vectors in a three-dimensional space for dynamics governing equations. Redefining a vector $\bar{s}$ such that $\bar{s} \in \mathbb{R}^3$ is in the form of a quaternion is carried out as shown below.

$$s_q = (s_0, \bar{s}) \quad \text{with } s_0 = 0 \tag{8}$$

Thus, the distinction is formed to differentiate the quaternion itself used for attitude representation, and variables in 'quaternion' form. Consider that when a vector $\bar{s}$ is converted into quaternion form, it will not be a unit quaternion. Lastly, the quaternion has explicit applications for changing reference frames—both in general and also for variables in quaternion format. The change of reference frame for a vector in quaternion form from the X frame to the Y frame is achieved via the following:

$$s_q^Y = q_{Y/X}^* s_q^X q_{Y/X} \tag{9}$$

2.4. Davenport's q-Method for Attitude Estimation

A solution for this estimation problem may be obtained by substituting the quaternion in place of the direction cosine matrix. The equation below is an identity for a direction cosine matrix as a function of q. Here, q_0 is the quaternion scalar component, $\bar{q}$ is the quaternion vector component, and $[\bar{q}\times]$ is the skew-symmetric matrix of $\bar{q}$.

$$C(q) = \left(q_0^2 - \bar{q}^{\mathrm{T}}\bar{q}\right) I_{3\times3} + 2\bar{q}\bar{q}^{\mathrm{T}} - 2q_0[\bar{q}\times] \tag{10}$$

Through some manipulation, this optimization problem can be parameterized in terms of q. By substituting Equation (10) into $G(q)$ for Equation (4) and isolating the quaternion, the system objective function $G(q)$ can be put into a compact format of $q^{\mathrm{T}}Kq$. K is a 4×4 matrix, and this provides a quadratic form which will allow for easy minimization [31] (or maximization, in this case).

$$\begin{aligned} G(q) &= \{q_0 \ \ \bar{q}\}^{\mathrm{T}} \begin{bmatrix} tr(B) & \sum_{i=1}^{n} \alpha_i \left[\hat{\mathbf{b}}_i \times\right] \hat{\mathbf{r}}_i \\ \left(\sum_{i=1}^{n} \alpha_i \left[\hat{\mathbf{b}}_i \times\right] \hat{\mathbf{r}}_i\right)^{\mathrm{T}} & (B + B^{\mathrm{T}} - tr(B) I_{3\times3}) \end{bmatrix} \begin{Bmatrix} q_0 \\ \bar{q} \end{Bmatrix} \\ &= q^{\mathrm{T}} K q \end{aligned} \tag{11}$$

A constraint $q^{\mathrm{T}}q = 1$ is then added to solve for the nontrivial solution of the optimization problem, as shown in Equation (12). This provides the optimal quaternion q_{opt} that defines the rotation between two frames. Afterwards, differentiating with respect to q and equating $q = 0$ solves for the maximum value of this quadratic form, shown in Equation (13).

$$G^*(q) = q^{\mathrm{T}}Kq - \lambda\left(q^{\mathrm{T}}q - 1\right) = max \tag{12}$$

$$\frac{dG^*(q)}{dq} = 2Kq - 2\lambda q = 0 \tag{13}$$

$$Kq_{opt} = \lambda_{max} q_{opt} \tag{14}$$

This yields a very straightforward solution where the optimal quaternion q_{opt} is exactly the eigenvector that corresponds to the maximum real eigenvalue λ_{max} for matrix K. Therefore, following this procedure provides q_{opt} to minimize the error in the attitude estimation system via the K matrix. This involves using summations of the B matrix—constructed with measurement vectors $\hat{\mathbf{b}}_i$ and known reference direction vectors $\hat{\mathbf{r}}_i$. Given the relation for Equation (10), the solution will provide the rotation matrix C that describes the rotation transformation from the Inertial frame to the Body frame.

For awareness, there are some nuances that can be added to the q-method attitude estimation process, the first being that $\hat{\mathbf{b}}_i$ and $\hat{\mathbf{r}}_i$ are not required to be directional unit vectors. Both $\bar{b}_i$ and $\bar{r}_i$ can be positional vectors instead. The results for attitude estimation are identical; the exact derivation of the Wahba problem and q-method would differ slightly due to a lack of simplification that comes with the unit vectors, but the core process and equations will remain the same. In practice, unit directional vectors are more commonly used [32] because knowledge of the precise distance for a reference body can be difficult, especially for satellite applications.

With the above assumptions in play, Davenport's q-method yields great results. However, we can develop a more generalized model that will continue to be applicable for satellite applications, but will not be limited by approximations or simplifications. The following section will expand on this knowledge to incorporate position estimation in tandem with attitude estimation to formulate a robust pose estimation method.

3. q-Method Pose Estimation Derivation

This section will now shift focus onto the newly developed q-method for pose estimation. Prior use of the q-method focused on spacecraft applications for attitude estimation, whereas this more generalized approach eliminates both the unit vector direction and $\overline{r} = \overline{r}^I_{\mathrm{D/I}} \approx \overline{r}^I_{\mathrm{D/B}}$ assumptions for the 'known' reference values of $\overline{\mathbf{r}}_i$. This indeed complicates the problem and makes it more difficult to estimate, but also incorporates inertial position $\overline{p}$ into the estimation; the final model shows promising compatibility for 6DoF drone pose estimation. Figure 2 illustrates how the q-method pose estimation will relate back to the global system definitions in Figure 1, differing from the original q-method. The objective is to estimate the attitude, C, as well as the satellite position, $\overline{p}$, in order to estimate the pose of the air vehicle. A list of variables is provided:

- $\overline{r}^I_{\mathrm{B/I}}$—Body position from Inertial Reference (Earth), to be estimated, $\overline{p}$.
- $\overline{r}^I_{\mathrm{D/I}}$—Known reference position from Inertial frame (Earth), $\overline{r}$.
- $\overline{r}^B_{\mathrm{D/B}}$—Measured position vectors in the Body frame, with error, $\overline{b}$.
- $\overline{r}^I_{\mathrm{D/B}}$—Unknown. Essentially requires prior knowledge. If we had this, position estimation would be easy. Let it be known as $\overline{d}$. We will use this to estimate $\overline{p}$ given that $\overline{d} = \overline{r} - \overline{p}$ by following the vector addition for the system definition.

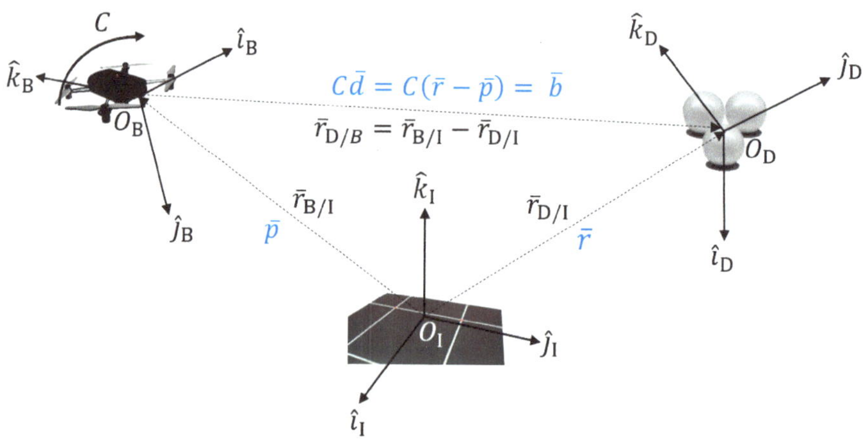

Figure 2. Definition of q-method pose estimation system.

We are able to take advantage of a simple substitution for $\overline{r}^I_{\mathrm{D/B}}$ by virtue of system definition commonality. The q-method takes $\overline{r}^I_{\mathrm{D/B}} = \hat{r}$. For the q-method pose derivation, let $\overline{r}^I_{\mathrm{D/B}} = \overline{d} = \overline{r} - \overline{p}$. The relation for the drone body measurements with error continues as $\hat{b}_i = C\hat{d}_i + \delta_i$. Substituting this equivalence into Equation (1), a slight modification is made to the original Wahba problem. This straightforward substitution unifies the estimation derivations, allowing for a similar system of equations. Throughout this derivation, all aforementioned variables will be kept identical, including the weighting factor α_i. Curiously, there exists other research expanding upon the Wahba problem to incorporate position [33,34], but not for the quaternion-based purposes used here.

$$\begin{aligned} L(C) = \delta_i^2 = \|\overline{b}_i - C(\overline{r}_i - \overline{p}_i)\|^2 &= \left(\overline{b}_i - C(\overline{r}_i - \overline{p}_i)\right)^{\mathrm{T}}\left(\overline{b}_i - C(\overline{r}_i - \overline{p}_i)\right) \\ &= \left(\overline{b}_i - C\overline{d}_i\right)^{\mathrm{T}}\left(\overline{b}_i - C\overline{d}_i\right) \end{aligned} \tag{15}$$

Further expanding the minimization cost function $L(C, d)$, and then grouping terms, gives

$$L(C,d) = \frac{1}{2}\sum_{i=1}^{n}\alpha_i\delta_i^2 = \frac{1}{2}\sum_{i=1}^{n}\alpha_i\left(\overline{b}_i^{\mathrm{T}}\overline{b}_i + \overline{d}_i^{\mathrm{T}}C^{\mathrm{T}}C\overline{d}_i - \overline{d}_i^{\mathrm{T}}C^{\mathrm{T}}\overline{b}_i - \overline{b}_i^{\mathrm{T}}C\overline{d}_i\right) \tag{16}$$

$$L(C,d) = \frac{1}{2}\sum_{i=1}^{n}\alpha_i\bar{b}_i^{\mathrm{T}}\bar{b}_i + \frac{1}{2}\sum_{i=1}^{n}\alpha_i\bar{d}_i^{\mathrm{T}}\bar{d}_i - \sum_{i=1}^{n}\alpha_i\bar{b}_i^{\mathrm{T}}C\bar{d}_i \tag{17}$$

Note that (17) possesses two terms that Equation (3) combines into one, due to unit vector measurements. Equation (3) also converts the problem into a maximization problem for simplicity, but the above form continues with the form $L(C,\bar{d})$ and $L(q,\bar{d})$, opting to keep the third term negative. Also, note the dimensions of q (4×1) and $\bar{d}$ (3×1); the following equations make use of the aforementioned 'quaternionized' vector format in Equation (8) when required.

The third term also still resembles that in (3), and subsequently will be equivalent to $\mathrm{tr}[C\beta^{\mathrm{T}}]$. Like B in Equation (5), β is defined as

$$\beta = \sum_{i=1}^{n}\alpha_i\bar{b}_i\bar{d}_i^{\mathrm{T}} \tag{18}$$

Similarly, the cost function is put into terms of $L(q,\bar{d})$ using Equation (10). Only the third term is a function of q (or C), and substituting $\bar{d}$ for $\hat{r}$ allows for the use of (11), as derived by Davenport. The cost function then becomes

$$L(q,d) = \frac{1}{2}\sum_{i=1}^{n}\alpha_i\bar{b}_i^{\mathrm{T}}\bar{b}_i + \frac{1}{2}\sum_{i=1}^{n}\alpha_i\bar{d}_i^{\mathrm{T}}\bar{d}_i - q^{\mathrm{T}}\kappa q \tag{19}$$

where the third term $q^{\mathrm{T}}\kappa q$ is expressed as

$$q^{\mathrm{T}}\kappa q = \{q_0 \quad \bar{q}\}^{\mathrm{T}}\begin{bmatrix} \mathrm{tr}(\beta) & \sum_{i=1}^{n}\alpha_i\left[\bar{b}_i\times\right]\bar{d}_i \\ \left(\sum_{i=1}^{n}\alpha_i\left[\bar{b}_i\times\right]\bar{d}_i\right)^{\mathrm{T}} & (\beta+\beta^{\mathrm{T}}-\mathrm{tr}(\beta)I_{3\times3}) \end{bmatrix}\begin{Bmatrix} q_0 \\ \bar{q} \end{Bmatrix} \tag{20}$$

With the cost function $f(x)$ now in terms of the variables $x = [q,\bar{d}]$ of interest, the quaternion norm constraint $q^{\mathrm{T}}q = 1$ is imposed as an equality constraint $g(x)$. The Lagrangian function [35] is applied to obtain the optimal solution that minimizes the objective function $f(x)$ (minimizing the error), therefore obtaining the optimal solution for q and $\bar{d}$. The Lagrangian function is defined as $\mathcal{L}(x,\lambda) \equiv f(x) - \lambda\langle g(x)\rangle$.

$$\mathcal{L}(q,d,\lambda) = \frac{1}{2}\sum_{i=1}^{n}\alpha_i\bar{b}_i^{\mathrm{T}}\bar{b}_i + \frac{1}{2}\sum_{i=1}^{n}\alpha_i\bar{d}_i^{\mathrm{T}}\bar{d}_i - q^{\mathrm{T}}\kappa q - \lambda(q^{\mathrm{T}}q - 1) \tag{21}$$

The necessary condition is implemented to minimize the system such that $\frac{\partial\mathcal{L}}{\partial\bar{d}} = 0$ and $\frac{\partial\mathcal{L}}{\partial q} = 0$. The partial differential equations used to solve the system are shown below.

$$\frac{\partial\mathcal{L}}{\partial\bar{d}} = 0 = \sum_{i=1}^{n}\alpha_i\bar{d}_i - q^{\mathrm{T}}\frac{\partial\kappa}{\partial\bar{d}}q \tag{22}$$

$$\frac{\partial\mathcal{L}}{\partial q} = 0 = 2\kappa q - 2\lambda q \tag{23}$$

Equations (22) and (23) can now be used to solve for the optimal position estimation $\bar{p}^*$ and optimal quaternion to describe the rotation q_{opt}. $\frac{\partial\mathcal{L}}{\partial q}$ fortunately gives a solution comparable to the q-method, making use of κ shown in (20) to find q_{opt} as the eigenvector for the corresponding maximum real eigenvalue. E.g., (22) will substitute $\bar{d}_i = (\bar{r}_i - \bar{p}^*)$ so that $\sum_{i=1}^{n}\alpha_i\bar{d}_i = \sum_{i=1}^{n}\alpha_i\bar{r}_i - \sum_{i=1}^{n}\alpha_i\bar{p}^*$. The end set of equations for q-method pose estimation are hereby formed.

$$\overline{p}^{*} = \frac{1}{N}\Big(\sum_{i=1}^{n}\alpha_i\overline{r}_i - q^{\mathrm{T}}\frac{\partial\kappa}{\partial\overline{d}}q\Big) \tag{24}$$

$$\kappa q_{opt} = \lambda_{max}q^{*} \tag{25}$$

One last simplification is made due to the term $\frac{\partial\kappa}{\partial\overline{d}}$ within Equation (24). The partial differential of $\kappa_{4\times4}$ w.r.t. a vector $\overline{d} \in \mathbb{R}^3$ yields a $4 \times 4 \times 3$ higher-order tensor. The evaluation of this term can be found in a prior publication [36] as well as validation of q-method pose estimation for simulated spacecraft applications. Thus, an intuitive equation for $\overline{p}^{*}$ exists by again referencing the system definition and taking a weight average for $\overline{p} = \overline{r}^{I}_{B/I} = \overline{r}^{I}_{D/I} - \overline{r}^{I}_{D/B}$. By definition, $\overline{r}^{I}_{D/B} = q_{I/B}\overline{r}^{B}_{D/B}q^{*}_{I/B} = q^{*}_{B/I}\overline{b}_i q_{B/I}$, where q^{*} is the quaternion conjugate.

$$\overline{p}^{*} = \frac{1}{N}\Big(\sum_{i=1}^{n}\alpha_i\overline{r}_i - q\langle\sum_{i=1}^{n}\alpha_i\overline{b}_i\rangle q^{*}\Big) \tag{26}$$

To summarize, the goal of q-method pose estimation is to find $\overline{p}^{*}$ and q_{opt}. In order to do so, one must first formulate κ, making use of β, which makes use of $\overline{d}$. If one of the optimal values ($\overline{p}^{*}$ or q_{opt}) is known, the other can then easily be calculated. However, a solution can be found using any valid numerical solving method. The next sections will investigate the results using a numerical solving scheme in parallel with physical sensors in a controlled lab environment.

4. Experimental Setup and Methodology

4.1. Experimental Overview and Setup

To further verify the functionality of the system of equations, the q-method pose estimation is incorporated with onboard measurements obtained from drone-camera experiments. Figure 3 illustrates how a Quanser drone used an Intel RealSense (R200) camera to measure the locations of five objects/beacons in 3D space. The beacons are placed at varying locations and depths upon a curved surface. The camera's depth-sensing feature was used to determine the location of each beacon in space relative to the camera/drone frame. Camera detection of each beacon is shown in Figures 4 and 5. These are used to provide positional estimates in a pictured frame, and then converted into the body-centered drone frame.

To validate correct estimates, the results were compared against true reference data. Vicon's motion capture system defined the locations of the five beacons as well as the drone. The Vicon system consists of eight high-resolution cameras that capture moving objects at a rate of 120 Hz. Each beacon and drone were defined as rigid bodies (objects) in three dimensions using reflective markers. Figure 4 shows two of the eight Vicon cameras mounted across the lab and test setup, forming the Inertial frame. This provides a full overview of the hardware used for the experimental setup across the Body and Inertial frames.

To test the method for orientation and position estimation in a physical setting, the drone operated on a static mount while detecting the beacons to obtain reference measurements. The first experiment used a stabilized, nominally oriented (no-tilt) drone to measure beacons, as shown in Figure 5 with the drone camera. A second experiment was performed where the drone was statically tilted at a roll angle of -15 degrees in the Body frame, as shown in the Figure 6, also with the onboard camera.

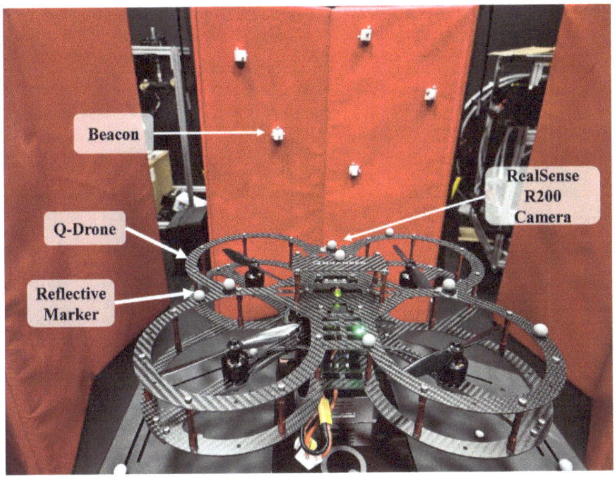

Figure 3. Quanser drone and five reference beacons.

(**a**)

(**b**)

Figure 4. (**a**) RGB imagery from drone camera. (**b**) Partial Vicon system within SDSU SPACE lab.

Two corrections to the experimental setup were required while post-processing data. First, the camera position placed on the drone needed to be taken into account, as it was +15 cm ahead of the true Vicon-defined drone centroid. This fix was performed by adding a +15 cm bias to the Vicon positional truth data during the pose estimation analysis and validation. Second, the camera measurements contained a constant bias outside of the camera position that needed to be considered. This was achieved by performing a static calibration in the same position as the experiment for both drone orientations. Afterwards, the mean bias of the camera for each individual reference marker was subtracted. Further details will be provided in the following section after the inputs.

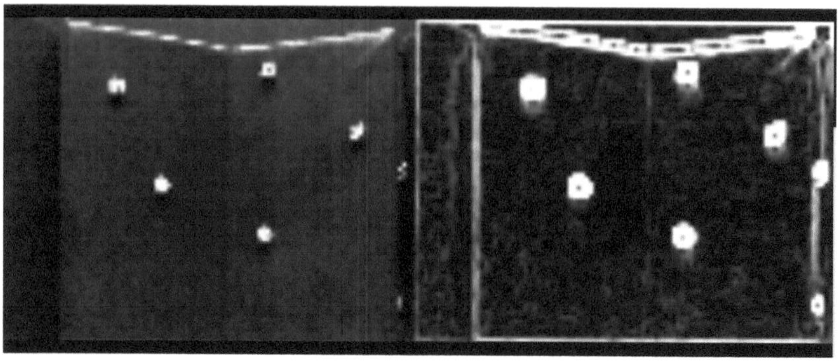

Figure 5. Beacon detection and location definitions.

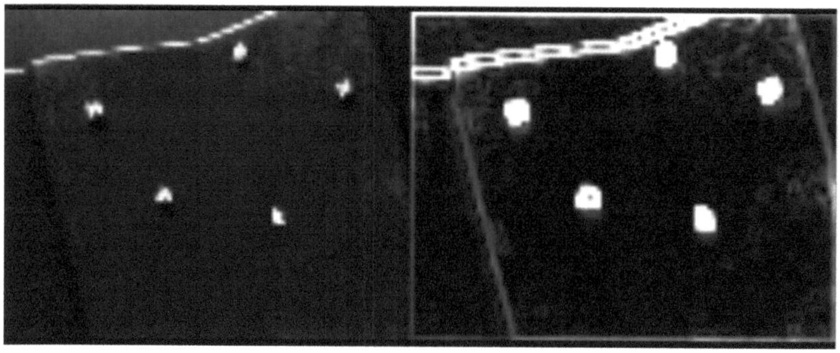

Figure 6. Reference beacons measured with 15° tilted drone.

4.2. Initial Conditions and Inputs

The two experiments were performed and the resulting output data were fed into MATLAB 2018a for post-processing, algorithm implementation, and analysis. Figure 7 provides an illustration of or pseudo-code for the validation process for both configurations. Figure 8 provides a visual representation of the true data within MATLAB using the Vicon system. A correction factor of +15 cm is added to account for the position of the camera on the drone with respect to the Body frame. This offset can be seen in the figure, as the camera centroid is offset from the drone body itself. Additionally, several hundred measurements were taken to characterize the average measurement bias of the camera sensor and correct the input reference measurements. The estimation algorithm and results come from a single instantaneous state—one each for the 0° roll and −15° roll.

The pose estimation using the previously outlined system of equations was accomplished with the aid of the built-in MATLAB numerical equation solver fsolve for this multi-variable system. The function uses a Trust-Region Dogleg method to solve for the optimal solution that minimizes the residual error of the system. The following is the exact input to calculate the residual $f(x)$ used within the solver based on Equations (25) and (26).

$$f(x) = 0 = \begin{bmatrix} \sum_{i=1}^{n} \alpha_i(\overline{r}_i - \overline{p}^*) - q\langle \sum_{i=1}^{n} \alpha_i \overline{b}_i \rangle q^* \\ (2\kappa q - 2\lambda q)/10 \end{bmatrix} \tag{27}$$

The second row of $f(x)$ is multiplied by 10^{-1} in order to better condition the system. Equating both sides of the equation to 0, this multiplication scalar does not make a difference to the solution. λ is included for the purpose of solving the system numerically. Thus, $f(\overline{p}, q, \lambda) = 0$ where $x = [\overline{p}, q, \lambda]$ to minimize the system.

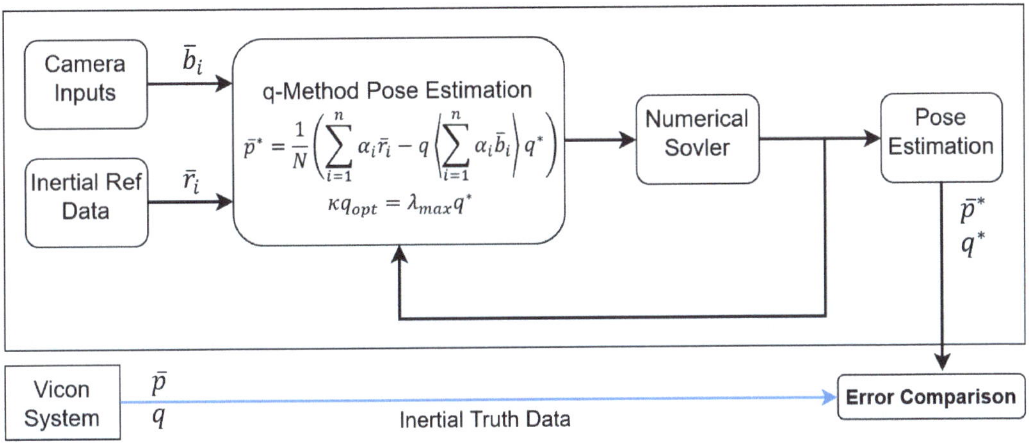

Figure 7. Procedural validation for q-method pose estimation.

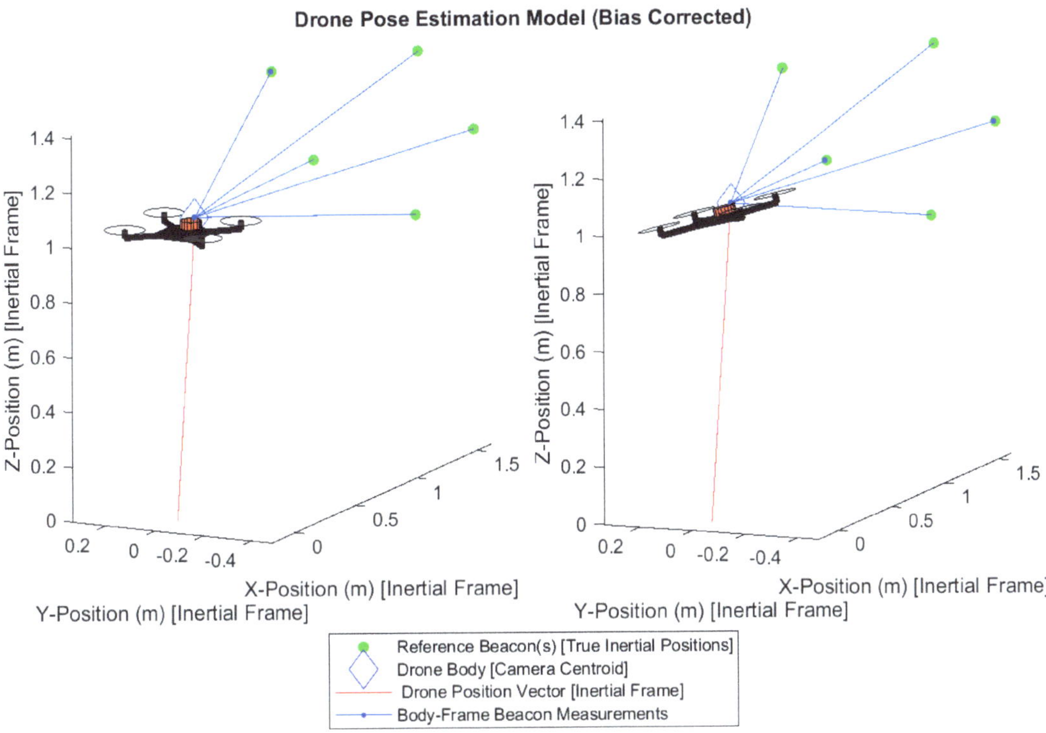

Figure 8. Experimental visualization. [Left: 0° roll. Right: −15° roll].

An initial guess x_0 is required to iteratively solve the function via numerical methods. As such, a nearby initial guess is arbitrarily selected for the quaternion q_0, position p_0, and λ of the system. For this experiment, let $p_0 = [0, 0, 1]$ m, $q_0 = [1; 0; 0; 0]$, and $\lambda_0 = \Sigma \bar{b}_i \bar{b}_i^{\mathrm{T}}$. This formulation for λ_0 is obtained through empirical observations of what generally works well while using this estimation model in conjunction with a numerical equation solver.

Table 2 provides the value of each parameter to initialize the system. The results generated hereafter make use of these provided inputs. Given N reference points for known reference vectors $\bar{r}^I_{D/I}$ and N body measurements $\bar{b}_i$ with measurement error σ_i, the objective is to estimate $\bar{p}^*$ and q_{opt} to be as close to the true values $\bar{r}^I_{B/I}$ and $q_{B/I}$ as possible.

Measurement errors and correction factors across each roll condition are all provided in Table 2 for transparency. Starting with the reference measurement error σ_i, a scalar percent error value is given for each on-board measurement of each reference vector in the Body frame. σ_i is computed using the experimental and true positional values of the reference beacons with respect to the camera given as $|\bar{r}^B_{D/B} - \bar{b}_i| / |\bar{r}^B_{D/B}| \times 100\%$. Mean measurement error $||\sigma_i||^2$ is the average error across all N measurements used in this procedure. This measurement error, while missing directional value, is moreso a simplified method to partially quantify the error in the system when comparing the accompanying algorithm results.

In addition, the average sensor error $\bar{\sigma}$ is the average error across time (700 data points) of the camera sensor for each reference beacon. A static collection of data identical to $\bar{b}_i$ is performed to verify that the snapshot state obtained in $\bar{b}_i$ is not drastically different from the average of the camera's ordinary noisy measurements. The same static collection of data was also used to detect the sensor's average experimental bias $\delta\bar{r}^B_{D/B}$ to correct the raw measurement vectors $\bar{b}_i^+$. Thus, the bias correction is made such that $\bar{b}_i = \bar{b}_i^+ + \delta\bar{r}^B_{D/B}$. Information for methods and error are disclosed, but procedures for absolute correction and minimization of sensor error are intentionally not incorporated here due to the desire to understand estimation performance with some degree of measurement error.

Table 2. Simulation initial condition inputs.

Parameter	Sym.	Unit	Input 0° Roll	Input −15° Roll
Num. Ref. Measurements	N	—	5	5
True Orientation	$q_{B/I}$	—	[0.9999; 0.0007; −0.0014; −0.0044]	[0.9913; −0.1309; 0.0003; 0.0004]
True Position	$\bar{r}^I_{B/I}$	[m]	[0.1454, −0.0038, 1.0788]	[0.1539, −0.0127, 1.0842]
Ref. Measurement Error	σ_i	[%]	[0.28, 0.29, 0.23, 0.05, 0.40]	[0.63, 0.64, 0.01, 0.27, 0.36]
Mean Measurement Error	$\|\|\sigma_i\|\|^2$	[%]	0.2487	0.3809
Average Sensor Error	$\bar{\sigma}$	[%]	[0.25, 0.24, 0.30, 0.33, 0.25]	[0.29, 0.40, 0.31, 0.42, 0.22]
Measurement Vectors (w/Error)	$\bar{b}_i$	[m]	$\begin{pmatrix} 1.317 & 0.326 & 0.241 \\ 1.390 & 0.186 & -0.080 \\ 1.472 & -0.201 & 0.336 \\ 1.339 & -0.490 & 0.102 \\ 1.464 & -0.192 & -0.261 \end{pmatrix}$	$\begin{pmatrix} 1.327 & 0.276 & 0.312 \\ 1.367 & 0.224 & -0.036 \\ 1.465 & -0.252 & 0.267 \\ 1.459 & -0.091 & -0.307 \\ 1.324 & -0.470 & -0.032 \end{pmatrix}$
True Reference Positions	$\bar{r}^I_{D/I}$	[m]	$\begin{pmatrix} 1.467 & 0.335 & 1.316 \\ 1.530 & 0.194 & 0.994 \\ 1.616 & -0.190 & 1.411 \\ 1.606 & -0.1821 & 0.814 \\ 1.485 & -0.481 & 1.178 \end{pmatrix}$	$\begin{pmatrix} 1.469 & 0.333 & 1.316 \\ 1.533 & 0.193 & 0.993 \\ 1.618 & -0.188 & 1.409 \\ 1.608 & -0.180 & 0.812 \\ 1.485 & -0.478 & 1.177 \end{pmatrix}$
Raw Measurements	$\bar{b}_i^+$	[m]	$\begin{pmatrix} 1.416 & 0.131 & 0.241 \\ 1.507 & 0.013 & -0.089 \\ 1.495 & -0.279 & 0.315 \\ 1.495 & -0.274 & -0.276 \\ 1.355 & -0.483 & 0.084 \end{pmatrix}$	$\begin{pmatrix} 1.404 & 0.173 & 0.142 \\ 1.507 & -0.008 & -0.140 \\ 1.548 & -0.171 & 0.353 \\ 1.478 & -0.314 & 0.213 \\ 1.363 & -0.456 & -0.198 \end{pmatrix}$
Camera Correction Factor	$\delta\bar{r}^B_{D/B}$	[m]	$\begin{pmatrix} -0.099 & 0.196 & 0.000 \\ -0.117 & 0.173 & 0.009 \\ -0.023 & 0.078 & 0.021 \\ -0.156 & -0.216 & 0.378 \\ 0.109 & 0.291 & -0.346 \end{pmatrix}$	$\begin{pmatrix} -0.077 & 0.103 & 0.170 \\ -0.140 & 0.232 & 0.104 \\ -0.083 & -0.081 & -0.087 \\ -0.019 & 0.222 & -0.520 \\ -0.038 & -0.014 & 0.167 \end{pmatrix}$
Weight Parameter	α	—	1.0	1.0

5. Results

The generalized q-method pose estimation formulation was able to successfully estimate both the position $\overline{p}$ and orientation q_{opt} of the drone for both attitude configurations. Both $\overline{p}$ and q_{opt} are the optimal values for the pose that minimizes the error of the system given the drone camera measurements with sensor noise. Figure 9 is designed to mirror the system definition (Figure 2). It overlays the estimated position $\overline{p}$ onto the true body position vector $\overline{r}^{I}_{B/I}$. Here, all inertial and truth data obtained from the Vicon system are used to validate the results of the pose estimation. The inertial system origin is the center floor of the laboratory in which the experiment is performed. True reference position vectors $\overline{r}^{I}_{D/I}$ are essential prior-knowledge of the Inertial frame or system, or wherever the drone intends to operate—also obtained by Vicon.

The position error ($\sim$1.8% and 1.56%) is calculated as $\|\overline{r}^{I}_{B/I} - \overline{p}^{*}\| / \|\overline{r}^{I}_{B/I}\|$, and the position angular error is defined as $\cos^{-1}\left((\overline{r}^{I}_{B/I} \cdot \overline{p}^{*}) / \|\overline{r}^{I}_{B/I} \cdot \overline{p}^{*}\|\right)$ to represent the angular error between the true vector and estimated position. The quaternion attitude error is not as intuitive to represent, but is found by the principal angle ϕ within the quaternion product $\delta q = q_{B/I} q^{*}_{opt}$ [15] where q^{*}_{opt} is the conjugate of the estimated quaternion. The equation for ϕ is given by $\phi = 2\cos^{-1}(q_0)$ from the prior quaternion definition. Observe that the quaternion error (principal angle error, $\delta\phi$) for both test cases is $\sim$0.453°and 0.716° respectively, between the true and estimated quaternion. The small positional error and small quaternion error demonstrate the reliability and accuracy of the q-method for pose estimation despite the relatively few reference measurements.

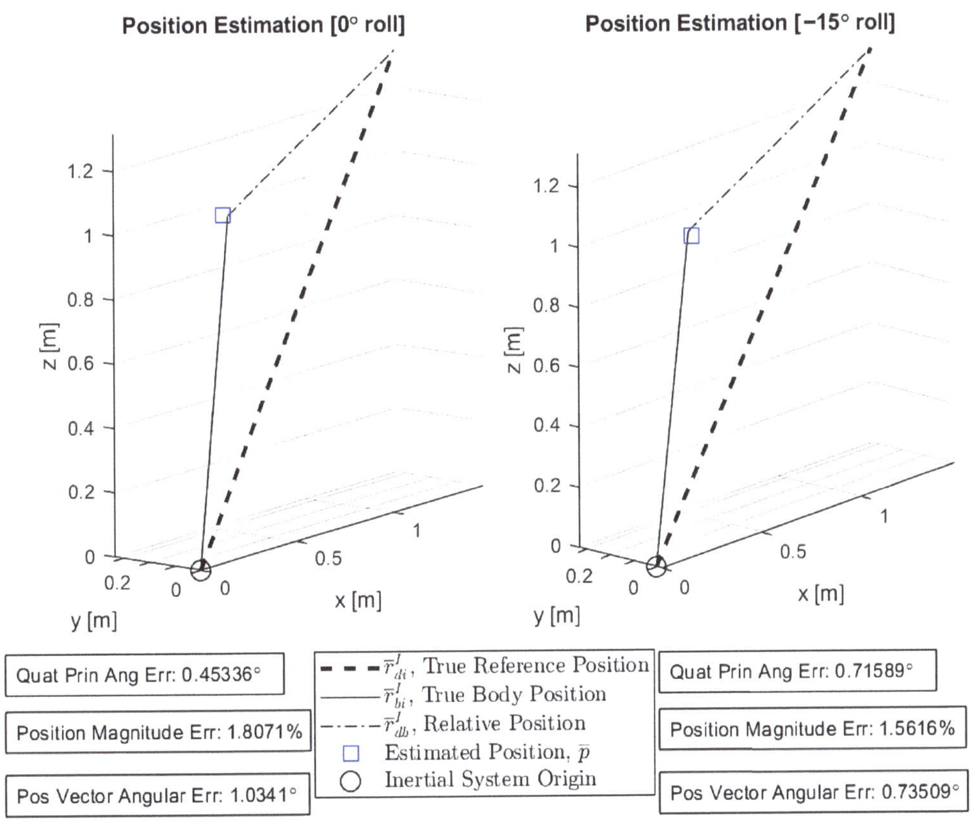

Figure 9. Pose estimation results.

The remaining figures detail the results of interest across each iteration of the numerical solver until convergence is reached. Iterative results for both the 0° roll and −15° roll cases are overlaid within Figures 10–12 for ease of presentation. Figure 10 displays the principal angle error $\delta\phi$ for an understanding of the aggregate attitude error. The bottom plot of the same figure shows the norm of the residual error $||f(x)||$ as the function fsolve converges to a stopping condition.

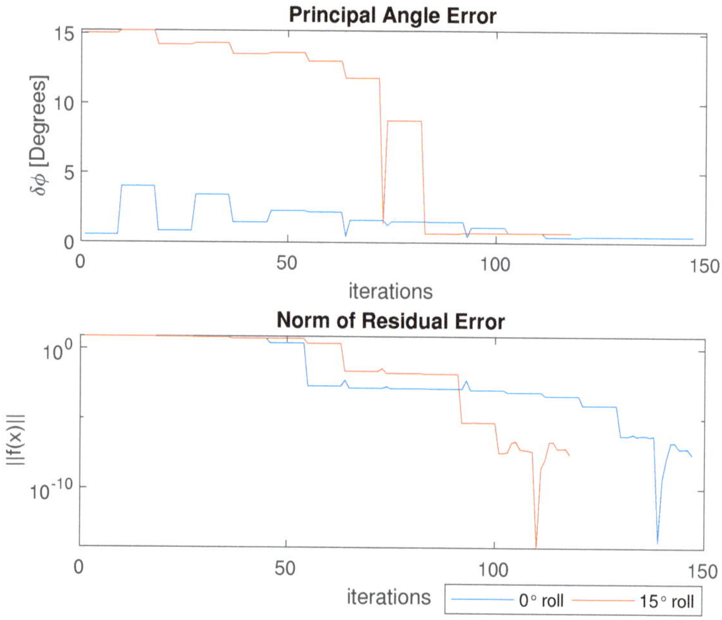

Figure 10. Principal angle error between true and estimated quaternions (**Top**) and norm of residual error for numerical solver (**Bottom**) across iterations.

Figure 11 itemizes each element of the attitude error $(\delta q_1, \delta q_2, \delta q_3, \delta q_4)$ between the true quaternion $q_{B/I}$ and the estimated quaternion q_{opt}. Lastly, Figure 12 shows the error for each element between the estimated and true positions. The deltas are not expected to be exactly zero; the expectation is not to find the exact true value, but rather to find the optimal estimation that minimizes the error of the system (introduced by the noisy measurements). Further performance enhancements could be made with the addition of more reference points. For an estimation model like this, the enhancements will directly lead to improved performance.

All figures and results presented show the advantage of the pose estimation q-method in its ability to solve for the state of the system with minimal error. In each instance, the numerical equation solver concluded its convergence and obtained a sufficient solution for the UAV. When compared to the inertial, Vicon truth data, the outcome for the predicted position and orientation is greatly comparable despite the error embedded in the attached camera measurement reference vectors.

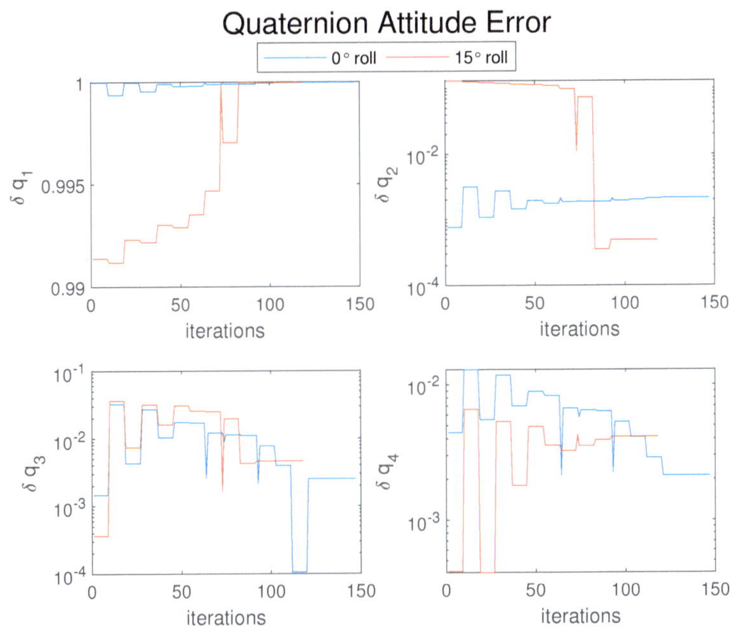

Figure 11. Attitude error between true and estimated quaternion elements $[\delta q_1, \delta q_2, \delta q_3, \delta q_4]$ across iterations for numerical solver.

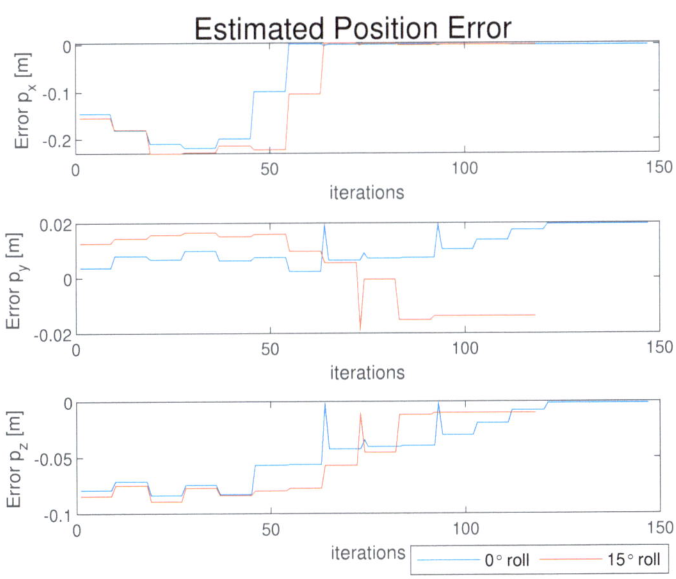

Figure 12. Error for estimated position elements $[p_x, p_y, p_z]$ across iterations for numerical solver.

Method Comparison and Evaluation

The generalization of the q-method focuses on adopting a solution to estimate both orientation and position simultaneously. The main aim of this paper is not to assess estimation performance or related metrics. Since the proposed algorithm solves the algebraic equation using an eigenvalue approach, it is not entirely fair to compare it with other methods that may involve additional computational steps beyond the scope of this approach. Nevertheless, tests were conducted to compare this method with the q-method (for orientation only) and the modified Optimal Linear Attitude Estimator, OLAE (for orientation and position) [37], showing nearly identical performance results. The same sensor data and initial conditions were used for the comparison.

OLAE is a single-point real-time estimator that utilizes the Rodrigues (or Gibbs) vector, a minimal-element attitude parameterization. The optimality criterion, which differs from Wahba's constrained criterion, is strictly quadratic and unconstrained. This method estimates both position and orientation based on vector observations obtained through vision-based camera technology. The generalized q-method also estimates the translation vector with the same accuracy and rate. Table 3 summarizes the results.

Table 3. Comparison of error results for different roll angles.

Roll Angle	Estimation Method	Position Error $\|\|\delta p\|\| [m]$	Angle Error $\delta\phi$ [deg]
0°	q-Method Pose	0.01965	0.4493
	OLEA	0.01898	0.4544
15°	q-Method Pose	0.01708	0.7159
	OLEA	0.01720	0.7202

The results show that both the generalized q-method and OLEA perform similarly in terms of position and orientation errors. At 0° roll angle, OLEA has a slightly smaller position error but a slightly higher orientation error. At 15° roll angle, the position errors for both methods are nearly identical, with OLEA having a marginally larger error. Both methods show an increase in orientation error at 15°, with the generalized q-method maintaining a slight advantage in orientation accuracy. Overall, the differences are small, with both methods offering comparable performance depending on the roll angle.

6. Conclusions

The q-method for pose estimation was designed to take the body-frame positional reference measurements and compare them to relative, known positions from an Inertial frame. Doing this, both the orientation and position of the body may be estimated to minimize the error introduced into the system via the measurements. This paper first provided a summary of Davenport's q-method for attitude estimation and then built on the existing model, while also acknowledging the assumptions and differences that went into the original model. The pose estimation model presented here went to great lengths to keep a consistent system relation between the Inertial, Body, and reference frames in a way that mirrors the q-method. It is via this common system definition that the pose estimation equations were derived by mirroring the original q-method derivation. From this newly developed model, the pose estimation results proved to provide a very accurate solution with relatively simple equations. The computational cost for the numerical methods could be a limitation of this method, but further work could be performed to develop closed-form solutions for this system of equations. Nevertheless, with ample processing power and computing speed, any conventional numerical method used in conjunction with the q-method for pose estimation will provide reliable results.

While results within this paper are promising, the experiment performed was for a single steady-state estimation. Additional estimation approaches like these offer an extra layer of redundancy and dependability for any autonomous aircraft or spacecraft. Further integration into a dynamic system would be required with parallel use of filters, GPS, dynamic models, and/or accelerometer and gyroscopic sensors. Any hardware or communication failures for these flight-essential features would require backup methods for state estimation; the generalized q-method pose estimation provides a backup measure preventing loss of the vehicle, provided prior environmental reference knowledge is utilized.

The estimation error itself is determined by the optimality of the Wahba problem, which is formulated based on the measurement noise. This paper does not aim to reduce the estimation error further, but rather focuses on generalizing the classical q-method to provide pose estimation solutions for coupled vehicle dynamics. Certain validation steps of this pose estimation method were omitted due to redundancy, but additional validation was performed to verify equivalency between the generalized q-method and the classical q-method. Noise reduction and elimination can be fourthly investigated according to the estimation tolerance defined by the problem itself. Preceding work has also previously been performed as part of the validation of the q-method pose estimation by investigating error sensitivity to initial conditions and numerical convergence through a Monte Carlo analysis [36].

Unlike Davenport's q-method, which will often make the far-away star assumption and use a star-catalogue with respect to earth, this pose estimation method generalizes the model to be more mathematically intuitive for general applications. The pose estimation model is unrestricted to just satellite attitude estimation, as demonstrated in this paper using a drone/UAV. This itself is a benefit. In addition, if the position is already known, these equations will function identically to the q-method. In essence, the integration of pose estimation methodologies with the q-method represents a significant advancement, promising more robust and adaptable solutions for spatial perception and analysis across various domains.

Author Contributions: Conceptualization, K.S. and A.B.Y.; methodology, K.S., A.B.Y. and M.H.; software, K.S.; validation, K.S.; formal analysis, K.S.; investigation, K.S., A.B.Y. and M.H.; resources, A.B.Y. and M.H.; data curation, K.S. and M.H.; writing—original draft preparation, K.S.; writing—review and editing, K.S., A.B.Y. and M.H.; visualization, K.S.; supervision, A.B.Y.; project administration, K.S., A.B.Y. and M.H. All authors have read and agreed to the published version of the manuscript.

Funding: This research received no external funding.

Data Availability Statement: The original contributions presented in this study are included in the article material. Further inquiries can be directed to the corresponding author.

Conflicts of Interest: The authors declare no conflicts of interest.

References

1. Crassidis, J.; Junkins, J. *Optimal Estimation of Dynamic Systems*; Advances in Applied Mathematics; CRC Press LLC: Boca Raton, FL, USA, 2024.
2. Forbes, J.R. Fundamentals of Spacecraft Attitude Determination and Control [Bookshelf]. *IEEE Control Syst. Mag.* **2015**, *35*, 56–58. [CrossRef]
3. Barfoot, T.D. *State Estimation for Robotics*; Cambridge University Press: Cambridge, MA, USA, 2017.
4. Fourati, H.; Belkhiat, D.E.C. (Eds.) *Multisensor Attitude Estimation: Fundamental Concepts and Applications*, 1st ed.; CRC Press: Boca Raton, FL, USA, 2016. [CrossRef]
5. Yang, Y.; Shen, Q.; Li, J.; Deng, Z.; Wang, H.; Gao, X. Position and Attitude Estimation Method Integrating Visual Odometer and GPS. *Sensors* **2020**, *20*, 2121. [CrossRef] [PubMed]
6. Wang, S.; Ahmad, N.S. A Comprehensive Review on Sensor Fusion Techniques for Localization of a Dynamic Target in GPS-Denied Environments. *IEEE Access* **2025**, *13*, 2252–2285. [CrossRef]
7. Zhang, A.; Atia, M.M. An efficient tuning framework for Kalman filter parameter optimization using design of experiments and genetic algorithms. *NAVIGATION J. Inst. Navig.* **2020**, *67*, 775–793. [CrossRef]
8. Liu, M.; Chen, Y.; Zhang, L.; Wang, Y. Attitude Estimation Algorithm of Portable Mobile Robot Based on Quaternion Complementary Filtering. *Micromachines* **2021**, *12*, 1373. [CrossRef]

9. Cheng, Y.; Crassidis, J.L. Particle Filtering for Attitude Estimation Using a Minimal Local-Error Representation. *J. Guid. Control Dyn.* **2010**, *33*, 1305–1310. [CrossRef]
10. Wang, S.; Clark, R.; Wen, H.; Trigoni, N. DeepVO: Towards end-to-end visual odometry with deep Recurrent Convolutional Neural Networks. In Proceedings of the 2017 IEEE International Conference on Robotics and Automation (ICRA), Singapore, 29 May–3 June 2017; pp. 2043–2050. [CrossRef]
11. Boughellaba, M.; Tayebi, A. Distributed Hybrid Attitude Estimation for Multi-agent Systems on SO(3). In Proceedings of the 2023 American Control Conference (ACC), San Diego, CA, USA, 31 May–2 June 2023; pp. 1048–1053. [CrossRef]
12. Davenport, P.B. *A Vector Approach to the Algebra of Rotations with Applications*; NASA TN D-4606; NASA: Washington, DC, USA, 1968.
13. Keat, J. *Analysis of Least-Squares Attitude Determination Routine DOAOP*; Technical Report CSC/TM-77/6034; NASA Goddard Space Flight Center: Greenbelt, MD, USA, 1977. [CrossRef]
14. Bani Younes, A.; Mortari, D.; Turner, J.D.; Junkins, J.L. Attitude Error Kinematics. *J. Guid. Control. Dyn.* **2014**, *37*, 330–336. [CrossRef]
15. Bani Younes, A.; Mortari, D. Derivation of All Attitude Error Governing Equations for Attitude Filtering and Control. *Sensors* **2019**, *19*, 4682. [CrossRef] [PubMed]
16. Bani Younes, A.; Turner, J.D.; Junkins, J.L. A Generic Optimal Control Tracking Solution for Various Attitude Error Parameterizations. In Proceedings of the 23rd AAS/AIAA Space Flight Mechanics Meeting, Advances in the Astronautical Sciences, Kauai, HI, USA, 10–14 February 2013; Volume 148.
17. Bani Younes, A.; Turner, J.D.; Majji, M.; Junkins, J.L. Nonlinear Tracking Control of Maneuvering Rigid Spacecraft. In Proceedings of the 21st AAS/AIAA Space Flight Mechanics Meeting, Advances in the Astronautical Sciences, New Orleans, LA, USA, 13–17 February 2011; Volume 140, pp. 965–981.
18. Euston, M.; Coote, P.; Mahony, R.; Kim, J.; Hamel, T. A complementary filter for attitude estimation of a fixed-wing UAV. In Proceedings of the 2008 IEEE/RSJ International Conference on Intelligent Robots and Systems, Nice, France, 22–26 September 2008; pp. 340–345. [CrossRef]
19. Van Der Ha, J.; Shuster, M. A Tutorial on Vectors and Attitude. *Control Syst. IEEE* **2009**, *29*, 94–107. [CrossRef]
20. Stanfield, K.; Bani Younes, A. Dual-Quaternion Analytic LQR Control Design for Spacecraft Proximity Operations. *Sensors* **2021**, *21*, 3597. [CrossRef] [PubMed]
21. Welch, G.; Bishop, G. *An Introduction to the Kalman Filter*; University of North Carolina at Chapel Hill, Department of Computer Science: Chapel Hill, NC, USA, 2006.
22. Titterton, D.; Weston, J. Strapdown inertial navigation technology—2nd edition—[Book review]. *IEEE Aerosp. Electron. Syst. Mag.* **2005**, *20*, 33–34. [CrossRef]
23. Särkkä, S. *Bayesian Filtering and Smoothing*; Cambridge University Press: Cambridge, MA, USA, 2013.
24. Risso, M.; Daghero, F.; Motetti, B.A.; Pagliari, D.J.; Macii, E.; Poncino, M.; Burrello, A. Optimized Deployment of Deep Neural Networks for Visual Pose Estimation on Nano-drones. *arXiv* **2024**, arXiv:cs.CV/2402.15273.
25. Farrell, J.L.; Stuelpnagel, J.C.; Wessner, R.H.; Velman, J.R.; Brook, J.E. A Least Squares Estimate of Satellite Attitude (Grace Wahba). *SIAM Rev.* **1966**, *8*, 384–393. [CrossRef]
26. Crassidis, J.; Andrews, S.; Markley, L.; Ha, K. Contingency Designs for Attitude Determination of TRMM. In Proceedings of the Flight Mechanics/Estimation Theory Symposium, Greenbelt, MD, USA, 15–18 May 1995.
27. Markley, L. Statistical Attitude Determination. In *Proceedings of the Encyclopedia of Aerospace Engineering*; John Wiley & Sons, Ltd.: Hoboken, NJ, USA, 2010. [CrossRef]
28. Diebel, J. Representing Attitude: Euler Angles, Unit Quaternions, and Rotation Vectors. *Matrix* **2006**, *58*, 1–35.
29. Koecher, M.; Remmert, R. Hamilton's Quaternions. *Numbers* **1991**, *123*, 189–220. [CrossRef]
30. Stanfield, K.S.; Younes, A.B. Dual Attitude Representations and Kinematics for Six-Degree-of-Freedom Spacecraft Dynamics. *IEEE Access* **2023**, *11*, 34349–34358. [CrossRef]
31. Shuster, M.D.; Oh, S.D. Three-Axis Attitude Determination from Vector Observations. *J. Guid. Control* **1981**, *4*, 70–77. [CrossRef]
32. Crassidis, J.L. Spacecraft Attitude Determination. In *Proceedings of the Encyclopedia of Systems and Control*; Springer International Publishing: Berlin/Heidelberg, Germany, 2021; pp. 2097–2104. [CrossRef]
33. de Ruiter, A.H.J.; Forbes, J.R. On the Solution ofWahba's Problem on SO(n). *J. Astronaut. Sci.* **2013**, *60*, 1–31. [CrossRef]
34. Cheng, Y.; Crassidis, J.L. A Total Least-Squares Estimate for Attitude Determination. In Proceedings of the AIAA Scitech 2019 Forum, San Diego, CA, USA, 7–11 January 2019. [CrossRef]
35. Rockafellar, R.T. Lagrange Multipliers and Optimality. *SIAM Rev.* **1993**, *35*, 183–238. [CrossRef]

6. Stanfield, K. Advanced Pose Developments for Aerospace Applications in Estimation, Modeling, and Control. Master's Thesis, School of Aerospace Engineering, San Diego State University, San Diego, CA, USA, 2024.
7. Mortari, D.; Markley, F.L.; Singla, P. Optimal Linear Attitude Estimator. *J. Guid. Control. Dyn.* **2007**, *30*, 1619–1627. [CrossRef]

MDPI AG
Grosspeteranlage 5
4052 Basel
Switzerland
Tel.: +41 61 683 77 34

Sensors Editorial Office
E-mail: sensors@mdpi.com
www.mdpi.com/journal/sensors

www.ingramcontent.com/pod-product-compliance
Lightning Source LLC
Chambersburg PA
CBHW041413210926
51850CB00013B/3263